The Vulnerable Andaman and Nicobar Islands

This first full-length book addresses disasters in the context of vulnerability of the Andaman and Nicobar Islands that comprise 572 islands in the Bay of Bengal. It looks at the disasters that the islands have experienced in the last 200 years and analyzes major disasters since colonization by the British.

Raising some critical questions, this book attempts to understand the overall profile of disasters – the facts, causes, damage, response and recovery – in the Andaman and Nicobar Islands. It discusses earthquakes, cyclones, tsunami and epidemics, as well as impacts of World War II, the penal colony and the post-Independence resettlement on the tribal population. The work will serve as a rich resource with its detailed tables, figures, maps and diagrams; appendices; and database ranging from travelogues, Census of India reports and fieldwork to Right to Information (RTI) petitions that collect hitherto unknown facts.

The book will be useful to students of geography, disasters and disasters management, climate and environmental studies, history, sociology, island and ocean studies, and South Asian studies.

Punam Tripathi is an independent researcher and has a PhD from the Department of Geography, Delhi School of Economics, University of Delhi, India. Previously, she was Assistant Professor, Salesian College, Siliguri. Her research interests and publications focus on the Andaman and Nicobar Islands. Her articles have been published in *Population Geography: A Journal of the Association of Population Geographers of India* and *Salesian Journal of Humanities and Social Sciences*.

The Vulnerable Andaman and Nicobar Islands

A Study of Disasters and Response

Punam Tripathi

Routledge
Taylor & Francis Group
LONDON AND NEW YORK

First published 2018
by Routledge
2 Park Square, Milton Park, Abingdon, Oxon, OX14 4RN

and by Routledge
605 Third Avenue, New York, NY 10017

First issued in paperback 2020

Routledge is an imprint of the Taylor & Francis Group, an informa business

British Library Cataloguing-in-Publication Data
A catalogue record for this book is available from the British Library

Library of Congress Cataloging-in-Publication Data
A catalog record for this book has been requested

ISBN 13: 978-0-367-73507-4 (pbk)
ISBN 13: 978-1-138-06236-8 (hbk)

Typeset in Sabon
by Apex CoVantage, LLC

This book is dedicated to my loving mother Sudha Tripathi. She has taught me how to build resilience and overcome vulnerability.

Contents

Figures

Note: All maps have been prepared by the author.

Tables

Preface

The Andaman and Nicobar Islands conjure an image of 'emerald green islands' surrounded by the deep blue ocean. The tranquil aquamarine beaches with white sand, mesmerizing sunsets and sunrises, exotic locales and the rich history of the islands have been used to lure tourists. But the Tsunami of 26 December 2004 presented a completely different picture. Entangled limbs of human and animal corpses were strewn across these pristine beaches post-Tsunami. Fallen houses, razed trees and upturned vehicles were mixed with the murk brought by the Tsunami. The unbearable stench of decaying bodies spread across these islands. It was characterized by death and destruction all across.

At this juncture came the book *Vulnerable India: A Geographical Study of Disasters*. The premise of this book is that there are no natural disasters and it is the vulnerability that is the root cause of the disaster. I was deeply inspired by this book and questioned, why did the Andaman and Nicobar Islands suffer in the Tsunami? Why did the islands become vulnerable to the Tsunami? Who had been most affected by the Tsunami – was it the non-tribal or the tribalpopulation? What was the response provided after the disaster? What are the other disasters that have occurred in the Andaman and Nicobar Islands? With this I began my research that finally took the shape of this book, *The Vulnerable Andaman and Nicobar Islands: A Study of Disasters and Response*.

The Andaman and Nicobar Islands present a critical area of study. It is an island territory of a country with a large mainland. Such a situation is not common. The islands of the world can be grouped into two categories on the basis of their jurisdiction. The first group consists of independent countries such as Sri Lanka, the Maldives, Fiji and those that belong to colonies of countries like New Caledonia in the southwest Pacific Ocean which is a colony of France. The development plans and policies for these countries cater to the needs of island ecology, even though small in size. In contrast, the second group consists of islands that are part of a country largely constituting a mainland. Such islands have been substantially ignored. Even though they're not a colony, the Andaman and Nicobar Islands fall in the second category. The development plans for the islands do not reflect its needs, rather it is a replica of the plans formulated for mainland India. A moot question is whether these mainlander policies have adversely affected and built the vulnerability of the islands.

The Andaman and Nicobar Islands are an integral part of India. It has a strategic geopolitical location at the centre of the Bay of Bengal. Its proximity to Myanmar, Indonesia, Thailand and Vietnam is advantageous from a military standpoint. The islands provide access to the Indian subcontinent, Southeast Asia and Myanmar, as well as to China through the land-based road. Its proximity to the Strait of Malacca, which is one of the busiest international trade routes with more than 65,000 vessels passing through it every year, is an added advantage. The Andaman and Nicobar Islands are also rich in resources. Of the total geographical area, 81 per cent is under forest cover. The forests of these islands range from tropical evergreen rainforests, deciduous forests, littoral forests and mangroves. They provide habitat for innumerable species. Situated between two major biodiversity hotspots, these islands exhibit extraordinary biodiversity. The islands are also surrounded by India's richest coral reefs that have the highest diversity among the four major reefs of India. The long coastline of the islands adds to the territorial waters of the country and extends India's Exclusive Economic Zone (EEZ).

The Andaman and Nicobar Islands are important from a historical perspective. It was the only penal colony of India. It is marked by the infamous Cellular Jail, popularly known as *Kalapani*. The ruins of the buildings on Ross Island bear testimony to the colonial history. It is one of the territories in India captured by the Japanese during World War II, the imprints of which are still found across the islands. The combination of natural beauty in terms of diverse forests, the azure sea, white beaches and rich coral diversity together with colonial architecture have made the islands an ideal tourist destination. This can be gauged from the increasing number of tourists both domestic and foreign. Adding to the significance of the Andaman and Nicobar Islands is its diversity of population. The islands are inhabited by six aboriginal tribes, most of whom still are hunter-gatherers. This population is juxtaposed along with the non-tribal population who came with the British, and later more were resettled under the government of India.

It is time that the Andaman and Nicobar Islands were brought to the forefront at all forums and policy making. Even though the islands are significant, these small islands located 1,200 km from the mainland India are victim to disasters all through their history. This book inquires into these disasters and seeks to understand the reason for the disasters and who are the vulnerable. It is the first of its kind of research on the Andaman and Nicobar Islands, where the issue of disasters is addressed through the lens of vulnerability. Tracing the history of disasters in the islands, the book identifies the disasters that have occurred in the islands and the vulnerabilities that led to it. It also puts forth the response to these disasters and in details assesses the post-Tsunami response in the Andaman and Nicobar Islands. Based on fieldwork, the book also documents and measures the recovery of the affected non-tribal and the tribal Nicobarese households. To accomplish this work, help and support came from many directions.

For this research I visited libraries in Delhi, Kolkata, Lucknow and Port Blair. In Delhi, the Ratan Tata Library, Central Reference Library and Central Science Library, Indian Institute of Economic Growth Library, Parliament Library, Planning Commission Library, Nehru Memorial Museum and Library, Indian Meteorological Department Library, Indian Institute of Planning and Administration Library, School of Planning and Architecture Library and National Institute of Disaster Management Library; in Kolkata, Asiatic Society Library and Anthropological Survey of India Library; in Lucknow, the Geological Survey of India Library; in Port Blair, the Archives, State Library and Anthropological Survey of India Library. Each serviced a different and detailed need. Appreciation is expressed to staff of these libraries.

What could not be archived through published material was sourced through officers and personnel in different government departments. I would like to thank Shri G. G. Saxena, Resident Commissioner of A&N Islands at New Delhi; Shri Rajat Sachar and Rizwan Uz Zaman of the Tsunami Rehabilitation Cell of the Planning Commission; and Ms. Andre Anurag (Director, A&N Islands, Ministry of Home Affairs), Shri Hasan Ali (Assistant Archivist of the State Archives) and Shri T. K. S. Ajayan (Assistant Director, Directorate of Disaster Management, Port Blair) for providing me with data regarding Tsunami Rehabilitation.

My thirst to seek more insights about the programmes and policies of rehabilitation led me to use the route of the Right to Information (RTI) Act. To this end detailed enquiries were channelled through RTI to different departments of the Andaman and Nicobar Administration: Andaman Lakshadweep Harbour Work, Andaman Public

Works Division, Department of Electricity, Department of Transport, Department of Social Welfare, Department of Education, Department of Environment and Forest, Department of Fisheries, Department of Animal Husbandry and Department of Tourism. The response and replies from all these were helpful to fill in the gaps of data much needed in this research.

Fieldwork was an important plank of this research as I was keen to know what the people of Andaman and Nicobar felt about the Tsunami, how they coped, what they did, how they responded and what were their hopes and aspirations. In this respect I must share that the families I interacted with were most welcoming and warm-hearted. Many shared a meal and also were ready to share their experiences.

In Car Nicobar Island the 'Captains', Mr Lamson Gedian (Second Captain of Arong Village) and Mr Reginald Watchful (First Captain of Kimius Village) were most helpful in arranging for a translator, as I am unfamiliar with the Nicobarese language. In this regard I am grateful to Loveleena (Arong village) and Prudence (Kimius village), without whom the fieldwork among the Nicobarese would not have been possible.

To visit Car Nicobar Island, a special permit called the tribal pass is required which is provided by the A&N Islands Administration. Thanks are due to Brigadier (retd.) Govind Singh Sisodia, Shri Shakti Sinha (Principal Secretary, Revenue/Disaster Management) and Shri Triwedi (Assistant Commissioner, Tribal Welfare) for facilitating the process of acquiring the tribal pass. At Car Nicobar Island, Shri S. C. Meena, Assistant Commissioner, Nicobar District was kind in extending my stay to complete my research. I acknowledge the hospitality of Nowry and her husband M. Dhanwar, Shri Vijay Bahadur Yadav, Sameer Dey and his entire family. It was the support and assistance of Swami Amartyananda (Secretary, Ramkrishna Mission, Port Blair), Mrs Mallika Syamchoudhury and Shri S. Syamchoudhury (Principal Chief Conservator of Forest), without which it would be difficult to complete this work.

I would like to convey kind appreciation to Ms Leela Solomon for editing the draft. Much needed breaks and support was provided by friends and family. Deeptima, Shabana, Kanchan and Niti Bhushan, all studying at foreign universities, made available the research, cheer and support that a researcher needs in different ways. Viola, Tanaya, Aniruddh, Kamini and Barsha were generous in their assistance. My family has rendered crucial help all through the work. Sincere word of appreciation is for my brothers, Vivek Pandey, Ravi Tripathi and Kshitiz Setu who supported me through this endeavour. The acknowledgement would remain incomplete if I do not express gratitude to Mrs Anju Gandhi, Mr Harvinder Gandhi and Saakshi Gandhi who provided me the comfort and warmth of a home in Delhi all throughout the research.

I sincerely appreciate and express my gratitude to Prof. Emeritus Gopal Krishan of Panjab University for academic advice and insights to my research problem. His encouragement kept me motivated to complete the work. I am indebted to Prof. B. Thakur and Prof. H. Ramachandran for their feedback and suggestions in the initial stage of my research. In the later part it was Prof. Surya Kant and Prof. S. Chattopadhyay who provided insight and expertise that greatly assisted in the research.

But, even with all this help I would not been able to accomplish this book without Prof. Anu Kapur. She not only introduced me to the study of vulnerability to disasters but provided never failing support to this research. Her astute and innumerable rounds of extremely detailed corrections and critical comments have shaped this research. Not only my Ma'am, but her parents have supported me in this research. Uncle and Aunty's love and motivation have helped me in completing this research. I remain forever indebted to them.

I would especially like to convey my appreciation to my husband Shaibal Sinha. He has encouraged and supported me throughout in this work. It is my mother, Mrs Sudha Tripathi, to whom I am indebted. Completion of this book would be impossible without her. She has not only supported this research financially but also assisted me in the fieldwork. Her valuable insight and contributions has enriched this work. It is her never-failing guidance and encouragement which gave me the courage to research on the vulnerability and response to disasters.

Last but not least I would like to add, in spite of care and concerns, yet I own any lapses that may have occurred along the way. Research is a journey and this book is my first step in this direction.

1 Identifying disasters in the Andaman and Nicobar Islands

The Andaman and Nicobar Islands (henceforth A&N Islands) were in the news again. The lone active volcano of the A&N Islands erupted on 23 January 2017. It suddenly began spewing lava and ash while the scientists from the National Institute of Oceanography, Goa were collecting sea floor samples in the Andaman Basin. The eruption ranked '2' on the Volcanic Explosivity Index, where '1' is the mildest and '8' the largest. It was a rare event for India. Eleven years earlier, another 'rare event' shook not only the A&N Islands but also the world. A tsunami generated by a 9.3 Mw earthquake on the Richter scale hit the A&N Islands at 6.18 a.m. (Indian Standard Time) on 26 December 2004. The epicentre of the earthquake was off the west coast of Sumatra Island (Indonesia). The Tsunami hit the coasts of all countries across the globe and killed nearly 283,000 people; millions were displaced and property of nearly USD 10 billion was damaged in Indonesia, Thailand, Sri Lanka, Somalia and India, among other countries. It became the world's deadliest disaster of the twenty-first century and remains so till today. In this Tsunami, more than 3,000 people were killed in the A&N Islands. But the tsunami was not the only disaster to occur in the A&N Islands. A historical enquiry reveals that the A&N Islands have experienced disasters in the past as a result of its vulnerability.

The A&N Islands present a unique area of study. A stretch of 1,200 km separates them from mainland India. The strategic location of the A&N Islands provides access to the Indian subcontinent, Southeast Asia, Myanmar and to China (China has established a naval base at the Coco Islands of Myanmar, which is less than 50 km from the northernmost point of the A&N Islands). The geographical proximity to Myanmar, Indonesia, Thailand, Vietnam and Singapore gives location-specific advantages from a military standpoint. The A&N Islands are in proximity to the Strait of Malacca, one of the world's most important shipping lanes, which connects the Indian Ocean to the Pacific Ocean. Over 65,000 vessels pass through the Strait of Malacca every year, carrying about one-fourth of world's traded goods, including oil (Ibrahim and Nazery 2007). The A&N Islands account for one-fourth of the total coastline and about 30 per cent of the total Exclusive Economic Zone (EEZ) of India. More than 148,000 tonnes of pelagic, demersal and oceanic fishery resources are estimated to be available in the 0.6 million km^2 of EEZ of the A&N Islands. The A&N Islands are also rich in forest resources and have a vibrant coral reef system. More than 80 per cent of its geographical area is under forest that supports a rich biodiversity. The A&N Islands are one of the 34 biodiversity hotspots of the world. They form a part of the Indo-Burma and Myanmar biodiversity hotspot.

The A&N Islands also have a unique socio-economic setup. They are inhabited by the aboriginal tribes that still live in the forests. Also, there are the descendants of the prisoners of Andaman Penal Colony, the settlers (refugees from East Pakistan that were settled under the resettlement plans), the opportunists (who flocked to the A&N Islands during 1960–70 in search of 'government jobs'), the officers of the Administration (who are frequently transferred) and the tourists (both domestic and international).

The A&N Islands: the area of study

The A&N Islands are a union territory of India. They lie towards the east of the Indian mainland in the Bay of Bengal (6° to 14° north latitude and 92° to 94° east longitude) and appear like emerald beads strewn in the blue sea. The average distance between the A&N Islands and the Indian mainland is 1,200 km. To the west of the islands lies the Bay of Bengal that separates them from mainland India, and to the east is the Andaman Sea which separates them from Thailand. In the north the A&N Islands are separated from Myanmar by Preparis North Channel and on the south the Great Channel separates them from Indonesia (Figure 1.1). In fact the A&N Islands are closer to Southeast Asia. The northern tip of Indonesia is at a distance of 150 km from the A&N Islands while Myanmar and Thailand are 600 km away to the east.

The British colonized the A&N Islands in the late eighteenth century and developed them into a penal settlement in the mid-nineteenth century, to house the prisoners of the Revolt of 1857. After a brief occupation by the Japanese from 1942 to 1945, the British again gained its command till 1947. In 1947, after Independence, the A&N Islands became part of the Indian Union and were administered by a commissioner appointed directly by the president. The States Reorganization Act of 1956 constituted these islands into a union territory administered by the central government (Ministry of Home Affairs) through the lieutenant governor, who is also the administrator.

The A&N Islands comprise two archipelagos containing a total of 572 islands, including all rocky outcrops and islets (Directorate of Economics and Statistics 2011). The two archipelagos, Andaman and Nicobar, are separated by the Ten Degree Channel, so-called because of the 10° north latitude which divides it. The channel is 139 km wide and 1,399 km deep. The Andaman Islands are the larger of the two archipelagos, and have 548 islands and cover 6,408 km^2, which is 78 per cent of the total area of the A&N Islands.

The Nicobar archipelago has 24 islands which cover 1,840 km^2, and comprises the remaining 22 per cent of the total area. The total length of the Andaman Islands is 467 km and of Nicobar Islands is 259 km. The average width is 24 and 58 km for the Andaman and Nicobar Islands, respectively. The maximum width of the A&N Islands is not more than 60 km in any part (Directorate of Economics and Statistics 2011).

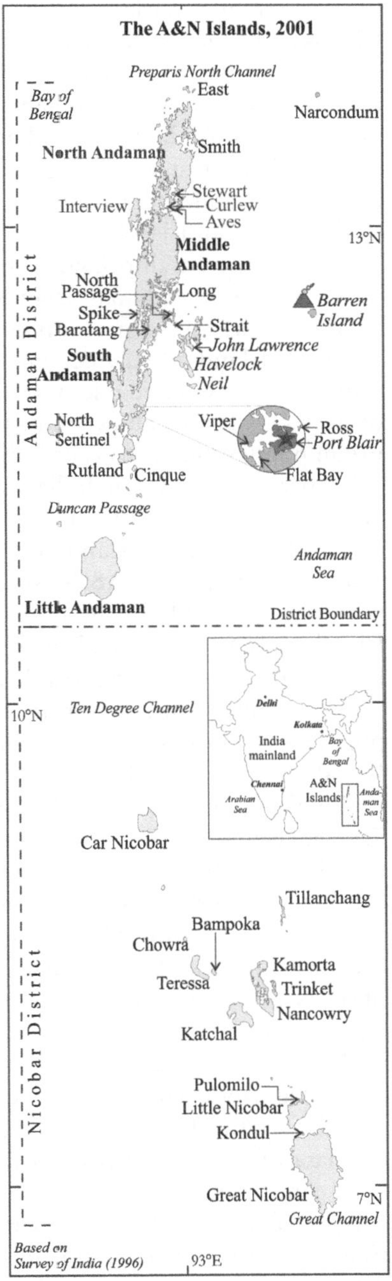

Figure 1.1 A&N Islands, 2001

Of the total 572 islands, only 38 are inhabited: 25 in the Andaman Islands and 13 in the Nicobar Islands. All islands of the Andaman archipelago, except Little Andaman Island, lie close to each other. Together they are known as Great Andaman Island. The Great Andaman Island and the Little Andaman Island are separated by the Duncan Passage. The northernmost is Landfall Island and the southernmost is Little Andaman Island. India's only volcano is in Barren Island which is part of the Andaman Islands.

The southern archipelago is the Nicobar Islands. They are smaller in size than the Andaman Islands. The northernmost is Car Nicobar Island and the southernmost is Great Nicobar Island. Administratively the A&N Islands are divided into three districts and nine *tehsils*. It has four census towns and one statutory town. Port Blair is the capital city of the A&N Islands. It accounts for more than 75 per cent of the total urban population (Census of India 2011a).

As per the Census of India (2011a), the population of the A&N Islands was 380,581, of which Andaman Islands accounted for 90 per cent and the Nicobar Islands accounted for the remaining 10 per cent. The population of the A&N Islands can be divided into two groups – non-tribal and tribal. The non-tribal population came to the A&N Islands in 1858, when the British set up the Penal Colony at Port Blair. It grew rapidly after Independence in 1947, after the families were 'resettled' under various schemes from 1947 to 1980. The non-tribal population consists of mainlanders who were 'settled' in the A&N Islands over the last 150 years. They account for 92.66 per cent of the total population. The tribal population constitutes the remaining 7.34 per cent of the total population. The tribal population can be further divided into tribes of the Andaman Islands and tribes of the Nicobar Islands. There are six tribes, four in the Andaman Islands and two in the Nicobar Islands (Table 1.1). The tribes of the Andaman Islands are the Sentinelese, Great Andamanese, Onges and Jarawas. The tribes of the Nicobar Islands are the Shompen and the Nicobarese. The Nicobarese account for nearly 97 per cent of the total tribal population in both the 2001 and 2011 censuses. The tribes of the A&N Islands are scheduled tribes. The last large-scale disaster experienced by the A&N Islands was the Tsunami. Therefore, for all purposes of this research, the data of the last census before the Tsunami, i.e. Census 2001 has been considered.

Apart from the strategic location, economic potential, ecological fragility and a complex socio-cultural setup, what gives credence to the A&N Islands as an area worthy to investigate issues of vulnerability,

Table 1.1 Tribal population in A&N Islands, 2001 and 2011

Tribe	*2001*			*2011*		
	Population	*Per cent of total tribal population*	*Per cent of total population of A&N Islands*	*Population*	*Per cent of total tribal population*	*Per cent of total population of A&N Islands*
Sentinelese (estimated)	39	0.48	0.04	15	0.05	0.00
Great Andamanese	43	0.14	0.01	44	0.16	0.01
Onges	100	0.32	0.03	101	0.36	0.03
Jarawas	270	0.87	0.08	380	1.36	0.10
Shompens	380	1.23	0.11	229	0.83	0.06
Nicobarese	30,000	96.95	8.42	27,168	97.25	7.14
Total	**30,832**	**100**	**8.69**	**27,937**	**100**	**7.34**

Source: Based on Census of India (2001a) and Census of India (2011b)

response and disasters is the fact that the A&N Islands have been victim to recurrent disasters. The main questions raised and addressed in this book are: Which are the disasters that have affected the A&N Islands? What is the spatial pattern of these disasters? What were the vulnerabilities that led to these disasters? Who were affected most in these disasters? How many people were killed in these disasters? Who were the ones killed? Why were those killed vulnerable? The key issue addressed is the role of vulnerability and response to disasters in the A&N Islands. Before the role of vulnerability can be assessed, it is imperative to identify the disasters experienced by the A&N Islands.

Identifying the disasters of the A&N Islands

The High Powered Committee on Disaster Management (HPC) lists 31 disasters that have affected or can have an impact on India (Ministry of Agriculture 2001). Not all of these are pertinent to the A&N Islands (Table 1.1a). Among these 31 disasters, there are five which at present are not applicable to the A&N Islands, namely (1) snow avalanches; (2) mine flooding; (3) mine fire; (4) chemical and industrial disaster and (5) nuclear disaster. The A&N Islands lie in the tropics and the highest point is 732 m (Saddle Peak in North Andaman Island); therefore, there is no possibility of snow avalanche. Till date, there are no mines or chemical factories, industries or nuclear power plants, so disasters by these can be ruled out.

Of the remaining list of 26 disasters, in HPC, research detailed from records in gazetteers, India Meteorological Department's (IMD) disastrous weather reports, disaster management plans and the published materials confirmed that 12 other types of disasters listed in the High-Power Committee's Report have not occurred in the A&N Islands. These are (1) serial bomb blasts; (2) festival-related disasters; (3) pest attacks; (4) cattle epidemics; (5) food poisoning; (6) tornadoes and hurricanes; (7) hailstorms; (8) cloudbursts; (9) heat waves and cold waves; (10) dam failures/dam bursts; (11) droughts and (12) oil spills. This brings the tally down to 14. These are (1) air, road and rail accidents; (2) boat capsizing; (3) cyclones; (4) earthquakes; (5) electrical disasters and fires; (6) floods and drainage management; (7) forest fires; (8) landslides and mudflows; (9) major building collapses; (10) sea erosion; (11) thunder and lightning; (12) urban fires; (13) village fires and (14) biological disasters and epidemics which have occurred in the A&N Islands. Among the 14 types of disasters, cyclones and earthquakes have occurred frequently in the A&N Islands.

Ruling out disasters associated with cyclones

The Andaman Sea and the Bay of Bengal are 'breeding grounds of the cyclones' that later move to the Indian mainland (Lal 1989). The latitudinal extent of the A&N Islands is 6° to 14° north, while the longitudinal extent is 92° to 94° east, and the A&N Islands are situated within the cyclone-forming region. The maximum width of the A&N Islands is less than 60 km; hence the cyclones that cross over the A&N Islands do not dissipate due to friction. To assess the frequency and time of cyclones originating or crossing over A&N Islands, a grid was laid on latitudes 5° to 15° north and longitudes 90° to 95° east. Within this grid, all cyclones that have been reported for a period of 120 years (1891–2010) were plotted. Such a grid was selected because cyclones within this range can affect the A&N Islands as minimum width of an average cyclone covers 2° latitude. A total of 175 cyclonic formations had crossed this grid (Table 1.2).

Not all cyclones were of similar intensity. Out of the 175, about 69 per cent were deep depressions, 19 per cent were cyclonic storms and 12 per cent were severe cyclonic storms. Neither very severe cyclonic storms nor super cyclonic storms have hit the A&N Islands. The reason for this low or near absence of severe cyclones is that the seas surrounding the A&N Islands are where cyclones are born, therefore they are unable to garner speed. However, the notion that 'seas surrounding the A&N Islands are only a "breeding ground" of a cyclone' is far from true. A glance at the records shows that the A&N Islands are visited at intervals by most severe storms. In this research, the term 'cyclone' is used for all three types of cyclonic formations – deep depression, cyclonic storm and very severe cyclonic storm.

In the last 120 years there is no evident pattern of increase or decrease in the number of cyclones. However, when the occurrences of the cyclones were classified on a monthly basis, October, November and December accounted for 66 per cent of the total 175 occurrences of cyclones, followed by 27 per cent in the months of April and May. The months of October, November and December are the inter-monsoon period between the South-West monsoon and the retreating monsoon, while April and May are months before the onset of the South-West monsoon.

The inter-monsoonal phase is conducive for the development of cyclones, as this period is marked by the shifting of the inter-tropical convergence zone (ITCZ), where the trade winds of both hemispheres converge. In April and May, the ITCZ shifts north of the equator and again shifts south during October through December. When the

Table 1.2 Type of cyclonic formations in and around A&N Islands (1891–2012)

Type	*Wind speed (km/hr)*	*No. in and around A&N Islands*
Depression	Up to 51	0
Deep depression	52–61	121
Cyclonic storm	62–87	33
Severe cyclonic storm	88–117	21
Very severe cyclonic storm	118–221	0
Super cyclonic storm	222 or more	0
Total		**175**

Source: Based on Indian Meteorological Department (1979, 1996, 2013)

ITCZ is further from the equator, there is a contrast between the two trade winds in terms of air temperature and moisture content. These differences in air masses result in the formation of centres of low pressure. These weak disturbances intensify into a cyclone in an environment that is conducive. Thus the presence of these weak disturbances is essential for the formation of cyclones, and in the Bay of Bengal these weak disturbances are formed during the inter-monsoonal period. When 21 severe cyclonic storms out of a total of 175 cyclones were mapped, it emerged that 16, or three-fourth of these cyclones, originated north of 10° north latitude. These cyclones too had distinct paths, 12 moved towards the north-west, two towards the north, five towards the north-east and one towards the east (Figure 1.2).

The tracks represent the variable wind direction along which the cyclones moved. All other types of cyclonic formations have similar pattern of origin and movement.

Though there are no very severe cyclones or super cyclones, other lower intensity cyclones are common in and around the A&N Islands which are strong enough to disrupt life on the islands. The life in the penal colony, which was set up by the British in 1858, was severely affected after the 1891 cyclone. This was noted by M. V. Portman, the officer in charge of the Andamanese (1879–1901):

> when the present Settlement experienced the cyclone of 2 November 1891, great surprise was exhibited, as it was supposed that the Andamans only *formed* the cyclones, and were out of the reach of their violence.
>
> (Portman 1899: 88)

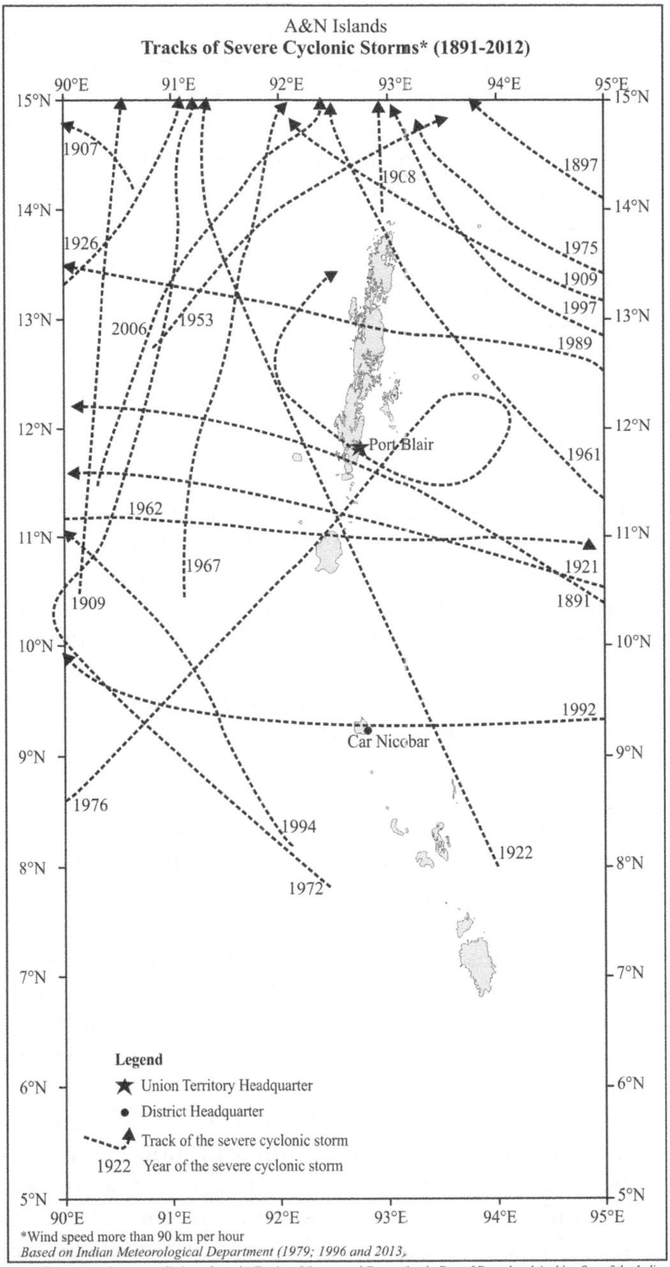

Note: The cyclone data was culled out from the Tracks of Storms and Depression in Bay of Bengal and Arabian Sea of the Indian Meteorological Department (IMD 1979 and 1996) and was updated from the cyclone e-atlas for the remaining period from 1997 to 2012. IMD on the basis of wind speed has classified cyclonic formations in six classes – depression, deep depression, cyclonic storm, severe cyclonic storm, very severe cyclonic storm and super cyclonic storm. If depressions intensified into a storm, the event has been listed as storm and not under depression. This was done to avoid duplication of the same event.

Figure 1.2 A&N Islands: tracks of severe cyclonic storms (1891–2010)

Another notable cyclone that was documented occurred in 1844. Describing the force of the wind, the captain of the ship *Runnymede*, who was in the wrecked ship during this cyclone, notes:

> The gusts were so terrific mixed with the drift and rain, that no one could stand on deck to man the pumps or clear the wreck . . . the severity of the wind was beyond description, there was nothing to which it could be compared, for unless present, no one could conceive the destructive power and force of wind, crushing everything before it as if it were a metallic body.
>
> (Captain Doughty in Portman 1899: 137)

Again, on 1 November 1891, a cyclone crossed over Port Blair settlement which led to damage:

> The RIMS *Enterprise* was driven on shore on the South Point Reef . . . the trees were either felled or badly injured, and the roads were blocked with fallen branches. In the jungle the debris was so thick and matted that even the Andamanese were unable to make their way along their accustomed paths. All the buildings at the Homes were levelled, the falling of the Barrack at Baja Jág-da killing an Andamanese girl. The Andamanese hospital at Haddo fell, killing an old blind man, and injuring another man.
>
> (Portman 1899: 677–8)

When the records maintained by the British (documented in Portman (1899)) were compiled, then out of the 11 cyclones listed, only six had registered deaths. The cyclone in 1792 killed more than 100 people. For all other events the deaths were limited to fewer than 10 persons Table 1.3.

Many ships were lost during the cyclones causing immense loss of life and property. It is a fact that one of the reasons for establishing the penal colony in the A&N Islands was the safe harbour of Port Blair to curtail the damages and heavy losses incurred by the merchant ships in the cyclones of the Bay of Bengal.

In the post-Independence period, three incidents of damage due to cyclone have been recorded (A&N Administration 2004). First was the cyclone of 24 November 1988 which caused flooding in Diglipur (North Andaman Island). The second cyclone crossed through Mayabunder (Middle Andaman Island) and neighbouring areas on 8 December 1988. The third, Cyclone Gay, made a landfall in Thailand on 4 November 1989, weakened to cyclonic storm over Bay of Bengal and crossed across Mayabunder on 6 November 1989. It moved east, gained speed and made landfall on 8 November 1989

Table 1.3 Damages ascribed to cyclones in the A&N Islands (1792–1893)

Year	*Location (with respect to A&N Islands)*	*Name of ship wrecked*	*Damage*	*Persons killed*
1792	Off the Cape Negaris (Cape in Myanmar, 193 km north of A&N Islands)	*Juno*	Ship sunk; food grain and other supplies drowned	More than 100
1844	Interview Island, west coast of Middle Andaman Island	*Briton* and *Runnymede*	Ships smashed in coral reef and wrecked	4 killed
1849	Island lying on west of North Andaman Island	*Emily*	30 crew members stranded	1 killed
1849	Narcondam Island, off east coast of North Andaman Island	*Flying Fish*	Ship capsized, crew swam to Narcondam Island; rescued after 18 days	
1859	Port Blair		Superintendent's bungalow and hospital unroofed; convict buildings prostrated; convicts huts in every station thrown down	2 killed
1883	Cocos Island (50 km north of A&N Islands)	*Scottish Chieftain*	Ship wrecked 4 miles north of Flat Island	
1885	North-west coast of Andaman Island	*Helen Pembroke*	Ship wrecked; crew sailed to Cocos Island	
1885	North-west coast of Andaman Island	*Fazel Kureem*	Ship wrecked; crew sailed to Cocos Island	
1889	Cinque Island	Country boat	Ship wrecked; crew swam to nearest Island	
1891	Port Blair		Collapse of many buildings that killed and injured convicts	Convicts killed – number not available
1893	Seas off North Andaman Island	Country boat	Crew drifted north to Burma with the cyclone, with north-east monsoon winds drifted south and reached Cuddalore coast (Madras Province)	Numbers unknown

Source: Based on descriptions in Portman (1899)

near Kavali in Andhra Pradesh as a super cyclonic storm with wind speed close to 240 km/hr. In this storm 69 people were killed and property worth Rs. 410 million was destroyed (Raghavan and Rajesh 2003). There were no deaths reported from the A&N Islands when Cyclone Gay crossed over.

The damages attributed to cyclones have been minuscule in the post-Independence period. The best technology for prediction of tracks of cyclones has been installed and in most cases the approaching cyclones are predicted, thereby advising the fishing boats not to venture into the sea. The ships have been made sturdy and are able to withstand the storms. As of the present, cyclones disrupt sailing schedules, uproot trees, snap communication lines, create waterlogging, damage crops and houses, but there are no human deaths. The A&N Administration has provisions to restore damaged infrastructure.

In case of the A&N Islands, cyclone-led disasters had been minuscule in the pre-Independence period. The vulnerability of the penal settlement was high and hence there were disasters associated with cyclones. But with the advent of technology and better management, cyclones do not lead to disasters in the A&N Islands in the present time and hence cyclones are ruled out. Besides the surrounding seas where cyclones blow, the A&N Islands experience frequent earthquakes as they are located on a subducting plate boundary (the Indian plate subducts below the Burma micro-plate) which causes earthquakes.

Ruling out disasters associated with earthquakes in the A&N Islands

The Bureau of Indian Standards has redrawn the seismicity map of India five times in the past 40 years (Kapur 2010). The A&N Islands on all five maps have remained in the very high seismicity zone. The A&N Islands were in Zone V of the seismic zone map of India (2002), which represents region of highest seismicity. It lies in the 'very high damage risk zone' (A&N Administration 2004) as it is close to one of the active plate margins of the world. The A&N Islands lie on the Burma micro-plate which is seismically active (Figure 1.3).

To the west of the Burma micro-plate is the subducting Indian plate which is moving obliquely in the north-east direction at 5 cm/year. It is marked by the presence of the 3,500 m deep Sunda-Andaman trench and a row of volcanic islands of the Sumatra and Java island arc system to the east of the A&N Islands (Malik et al. 2006). The arc continues northwards as marked by the presence of two volcanic islands, Narcondam Island and Barren Island. Both of these are part of the Andaman Islands and lie to the east of Middle Andaman Island. While Narcondam Island is an extinct volcano, Barren Island is an active one, where the latest eruption was on 23 January 2017.

The Andaman Sea forms a back-arc basin; this is also an indicator of this subduction. The Andaman Sea lies between the A&N Islands and Thailand. Because of this, the region also has active and inactive faults marking the sites of plate movement. The most prominent fault in this section is the West Andaman Fault (WAF). This fault extends from Sumatra in the south-west up to the Burma micro-plate in the north (Malik et al. 2006). The other faults are the Sumatra Fault system, Seuliman Fault and the Sagaing Fault.

Given the tectonic characteristics, earthquakes and volcanic activities are common in and around the A&N Islands. One can find reference to earthquakes in the folklore of the tribes of the A&N Islands. The earliest recorded earthquake in the A&N Islands was the earthquake of 31 December 1881 (Kayal 2008). The documentation of earthquakes began only after the British colonized the islands in 1858. During the colonial period, significant earthquakes were reported by the British and details of the events

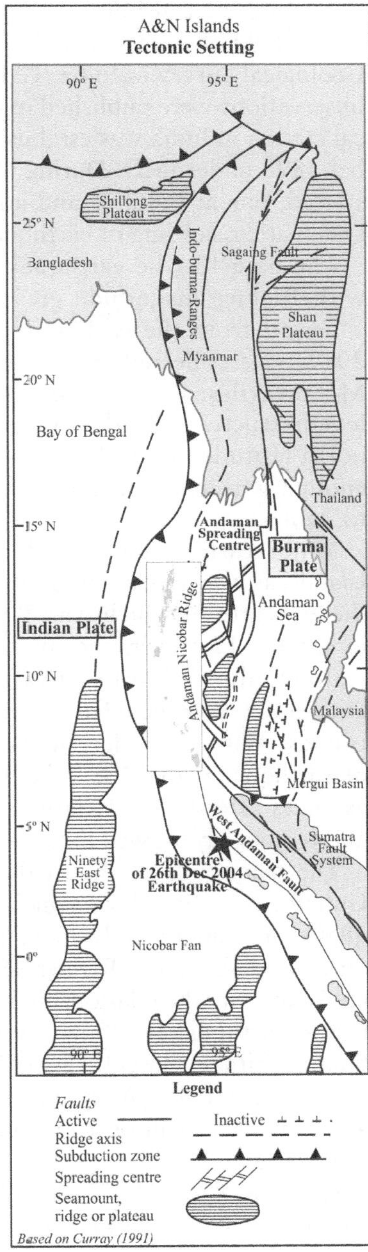

Figure 1.3 A&N Islands: tectonic setting

were published in journals. Before seismological observatories were set up, field observations of large earthquakes were carried out by the Geological Survey of India (GSI), which was established in 1851, and observations were published in the GSI memoirs. The first seismological station in India was established in Calcutta (Alipore) on 1 December 1898 under IMD. During 1898–99, two more observatories were started, one at Bombay and another at Kodaikanal, after which the systematic recording of earthquakes began.

The data for the earthquakes in and around the A&N Islands, with Richter magnitude greater than or equal to 5 ($M \geq 5$), for a period from the earliest recorded earthquake till 31 December 2012, was collected from the IMD's seismological division. Only $M \geq 5$ earthquakes were selected because earthquakes of $M < 5$ are less destructive. The data was collected for the area between 5°–15° north latitude and 90°–95° east longitude, as this covers area large enough within which the small and medium earthquakes can lead to damage.

Major earthquakes outside this grid too can impact the A&N Islands, as was the case in the 26 December 2004 earthquake ($M = 9.1$), the epicentre of which was 3°19′12″ north latitude and 95°51′ east longitude. However, the average return period of $M \geq 8$ in this region is about 80 years (Srivastava and Chaudhury 1979). An enlarged grid would have included the earthquakes that do not impact A&N Islands. Within this latitudinal and longitudinal span, the number of earthquakes of $M \geq 5$ from 1881 to 2011 was 424. This figure excludes aftershocks of large earthquakes, as these lead to spatially and temporally skewed data.

The magnitude of the earthquakes in A&N Islands confirms the Gutenberg-Richter law, which states that lower magnitude earthquakes are significantly more prevalent than the higher magnitude ones (Gutenberg and Richter 1944). The Gutenberg-Richter law (GR law) expresses the relationship between the magnitudes and total number of earthquakes in any given region and time period of at least that magnitude. Gutenberg and Richter (1944) proposed the following linear relation: $\log_{10} N(x) = a - bx$. For example, for every magnitude 4.0 event there will be 10 magnitude 3.0 quakes and 100 magnitude 2.0 quakes.

From 1858 to 2011, a total of 424 earthquakes of $M \geq 5$ have been recorded. Of these 424 recorded earthquakes, 83 per cent are of magnitude $6 > M > 5$ (M greater than 5, but less than 6), while another 16 per cent are $7 > M > 6$ (M greater than 6, but less than 7). Only five earthquakes, which is 1 per cent of the total, are greater than $M = 7$, and only one earthquake has been greater than 8. Thus, it is evident that large numbers of earthquakes experienced in the A&N Islands are moderate to strong.

The number of earthquakes appears to increase over the 131-year period. However, this is misleading interpretation as earlier recording of earthquakes was not advanced and therefore only very high magnitude earthquakes ($M \geq 7$) were recorded. Even the seismological observatory set up in Calcutta in 1898 was not sensitive enough to record low magnitude earthquakes of the A&N Islands. As a consequence there is no mention of any earthquake in the 33-year period following the December 1881 earthquake. Hence an annual trend or a decadal trend analysis of earthquakes is unreliable.

Earthquakes in the A&N Islands do have a spatial pattern where 76 per cent of the total 424 epicentres are concentrated between 93° and 95° east longitude (Figure 1.4). The epicentres form a continuous belt on the east of the A&N Islands marking the subduction zone, where the Indian plate is subducting below the Burma micro-plate. There is no distinct north-south divide between the occurrences of earthquakes when 10° north latitude, which separates the Andaman from the Nicobar Islands, is taken as the line of division. Earthquakes occurring north of 10° north latitude are slightly higher in number than the earthquakes occurring south of 10° north latitude, and account for 55 per cent of the total.

Besides occurrence and location, what is more significant in case of earthquakes, is their magnitude, intensity and depths. Magnitude, intensity and depth measure different characteristics of the earthquakes. Magnitude measures the energy released at the source of the earthquake. It is measured by the moment magnitude scale (a modified version of the Richter scale). It is a logarithmic scale, i.e. an earthquake which registers $M = 5.0$ has a shaking amplitude 10 times that of earthquake $M = 4.0$. Intensity measures the strength of the shaking produced by the earthquake and the associated damage at a certain location. Intensity is determined from effects on people, human structures and the natural environment. The intensity is measured by the modified Mercalli intensity scale. Intensity is determined by the depth of the earthquake. The recorded earthquakes around the world have depths ranging from near the earth's surface to about 700 km deep. The strength of shaking of an earthquake diminishes with increasing distance from the earthquake's source. Thus, shaking at the surface of the same magnitude earthquake at 500 km deep will be less compared to the earthquake at 20 km depth. Based on depths, earthquakes are classified as shallow focus earthquake with depth less than 70 km from the surface of the earth; intermediate

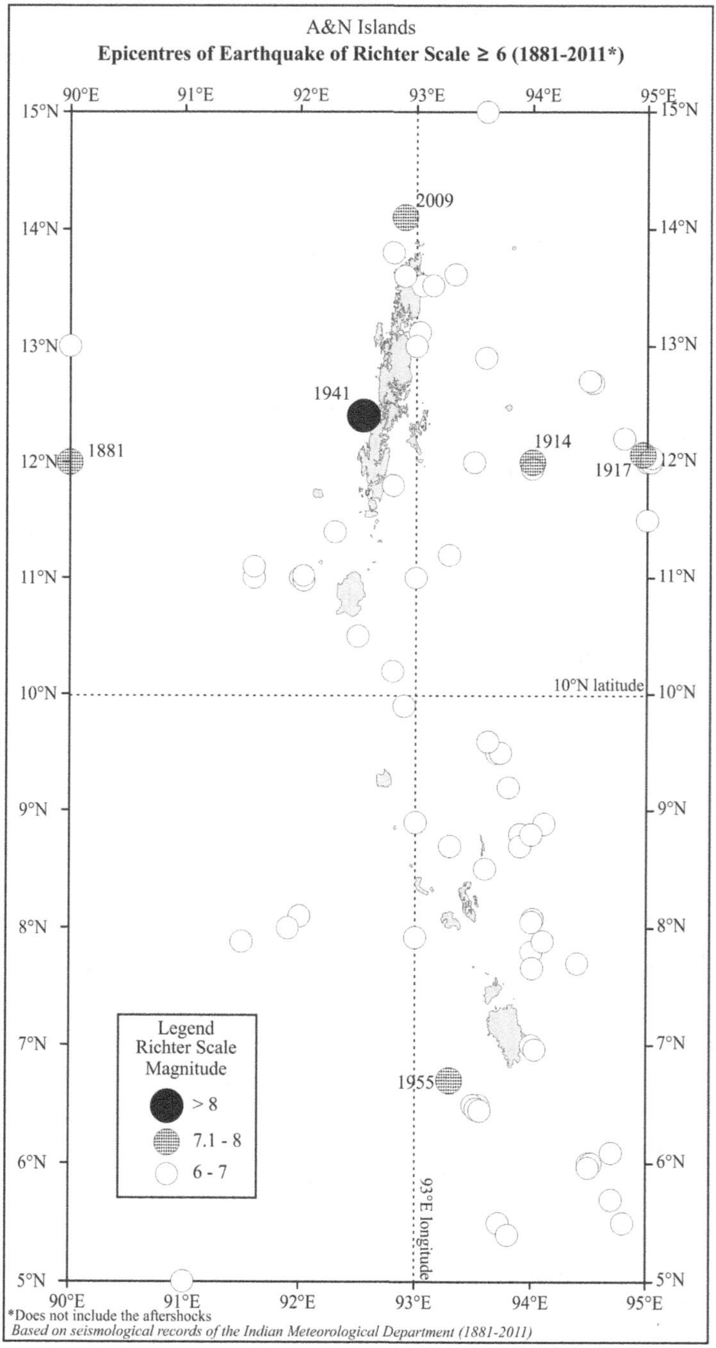

Figure 1.4 A&N Islands: epicentres of earthquake of Richter scale ≥ 6 (1881–2011)

focus earthquakes with depths 70–300 km; and deep focus earthquakes with depths greater than 300 km.

The damage associated with an earthquake is the function of both magnitude and depth. Of the 424 recorded earthquakes in the A&N Islands, the depths of 81 earthquakes were not known. Of the remaining 343 earthquakes, 89 per cent per cent had shallow focus, and the remaining 11 per cent had intermediate focus. Thus, there were no deep focus earthquakes in the region. Even though the shallow focus earthquakes abound in the A&N Islands, the damage associated with these is less as most of these are low magnitude earthquakes.

Six among these 424 earthquakes had a magnitude of $M \geq 7$, four of which occurred in the colonial period. Of these four earthquakes, details about damage and those killed could be gathered for only two of them. The first is the earthquake of 31 December 1881, of magnitude $M = 7.8$. Its epicentre was located 200 km north-north-west of the Andaman Islands. R. D. Oldham, famous geologist, notes in his field observations:

> On . . . 31 December 1881 a severe earthquake was felt over a large portion of the Indian Peninsula and Bengal, affecting also the Burmese coast and causing much damage in the Andaman and Nicobar islands . . . Port Blair is the only place where any damage was done to masonry buildings, and it is to be regretted that the damage should be so little instructive as is the case. The infantry barrack . . . were severely cracked . . . In the Car Nicobar extensive damage was done to the cocoanut groves and the huts of the natives, . . . On the seashore the trees were left standing, while further inland they were overthrown.
>
> (Oldham 1884: 47)

The second earthquake for which damage records are available is in 1941. The epicentre of this earthquake of magnitude $M = 8.1$ was 96.7 km north-north-west of Port Blair. The earthquake caused widespread damage in Middle Andaman and South Andaman. Most masonry structures in Port Blair were badly affected. All the buildings and roads on Ross Island, the administrative centre of the British, were damaged. The Cellular Jail, a three-storey building with 696 solitary cells, was destroyed, and the chief warden of the jail was killed (A&N Administration 2004). From 1947 to December 2004, even though 241 earthquakes were recorded, details of damage are available only for two of them – first, the earthquake swarms in 1982–83 and the second in 2002. From December 1982 to January 1983, the Great Nicobar Islands experienced a series of tremors which created panic among the settlers. Even though no damage was reported, some of the settlers left the place and came to Port Blair.

A second earthquake with associated damage occurred on 14 September 2002, located 23.6 km south-south-east of North Andaman Island. Its magnitude was M = 6.5 and it lasted for two minutes. While the tremors were felt across the A&N Islands, including the Nicobar Islands, damages were reported from the North and Middle Andaman Islands. The Ariel Bay jetty (in North Andaman Island) was damaged and a few shops near it collapsed. The public buildings most affected were a church, an orphanage, a bank, a girl's hostel, a postal building, teacher's quarters, shops and the school. The lighthouse on East Island was damaged and 50 kg of mercury used in the beacon spilled as a result (A&N Administration 2004). This spill was a health hazard to the local people as well as to the fragile ecology of the entire region. Environmental organizations expressed apprehension over the resultant toxicity that threatened the flora and fauna and the amount of mercury that entered the food chain. However, no action was taken to remove the spilled mercury from the sea.

Thus, it can be inferred that even though the magnitude of the earthquakes was high, there were negligible deaths reported in these incidences indicating low vulnerability to the earthquakes. A feature associated with earthquakes generated at a subducting plate boundary is tsunami. Occasionally some of these earthquakes have generated tsunami. This earthquake-tsunami association happened in the earthquakes of 1847, 1868, 1881, 1883, 1935, 1941, 2002 and 2004. However, no deaths or large-scale damage is ascribed to the earthquakes in the A&N Islands prior to the 26 December 2004 earthquake of M = 9.3.

For this research, a disaster is 'an event that causes large-scale *killing* and destruction within short period of time'. Given this definition, within a historical enquiry, the type of disaster listed in the HPC that has occurred in the A&N Islands is epidemics. Ironic as it may seem, the two other disasters which have affected the A&N Islands, World War II and the Tsunami of 26 December 2004, are not listed in the HPC report.

Identifying three main disasters

From the time of recorded history the A&N Islands experienced three types of disasters – the epidemics in the late nineteenth century, World War II (1942–45) and the Tsunami (26 December 2004). In each of these three events, large-scale killing, damage and destruction were recorded.

When did these three disasters occur? What were the characteristics of these three disasters? What were the impacts of these three disasters?

What is the spatial pattern of damage due to these three disasters? Who were the most severely affected in the disaster? Were they the 'most' vulnerable? What vulnerabilities of the A&N Islands led to these disasters? What were the responses to these disasters? Has the response to the series of disasters made the A&N Islands resilient or more vulnerable? These are some of the questions which this research addresses.

Appendix

Table 1A.1 Identifying ‘disasters’[×] in A&N Islands as per the High Powered Committee on Disaster Management’s Report (2001)

S. No.	*Disaster*	*Occurrence of event*	*Disasters associated with it*
I. Water and climate-related disasters			
1	Floods and drainage management	☑	No
2	Cyclones	☑	No
3	Tornadoes and hurricanes	☒	No
4	Hailstorms	☒	No
5	Cloudbursts	☒	No
6	Heat waves and cold waves	☒	No
7	Snow avalanches	☐	NA
8	Drought	☒	No
9	Sea erosion	☑	No
10	Thunder and lightning	☑	No
II. Geologically related disasters			
11	Landslides and mudflows	☑	No
12	Earthquakes	☑	No
13	Dam failures/dam bursts	☒	NA
14	Mine fires	☐	NA
III. Chemical, industrial and nuclear-related disasters			
15	Chemical and industrial disaster	☐	NA
16	Nuclear disaster	☐	NA
IV. Accident-related disasters			
17	Forest fires	☑	No
18	Urban fires	☑	No
19	Mine flooding	☐	NA
20	Oil spill	☒	No

S. No.	*Disaster*	*Occurrence of event*	*Disasters associated with it*
21	Major building collapse	☑	No
22	Serial bomb blasts	☒	NA
23	Festival-related disasters	☒	NA
24	Electrical disasters and fires	☑	No
25	Air, road and rail accidents	☑	No
26	Boat capsizing	☑	No
27	Village fires	☑	No
V. Biologically related disasters			
28	*Biological disasters and epidemics*	☑	*Late nineteenth century, epidemics*
29	Pest attacks	☒	NA
30	Cattle epidemics	☒	NA
31	Food poisoning	☒	NA
	*War**		*1942–45, World War II*
	*Tsunami**		*26 December 2004, Tsunami*

☑ – Occurred; ☒ – Not Occurred; ❏ – Not Applicable

† 'Disaster' is defined as an event that causes large-scale killing and destruction within a short period of time.

* Not in the list, but occurred in the A&N Islands.

NA: Not applicable in the case of A&N Islands.

Source: Based on Ministry of Agriculture (2001); Directorate of Economics and Statistics (2011); Indian Meteorological Department (1977–2007); and A&N Administration (2012)

2 Epidemics

A disaster of the nineteenth century

One of the earliest recorded disasters of the A&N Islands was the epidemics of syphilis which exterminated more than 90 per cent of the tribal population of the Andaman Islands in the late nineteenth century. It was a disaster of 'mega scale'. The British colonized the A&N Islands in 1858 and simultaneously began to document the islands. It is from these records that the disaster of epidemics has been pieced together.

Here is how it all began. The East India Company set up a colony in the Andaman Islands in 1789 at present-day Port Blair which was then called Port Cornwallis. It was closed down after seven years in 1796 because of high mortality due to sickness and insalubrious climate. The islands were not the focus of the British for the next 62 years. In 1857 they again decided to set up a penal colony in Andaman Islands.

There are five main reasons why the British took possession of the Andaman Islands in 1857. First, the British wanted to expand their territories for exploitation of resources. The Indian mainland, Burma and Sri Lanka were already British colonies. The neighbouring Nicobar Islands were in possession of Denmark, and the Andaman Islands were the 'un-colonized' territory left in the region. The Andaman Islands were a potential resource-rich area that would add to the territories of the British.

Second, the Andaman Islands had a strategic location in the Bay of Bengal because of its proximity to the trade route and trading centres. The Ten Degree Channel and the Strait of Malacca are two important lanes in the shipping route which link the Indian Ocean with the Pacific Ocean. Colonizing the Andaman Islands was definitely an advantage for the British, as this would have consolidated political control over South Asian countries, and the British could thus control trade in the Andaman Sea and the Bay of Bengal.

Third, piracy was extensive in the Bay of Bengal and the Andaman Sea during the nineteenth century. The Andaman Islands were the headquarters for the pirates of Malaysia, who used the well-sheltered harbours of the Andaman Islands as hiding places throughout the year (Portman 1899). The expansion of the British Raj on the Indian mainland, especially after setting up ports at Calcutta and Madras, led to an increase in trading vessels in the sea. Pirates regularly attacked these vessels, hitting the British commercial trade. To control this, the British wanted to set up a colony in the Andaman Islands. Further, the British trading vessels were also attacked by tribes of the Andaman Islands. The Andaman Sea and the Bay of Bengal remain rough for most of the year. In the monsoons, trading vessels took refuge on the Andaman Islands but were often attacked by tribal Andamanese. Attacks in the early to mid-1800s, including in 1819, 1824, 1825, 1839, 1844, 1852, 1855, 1856 and 1857, have been documented (Portman 1899). The frequency of attacks increased in the late 1830s. Therefore there was a need for a safe harbour where ships could berth in rough weather and also provide protection from pirates. It became imperative for the British to set up a colony in the Andaman Islands to guard against Malayan pirates at sea and tribal Andamanese on shore.

The fourth reason that accelerated colonization of the Andaman Islands was the implementation of the Indian Penal Code. Drafted by Lord Macaulay in 1837, it came into effect in 1860 to control the increasing resentment against the growing British Empire in India. In the Indian Penal Code, Macaulay proposed that the criminals be removed from society and from all who are acquainted with them, forever, to a distinct country of which they know nothing and from which they could never escape (Sen 2000). A penal colony is an institution where prisoners are exiled (often located on an island from which escape is difficult or impossible). Before the setting up of the Andaman Penal Colony, Indian convicts were sent to other penal colonies of the British (Figure 2.1).

Around the Andaman Islands there were seven such colonies. Transporting of prisoners and managing the affairs of these penal colonies was difficult as these were either very far from Calcutta, such as Penang, Malacca, Singapore and Bencoolen, while Arrakan, Moulmein and Tenasserim were difficult to reach because of mountainous terrain. Therefore the penal colony at the Andaman Islands was a better option. The British were still contemplating the decision to set up a penal colony, when the final jerk came with the 1857 Revolt, which was the fifth and final reason for the setting up of the Andaman Penal Colony.

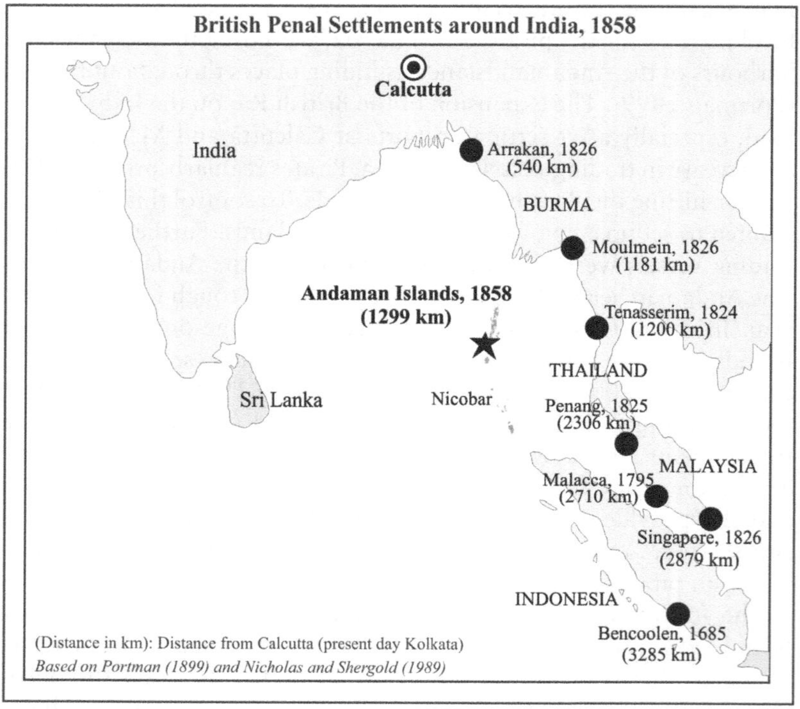

Figure 2.1 British penal settlements around India, 1858

The 1857 Revolt started on 10 May 1857 in Meerut and is said to have ended with the fall of Gwalior on 20 June 1858. It is also known as 'The Mutiny', since this revolt started as a mutiny of sepoys of the East India Company's army in northern India largely in the states of the upper Gangetic plain and central India; it lasted for nearly a year. It emerged as a threat to British rule in India. The British wanted a punishment for the rebels that would deter such uprisings in future. Coupled with it was the prevalent notion among the Hindus in the nineteenth century that crossing the seas entailed the loss of caste status (Sen 2000). Therefore the British saw transportation as a more painful deterrent for Indians. The Andaman Islands were projected as an unknown place inhabited by 'cannibals'. In fact, in the world map prepared by Ptolemy in the second century AD, the Andaman Islands were referred to as an 'Island of Cannibals' (Portman 1899). An extract from Ritchie's *Survey of the Andaman Islands* in 1771 describes Andaman islanders as cannibals who 'eat one another, or at least eat those of their enemies whom they took or kill in battle'. The 'fear of the place' was a reason to transport the rebels to the Andaman Islands. Other colonies were 'known' and hence not 'fearful'.

Thus, colonizing the Andaman Islands served five main purposes for the British: first, it increased the colonial territory of the British; second, it provided the British with complete control over the North Indian Ocean and Southeast Asia; third, it prevented trading losses; fourth, it would be an 'area' for transporting rebels; and fifth, it was to discourage revolts such as 1857 on the mainland in future.

Colonial settlements

During the Mutiny, in November 1857, the government appointed the Andaman Committee to recommend a site for penal settlement in the Andaman Islands. The committee recommended Old Port Cornwallis as the best site for the penal colony. This was the site of the earlier British settlement that was set up in 1789.

With the expansion of the British in India during the eighteenth century, traffic in the Bay of Bengal increased. Three different surveys were carried out in the Bay of Bengal during 1777 and 1788, and all three recommended the Andaman Islands for colonization. As the company was expanding its territorial limits over the mainland, it did not pay much attention to these reports. However, as the cases of piracy and killing of the shipwrecked crew increased, the need for safe harbour was felt.

In 1788 Lieutenant Archibald Blair of the Bombay Marine (the East India Company's navy) was entrusted with the task to search for a site for settlement in the Andaman Islands. After an intense three months survey of the entire Andaman Islands, Lieutenant Blair recommended harbour to the south-east of South Andaman Island as the best site to locate a settlement. The settlement came up in 1789 at the site of present-day Port Blair and was then named as Port Cornwallis, in honour of the then governor general of India, Lord Cornwallis. But in 1792 the settlement was shifted to a larger harbour (present-day Diglipur) in the North Andaman Island to accommodate the increasing number of merchant vessels. This new site was renamed as Port Cornwallis and the earlier one was called Old Port Cornwallis. This new settlement was plagued with sickness and in the monsoon of 1795, 50 people were killed. Therefore in 1796, this settlement was closed and people were shifted back to Penang and Calcutta. The Andaman Islands again were cut off from the rest of the world for 60 years. Due to the unsuccessful colonization attempt by the British, none of the other European powers ventured to colonize it. After the 1857 Revolt, the British government set up the Andaman Committee under the chairmanship of Dr F. J. Mouat to assess the feasibility of a penal colony at the Andaman Islands. The committee after a detailed survey recommended Old Port Cornwallis as the best site for the penal colony, and it was renamed as Port Blair, after Lieutenant Archibald Blair, who established the first colony there in 1789.

On 10 March 1858, Dr Walker, the first superintendent of the Andaman Penal Colony, arrived with 200 convicts of the 1857 revolt, 50 naval brigade men, two doctors and one overseer, giving the Penal Colony a total population of 254. When the British arrived they were quick to carve out and create a 'settlement' which was suited to their ways and means. The British needed roads, tram lines, barracks, a prison, a gallows, a court, bungalows, a secretariat, an army office, a mess, a magazine, polo grounds, a swimming pool, a church, a cemetery and other necessities and luxuries. For this they selected Ross Island, Chatham Island, Viper Island, and parts of South Andaman Island (Figure 2.2).

The British occupied Ross and Chatham Islands for themselves. Ross Island became the headquarters, with government offices and residences for British officers. The main jetty, where ships from the mainland anchored, was constructed at Chatham Island. Opposite to Ross Island, on South Andaman Island at Atlanta Point and Haddo, the convict barracks were constructed. By the end of 1858 Port Blair

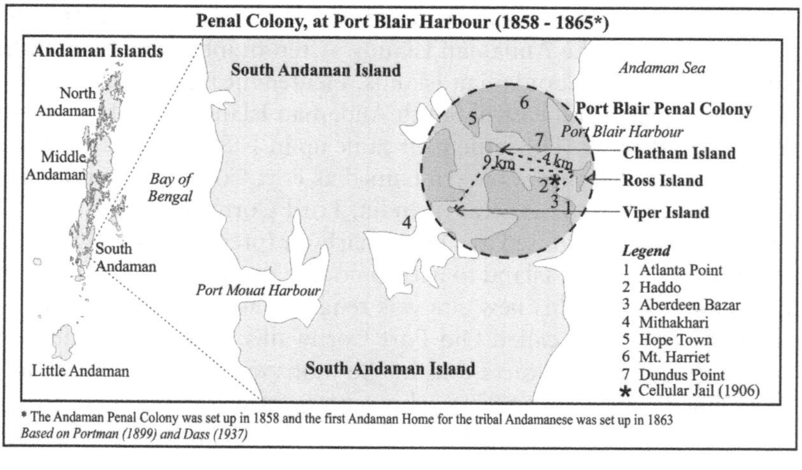

Figure 2.2 Penal colony at Port Blair Harbour (1858–65)

had many barracks and was like an open jail guarded from the sea from all sides (Roy and Choudhary 2002). As more convicts arrived, the population grew and the settlement expanded. In 1863 Viper Island, Aberdeen Bazaar, Mitha Khari, Hope Town and Mount Harriet were cleared. Roads and buildings came up across the settlement and a court, jail and gallows were constructed in Viper Island for convicts who revolted.

The Ross Island, Chatham Island, Viper Island and the cleared parts of South Andaman were known as the Port Blair penal settlement. In 1863 two divisions existed: the administrative headquarters which was at Ross Island, and the convict stations spread across South Andaman Island, Chatham Island and Viper Island. The distance between Ross Island and the nearest point on South Andaman, the Atlanta Point, is only 0.5 km, while Chatham and Viper are at distances of 4 km and 9 km, respectively.

With changing needs of the settlement, the penal colony expanded, and so did the administration. From the modest beginning of one superintendent, 50 naval brigades men, two doctors and one overseer in 1858, the administration grew to 12 departments in 1942. These were Jail, Revenue, Treasury, Law and Order, Supply, Forest, Marine and Shipping, Medical, Education, Postal, Public Works Division (PWD) and Transport. Another special department was the Andamanese Affair which was headed by the officer in charge of the Andamanese. These departments were headed by British officers and had further sub-divisions to exercise control over the people. The entire Administration was headed by one chief commissioner, one deputy commissioner, two assistant commissioners (one for settlement and one for revenue), *tahsildars, patwari, choudries* and *chowkidars*. The administrative offices functioned through a set of buildings and structures which further solidified the power of the British. But the forest and land where the British set up their penal settlement were not 'empty' and 'vacant', but were home to the tribal Andamanese. Portman (1899) noted 'when we reoccupied the Andamans in the latter year, the Aka-Beá-da were far the stronger and more numerous tribe, and had occupied the whole of the land round the Harbour of Port Blair' (Portman 1899: 703–4).

Original inhabitants of Andaman Islands

In 1858, when the British arrived to set up the penal colony, the Andaman Islands were occupied by people of the Negrito race who were hunter-gatherers. The British named them the 'Andamanese tribe' (presently known as the Great Andamanese). On the basis of the study of kitchen middens of the Andaman Islands, Cipriani (1955) concluded that the Andamanese tribe had occupied the islands for 5,000–6,000 years. Kitchen middens are accumulations of refuse, mostly from kitchens. They consist of animal bone, pottery, faeces, shell, botanical material and debris dumped near the site of settlement. The middens are studied by archaeologists. These are a rich source of information regarding the diet pattern of a population, and are the best source to authenticate studies about people in the remote past (Cooper 2002).

Radcliffe-Brown (1922), on the basis of genetic study, concluded that the tribal Andamanese came from the Philippines via Indonesia, Malaysia and Myanmar. With time, they moved to different parts of the Andaman Islands. Gradually contact between tribes was lost; isolation led to differences in the groups, and different groups evolved. In 1858, the British identified two groups: the Great Andamanese, who occupied the Great Andaman Islands (includes North Andaman, Middle Andaman, Baratang and South Andaman), and the Onges-Jarawas, who occupied the Rutland, Little Andaman, interiors of South Andaman and North Sentinel Island.

When the British arrived, various tribal groups of the Andamanese were spread across the entire island and had a 'sizeable population' of approximately 8,000 (Portman 1899). There are several evidences that confirm the 'sizeable tribal population'.

First, each time a ship anchored on any of the islands, the tribal Andamanese were seen. From larger islands such as the North Andaman, Middle Andaman, South Andaman, Little Andaman and Interview Island to smaller islands like Stewart Sound, Havelock, Rutland and Cinque, the tribal Andamanese were spotted. The presence of tribal Andamanese in the small island of North Sentinel (west of South Andaman Island) even today is proof of the size and spread of the Andaman tribal population.

A second evidence comes from the settlement sites discovered when the British explored not just the coast but also the forests of Andaman Islands. Tribal Andamanese settlements were of two types: a communal hut shared by all families, or an assemblage of nearly 10 huts. Subtribes living near the coast preferred communal huts, while sub-tribes living in the forest constructed separate huts. A communal hut was a large structure – 60 ft in diameter and 20–30 ft in height – and accommodated 20 families of three to four persons each. One communal hut supported nearly 80 persons. The individual huts too supported a large population of nearly 150 people. Based on the study of kitchen middens, Cooper (2002) estimated that on an average there were two to five kitchen middens per square kilometre in the Andaman Islands. The total area of Andaman Islands is 6,408 km^2. Therefore based on Cooper's calculation, there would have been 3,204 kitchen middens in all of the A&N Islands. If one settlement site supported 80 people, then the population density would be close to 40 persons per square kilometre before the arrival of the British in 1858. This is similar to the present-day population density of A&N Islands, which is 46 persons per square kilometre (Census of India 2011a).

A third clue to the large number of tribal population is the presence of a variety of tribes. Some sub-tribes were friendly while others were hostile: 'when the present settlement was first opened, we were not aware that there were different tribes of Andamanese, speaking different languages, nor did we know of the divisions of the race into "ar-yauto" (coast dwellers) and "eremtaga" (forest dwellers)' (Portman 1899: 704). The Great Andamanese were divided into 12 sub-tribes and the Onges into four sub-tribes at the time when British occupied the A&N Islands in 1857 (Figure 2.3).

This division indicates that there was a substantial population which could be organized into sub-tribes and local groups. Each of the tribes had demarcated territory for hunting and gathering. Infringement of another tribe's territory led to conflicts. The average group owned a territory of around 40 km^2 (16 sq. mi). The need for division of resources also indicates sizeable tribal population.

A fourth inference about extensive tribal Andamanese population can be drawn from the Nicobar Islands, which is the closest neighbour

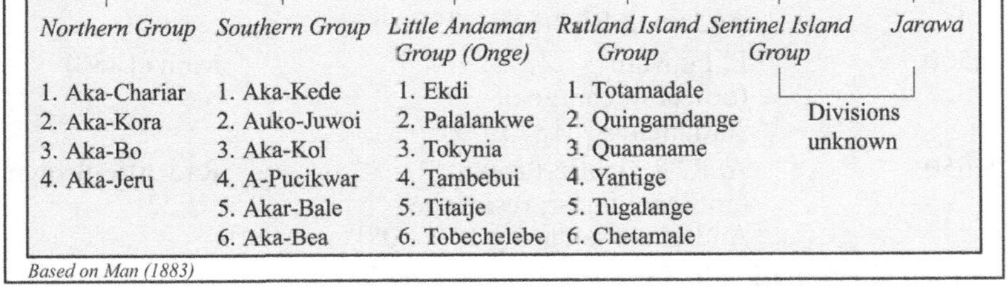

Figure 2.3 Sub-tribe division among the Andamanese, 1858

of Andaman Islands and shares many commonalities. In 1901, the Nicobar Islands supported a population of 6,000. It is an indication that the Andaman Islands, which are 3.5 times larger than the Nicobar Islands, could also have supported a large population.

The fifth evidence is the writings of the anthropologists and the administrators in charge of the tribal Andamanese, who estimated the population, which ranges from 3,500 to 15,000 (Table 2.1).

One of the records states: 'the Islands (Andaman) were *thickly* peopled, before our advent, and each tribe had to keep to its own country' (Portman 1899: 15; emphasis added). Therefore, it is evident when the British in 1858 colonized the Andaman Islands, the tribal Andamanese were large in number and spread across its length and breadth. Colonization brought the tribal Andamanese in close contact with the non-tribal British and the convicts. An important outcome of this contact and reason for epidemic on the scale of a disaster was the setting up of the institution of the 'Andaman Home'.

Table 2.1 Population of Andamanese in 1858

Population in 1858	*Estimate provided by*	*Source of data*
15,000	Dudhnath Tewari (runaway convict, spent a year with Andamanese in 1859)	Portman (1899)
8,000	M. V. Portman (officer in charge of Andamanese, 1879–1901)	Portman (1899)
5,000	Richard Temple (chief commissioner of A&N Islands, 1894–1904)	Imperial Gazetteer of India (1909)
5,000	F. J. Mouat (head of commission for finding suitable site for penal colony in Andaman Islands in 1857)	Mouat (1863)
3,500	E. H. Man (officer in charge of Andamanese, 1875–79)	Man (1883)
5,650	A. R. Radcliffe-Brown (anthropologist, researched Andamanese during 1906–08)	Radcliffe-Brown (1922)
Estimated population in 1858 = approximately 7,025		

Source: Compiled by the author

The Andaman Home

The growth of the penal colony in their territory, deforestation and the rampant attacks on their settlements caused discontent among the tribal Andamanese. They retaliated by attacking convicts who ventured into the forest, either to collect forest produce or to escape from the settlement. In 1863 one of the naval brigades men was attacked and killed by a tribal Andamanese, when the former tried to molest a tribal Andamanese woman. The British captured the killers and put them in confinement on Ross Island near the bazaar. The family members of these convicted Andamanese were allowed to visit and stay with the prisoners and were loaded with gifts of coconut and bananas when they returned back to the forests. In time more and more Andamanese came to visit the prisoners in greed of the 'gifts'. This was the origin of the Andaman Home (henceforth Home(s)). It was a collection of huts enclosed by a bamboo fence, a free asylum to which any tribal Andamanese could be admitted. With time several such Homes came up in the penal colony. These Homes were set up to teach benefits of a sedentary life and agriculture to tribal Andamanese.

The first Home was set up in 1863 and by the end of 1890 there were approximately 15 such Homes; 12 of these were located in South Andaman Island. One each was located at Ross Island, Viper Island and Rutland Island. These Homes were located at the margins of the settlement on the border of the forest so that the tribal Andamanese did not visit the main Port Blair settlement. The size of the Homes varied, and in the 1860s most housed close to 100 tribal Andamanese each. Parallel with the institution of the Homes, the British established the Andaman Orphanage: 'to wean the adult savage from his roaming, desultory life is nearly an impossible task; our only chance of being successful is by turning our attention to the children' (*Man's Annual Report* (1867–68) in Portman 1899). The place was called the Andaman Orphanage, set up in 1869, where the British administration kept the offspring of the tribal Andamanese parents.

Initially the tribal Andamanese were free to visit and stay at the Homes. However, later they were compelled to live in confinement, and forced to wear clothes, cultivate land and learn English and Hindi. In one of the letters of 30 June 1863, two months after the setting up of Homes, Colonel Tytler writes:

> we are in reality laying the foundation stone for civilising a people hitherto living in a perfectly barbarous state, *keeping these people*

> *in our custody as hostages*, for it undoubtedly secures the better behaviour of these inhospitable people towards our Settlement.
>
> (Tytler in Portman 1899: 378)

In the Homes, the tribal Andamanese were exploited in different ways. The British desire to control the A&N Islands and the tribal Andamanese led them to invest in surveys and preparation of maps and charts. The topographical, hydrographical, forest, flora and fauna and the anthropological survey of Andaman Islands were conducted in 1870s and 1880s. For these surveys and other explorations, the British administration used tribal Andamanese in Homes as 'navigators' because they could swiftly make way through the dense forest and sail with ease along difficult coral banks of the islands. Tribal Andamanese knew the place well and they could 'befriend' the tribes which were not friendly to outsiders.

Moreover, the British required forest products, especially bamboo and thatch for the houses of the convicts. They could not trust convicts for this purpose, as convicts used to escape whenever they were assigned work in the forest. Therefore the tribal Andamanese were best suited to fulfil this requirement. Besides assisting in surveys and explorations, the British 'used' the tribal Andamanese as 'domestic servants'.

The proven criminal background of the convicts meant that the British distrusted convicts of the penal settlement to look after their wives and children. A better option was to employ tribal Andamanese. Tribal Andamanese women were employed as nannies because they were affectionate towards children. The men were also employed as domestic servants and engaged in making boats, looking after the gardens, grazing the cattle, rearing pigs and other chores. In the penal settlement, where the population of convicts outnumbered the British, the latter felt secure having the tribal Andamanese close to themselves. Moreover, the most important task in which the tribal Andamanese assisted the British was in 'catching' the runaway convicts. In 1870s, on an average, tribal Andamanese caught 15–20 escaped convicts. The British and Indian guards were unfamiliar with routes and movement through these dense forests. Since the tribal Andamanese lived in and used the forests, they were most suited for the job.

Therefore, by setting up the Homes the British achieved three merits. First, it reduced the daily conflict and attacks by tribal Andamanese on the British settlements. Second, the attacks on the British vessels trading the shores of the islands stopped, and finally, the tribal Andamanese proved useful in surveys, gathering resources, as servants and catching the runaway convicts. But there was a serious flip side which was brewing. While the tribal Andamanese were being 'used' by the British, the Homes became a breeding ground of disasters. The tribal Andamanese were hunter-gatherers. They were put (forced almost) close to the non-tribal population. Within 50 years of the 'contact', 90 per cent of the tribal Andamanese had died. This is the first recorded disaster of the A&N Islands.

Epidemics among the tribal Andamanese

The epidemics of syphilis killed more than 90 per cent of the tribal Andamanese population (Census of India 1931). While syphilis killed the tribal Andamanese, those who managed to survive became vulnerable to other diseases. The population of the tribal Andamanese succumbed to one disease after another within 20 years of setting up of the Andaman Home (Figure 2.4).

From the historical record it is evident that the population of the tribal Andamanese was successively declining after the first case of syphilis was noticed in January 1875. This was within 12 years after opening

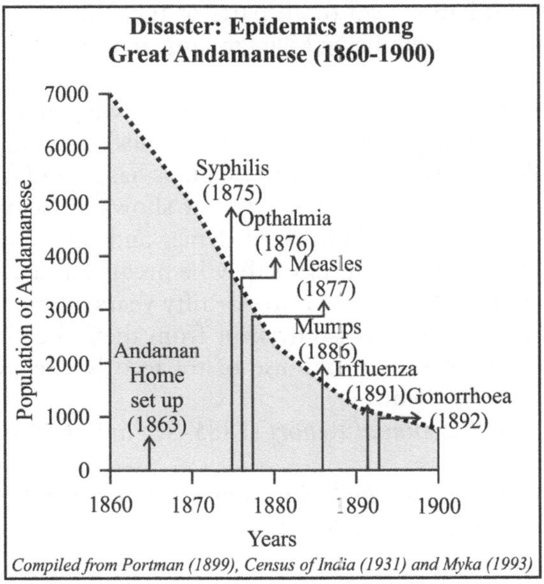

Figure 2.4 Disaster: epidemics among the Great Andamanese (1860–1900)

of the first Home: 'In February 1876 . . . the inmates of the Viper Home . . . were suffering from syphilis' (*Barwell's Annual Report* (1875–76) in Portman 1899: 604). By the end of the nineteenth century, even the children were carriers of hereditary syphilis. The condition was such that some of these poor people were in a shocking condition: 'being covered all over with sores like small-pox pustules, and smell from their bodies was so offensive that they could not be allowed on board the steamer, but were towed in a boat some distance astern [behind a vessel]' (*Cadell's Annual Report* (1884–85) in Portman 1899: 663).

In July 1876, ophthalmia broke out which rendered many tribal Andamanese completely and partially blind. Measles also killed:

> the disease [measles] gradually spread throughout the whole of the Great Andaman, affecting all but the Jarawa tribes, and was nearly a year in doing so. The actual number of deaths from the measles may not have been more than 15 per cent of the whole of the population, but the deaths from the sequel of the disease were many more. Half, if not two-thirds, of the whole of the Andamanese in the Great Andaman died from its effects.
>
> (Portman 1899: 614)

By the mid-1880s, i.e. within 17 years of establishment of Home, syphilis and measles had spread to all parts of the Andaman Islands and tribal Andamanese were being killed in large numbers:

> The friendly tribes of Andamanese are, I regret much to report, *rapidly dying out*. Great ravages were caused among them by an outbreak of measles in 1876, and syphilis has spread from one end of the Island to the other. Every year shows a decreasing population, the old and middle aged dying, and no children coming up to fill their places. It may safely be predicted that the friendly tribes will be extinct come thirty or fifty years hence. Intoxicating drinks have been successfully kept from them, but it is thought that over indulgence in the use of tobacco has tended to cause sterility among them.
>
> (*Cadell's Annual Report* (1885–86) in Portman 1899: 662)

To make matters worse, in 1886 mumps broke out followed by the epidemics of influenza in 1890–91:

> During the year 1890–91 the principal event was an outbreak of Russian influenza which spread rapidly throughout the Islands,

> and the Andamanese, with their constitutions already weakened by syphilis, had no strength to fight the disease, but died in large numbers.
>
> (Portman 1899: 673)

The year after, in 1892, there was an attack of gonorrhoea, and this killed the remaining hope for the survival of the tribal Andamanese. The onslaught of one disease after other did not give any chance to the tribal Andamanese to recoup and revive the population. By the end of the 1880s, the population of the tribal Andamanese had declined by 83 per cent and by the turn of the century it declined by 89 per cent. In 1899, the situation was so bad that

> all people on the Rutland Island and Port Campbell are dead, and very few remain in the South Andaman and the Archipelago Island. The children do not survive in the very few births which occur, and the present generation may be considered as the last of the aborigines of the Great Andaman. Even these have their constitutions to a great extent undermined by hereditary syphilis, and are unable to endure much exposure.
>
> (Portman 1899: 676)

Though the entire Great Andaman was affected by the epidemics of syphilis, measles, ophthalmia, gonorrhoea and Russian influenza, the disease did not spread to all islands at the same time. The spatial pattern of epidemics confirms that the tribes which came in contact with the British were the first to be killed.

Spatial character of epidemics

The epidemics of syphilis affected the Andaman Islands and not the Nicobar Islands. Syphilis was introduced in Home on Viper Island. Measles started on Ross Island among the convicts (who probably brought it from mainland India) and spread to the children staying in the Andaman Orphanage on the island. The Homes were the breeding sites of diseases. A distinct relation exists between the Homes and the percentage of population killed (Figure 2.5). As Homes were concentrated in South Andaman, it was here that 95–100 per cent of the population was killed. There were no Homes in North Andaman Island and thus the sub-tribes of this part registered a 91–95 per cent decline compared to near 100 per cent.

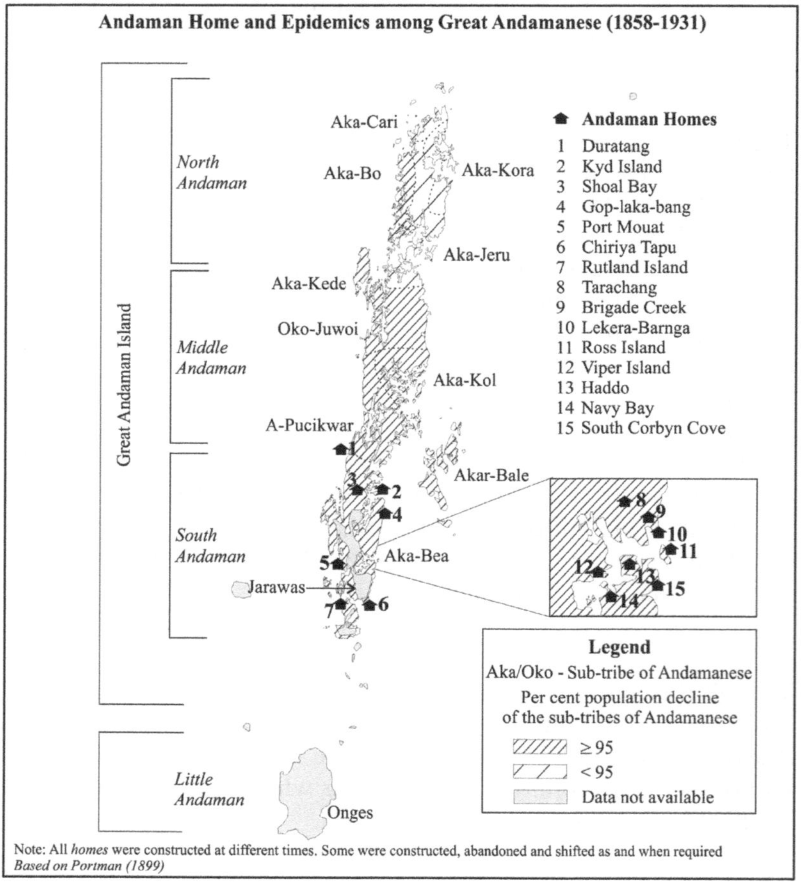

Figure 2.5 Andaman Home and epidemics among Great Andamanese (1858–1931)

By end of 1870s, the high rate of killings was apparent:

> contact with civilization has been marked with the *usual* lamentable result of reducing the aboriginal population; indeed, the death rate, among those *within the area of our influence*, during the past twenty years has so far exceeded the birth-rate . . . the race, at least that portion of it which inhabits Great Andaman, will be well nigh extinct.
>
> (Man 1883: 3)

Table 2.2 More deaths than births in Andaman Homes (1883–90)

Year	*Number of*	
	Births	*Deaths*
1883	7	11
1884	11	28
1885	7	23
1886	2	13
1887	2	13
1888	0	31
1889	2	13
1890	0	20

Source: Based on annual reports of the chief commissioner, in Portman (1899)

A sample from Annual Reports of the Homes between 1883 and 1890 shows that in these eight years, there were hardly any births but many deaths. In fact, in the years 1888 and 1890 there were no births but only 31 and 20 deaths, respectively (Table 2.2).

By 1890, all tribes in South Andaman Island had been killed: 'there appears to have been considerable mortality from the disease' (*Annual Report* (1890–91) in Portman 1899: 673). The islands were becoming depopulated. By the end of 1890, the islands of Rutland, Havelock, Neil, John Lawrence, Tarmugli, Steward, Sound and other islands had no tribal Andamanese population left. The situation became so grave that a stage was reached where not a single tribal Andamanese was available or surviving to work for the British.

A distinct relation exists between the growth of the penal settlement and the killing of the tribal Andamanese. All sub-tribes of the Andamanese did not come in contact with the British at the same time. The ones who first came in contact were rendered vulnerable compared to the others. The tribes living in South Andaman were the first to interact with the British and they were first affected by the epidemic and were the first sub-tribe that were killed. As the contact of the British with the sub-tribes inhabiting the northern parts, such as Aka-Cari, Aka-Kora, Aka-Bo and Aka-Jeru, occurred later, killing among them was delayed. The population of tribes that had no contact with the British were not affected at all. This includes the tribes of Jarawas, Sentinelese and Onges.

Even though the British took possession of the Nicobar Islands in 1866 and set up a penal colony there, the tribal Nicobarese population remained unaffected, because the tribal Nicobarese were not forced into Homes or the Orphanage. The interaction of the British and the Indian convicts with the tribal Nicobarese was limited, and therefore, the disease did not spread among the tribal Nicobarese.

Towards the end of the nineteenth century, two trends of population were observed in the A&N Islands. On one side, the population of the tribal Andamanese was declining rapidly; on the other the non-tribal population of the penal colony grew from 254 in 1858 to more than 20,000 in 1941, which was the last census taken before World War II (Figure 2.6).

The population swelled not only with the number of convicts, but also due to the arrival of Moplahs, Bhantus and Karens. Moplahs and Bhantus were categorized as 'convicts', but under different circumstances. In 1922, from the Malabar Coast of present-day Kerala, 1,885 Moplahs were transported into the Andaman Islands because they rebelled against the local landlords who were being supported by the British administration. Their revolt was classed as rebellion against the British and hence they were sent to the Andaman Penal Colony for life imprisonment.

The Bhantu tribes of Central India were notified as a criminal tribe and were convicted under the Criminal Tribe Acts in 1871, 1910 and 1920. Therefore, during 1924 to 1926, 285 Bhantus were transported to the Andaman Islands. To fulfil the requirement of labourers, 66 Karens

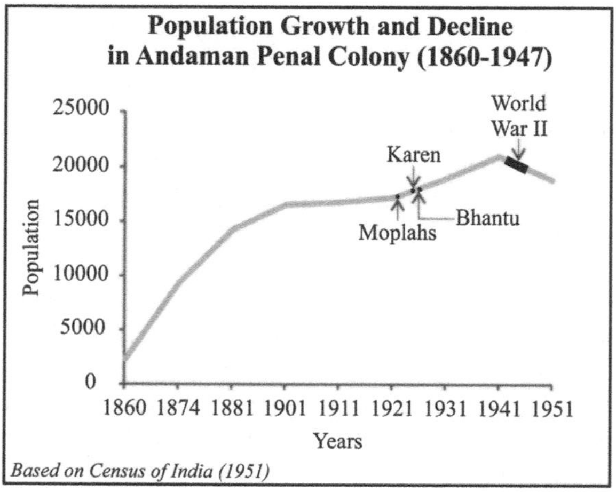

Figure 2.6 Population growth and decline in Andaman Penal Colony (1860–1947)

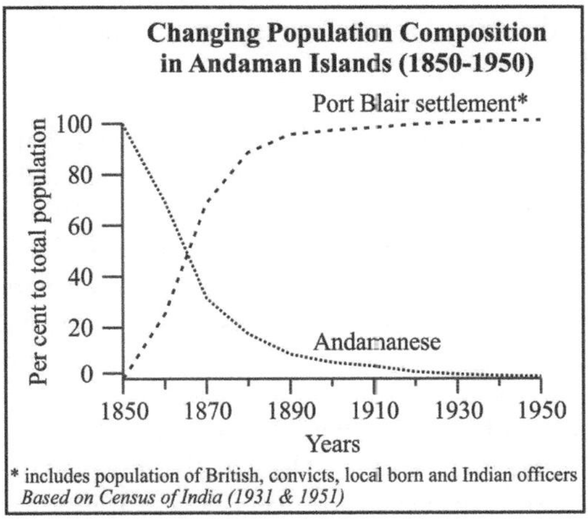

Figure 2.7 Changing population composition in Andaman Islands (1850–1950)

from Burma (present-day Myanmar) were brought and settled in the Andaman Islands. The three groups were later joined by their relatives, thereby increasing the population. This led to a change in the tribal and non-tribal population composition (Figure 2.7).

In 1858, 100 per cent of the population of the Andaman Islands was tribal Andamanese. By 1870 the population of tribal Andamanese was less than 50 per cent of the total population. By 1890, the tribal population was less than 10 per cent and continued to decline. In the 1930s, the Andaman Islands were occupied by the non-tribal that included the British, the convicts, their families and the officers from mainland India, and there were virtually no tribal Andamanese left. Only the Jarawas followed by Onges and Sentinelese remained, making up the Andaman tribal population in 1931. It was the vulnerable population of the tribal Andamanese that were killed while the non-tribal population not only survived the epidemics but also grew rapidly.

When diseases like syphilis, measles, ophthalmia, Russian influenza and gonorrhoea were widespread in the Andaman Islands and began to take a heavy toll, the British documented the reasons for the rampant spread of disaster. They attributed the spread to the tribal Andamanese and classed them as 'unhealthy' and 'unhygienic'. But this is far from the truth, for the records prove that prior to British arrival the tribal Andamanese in fact were a 'healthy' race.

Andamanese: evidences of a healthy tribe

The health of the tribal population can be gleaned from the writings by the British:

> white teeth, bright sparkling eyes, and very well shaped figures . . . epidemics and endemics are absent from the Andaman Islands, unless malarial fevers be classed as the latter . . . there is no marked fever or other disease present, as there are absolutely no signs or symptoms of an abnormal conditions . . . Andamanese by nature are far sighted and there are very few cases of natural deformities.
>
> (Portman 1899: 29)

The adeptness at physical activities is another clear indicator of the health of the tribal Andamanese. The British described the tribal Andamanese as expert swimmers and divers. Their ability to swiftly move through the dense forests was appreciated by the British and the convicts alike. Most significant is the fact that the tribal Andamanese had occupied the Andaman Islands as far back as 5,000 years. Had the tribe been unhealthy, they could never have survived so long. It is clear that it was the British who rendered the population vulnerable by forcing them into Homes and the Orphanage.

Their vulnerability had been rendered in four different ways: first was the 'power' of the British compared to the tribal Andamanese; second, the expansion of the British colony led to shrinking of resources of the tribal Andamanese; third was the master-servant relation of the British with the tribal Andamanese; and the fourth was the construction of Homes.

To boast their power, the British treated the tribal Andamanese harshly. They were invariably referred to as 'savages', 'degenerated negroes', 'wild animals', 'a race of treacherous, cold-blooded murderers', 'wretched outcasts' and 'barbarous people'. The British were cruel in their treatment of the tribal Andamanese and their repressive measures were excessive and terrified the Andamanese. It was to demonstrate this superiority that the British set up their 'symbols' of power in the penal settlement.

The buildings, structures and infrastructures raised by the British were alien to the tribal Andamanese. The land use of Ross Island changed from a dense evergreen forest with thick undergrowth to one of manicured gardens, orchards, tennis courts, cricket grounds, swimming pools and buildings (Figure 2.8). Ross Island could be reached in 15 minutes using the ferry from South Andaman Island (the area

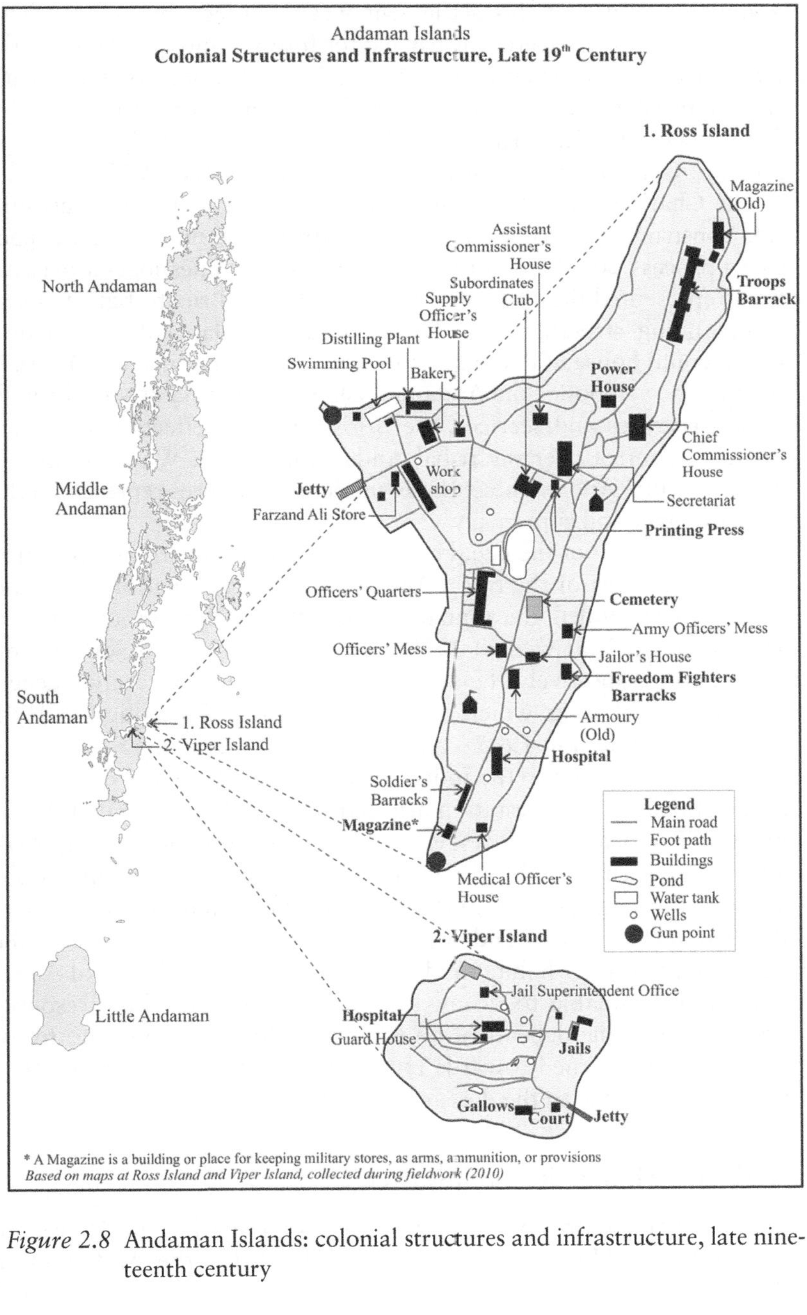

Figure 2.8 Andaman Islands: colonial structures and infrastructure, late nineteenth century

inhabited by the convicts), which is approximately 4 km away. The dominant feature of this settlement was Aberdeen Bazaar. In these locations tramways were constructed for forestry and quarrying work. Jetties were constructed at Chatham Island, Atlanta Point in South Andaman Island, Ross Island, Viper Island and Hope Town. Larger ferries connected Ross Island to Middle Andaman Island, while ships sailed to Calcutta and Madras on the mainland (Dass 1937).

The Chatham Saw Mill was an active symbol of power. This mill was constructed in 1883 and is Asia's largest saw mill (with a capacity to process 20,000 m^3 of timber). Heavy machines logged massive trees represented the control of resources by the British. But the most commanding structure was the infamous Cellular Jail, a structure which could house 668 inmates and which was constructed in 1910. It became the symbol of the Andaman Islands and remains so. Through these structures and services, the British built an effective and commanding control over the tribal Andamanese land. When compared with the tribal Andamanese, these 'modern' structures represented the British power and supremacy.

The expansion of the penal colony in different directions consumed the resource area of the tribal Andamanese. As population grew and the colony expanded, the annual average rate of deforestation between 1869 and 1929, as well as 1930 and 1950, reached 15,300 m^3 and 49,700 m^3, respectively. This is also evident from the revenue generated from lumbering, which increased from Rs. 15.8 million in 1890 to Rs. 58.3 million in 1905 – a 250 per cent increase in 15 years. The timber was exported to the Indian mainland, England and other colonies of England. Deforestation affected the health of the tribal Andamanese. The clearing of forest exposed them to sunstroke: 'Cold, bleak air of the clearings in Port Blair affected the Andamanese injuriously, and was one of the causes of their diminishing numbers' (Homfray in Portman 1899: 551). Land was the most significant resource of the tribes as it was their habitat and source of food. Since the tribal Andamanese were unable to do anything about their shrinking resource area, they became dependent on the British for food. With no forest available, they turned towards Homes for food and shelter and they became vulnerable to the disaster of epidemics.

Andaman Home: door to disaster

The German anthropologist Von Eickstedt remarked in 1926 that the Andaman Home was the door of death for the Andamanese (Dass 1937). Homes and the Andamanese Orphanage were 'new and alien institutions' to the way of life of tribal Andamanese. There are five reasons why the Homes were the door to disaster for the tribal Andamanese.

First, the Homes imposed sedentary life. Before the arrival of the British, the tribal Andamanese were nomads. This practice had two advantages – first, it helped restore depleted resources of the scarce land area and second, it helped maintain hygiene. The tribal Andamanese were not used to living in permanent huts or wearing clothes. Thus, the Homes became filthy and unfit for habitation.

Second, the Homes were a place where the British housed the different sub-tribes of the tribal Andamanese. Before the arrival of the British, the tribal Andamanese were friendly with the neighbouring tribes but in conflict with the tribes far away. In the Homes, hostility among sub-tribes dissolved and marriages began to take place. The interaction among different tribal groups became a 'channel' for the spread of diseases to all parts of Andaman Islands, through marriages and intermingling of the sub-tribes.

A drastic change in the diet of the tribal Andamanese in the Homes is a third means of vulnerability. Prior to settling in homes, the tribal Andamanese diet included pork, dugong, turtle, fish, wild fruits and honey. This food was rich in protein and low in carbohydrates. In the Homes, tribal Andamanese were fed coarse rice and pulses of low quality. The same was fed to the elephants as it was deemed unfit for the convicts (Myka 1993). Mr Homfray, the officer in charge of the Andamanese from 1864 to 1874, noted in 1866 that the deaths of 12 adults of a Home can 'be attributed partly to the locality selected for their Home being recently cleared, and therefore, unhealthy land, and to excess in grain diet, their more accustomed food being pig's flesh and fish' (Portman 1899: 540).

A fourth reason was their interaction with the convicts. The latter were appointed as guards to the Homes. They introduced the tribal Andamanese, adults and children alike, to smoking tobacco, chewing betel nuts and drinking liquor. These 'new habits' weakened immunity and the tribal Andamanese succumbed to diseases.

The physical and sexual exploitation of the tribal Andamanese by the convict guards is the fifth and most important reason of vulnerability. Because the tribal Andamanese caught escaped convicts for the British, the convicts regarded the tribal Andamanese as 'enemies'. Therefore, the convicts treated the latter harshly: 'the Andamanese required to be protected against us, quite as much as we required to be protected against them [convicts]' (Portman 1899: 481). The convict guards not only treated tribal Andamanese with contempt, but also sexually exploited them. Syphilis, the disease which killed the tribal Andamanese, originated in the Homes. The origin of syphilis epidemics among the tribal Andamanese is traced to a convict guard:

> Convict Shera, the senior Petty Officer at the Homes, was proved to be the chief, if not sole, offender, and was remanded to section in the labouring ranks (he subsequently died of syphilis). Further enquiries were made regarding the syphilis among the Andamanese. It was found that a woman from the west coast of the South Andaman had a child aged three years, the father of which was a convict, and the Andamanese admitted that they had been afraid to bring their sufferings to notice because they disliked the restraint they were necessarily subjected to in the hospital and because they were prevented from telling by the convicts, who were afraid of being punished.
>
> (Barwell in Portman 1899: 604)

Thus, the Homes were 'centres' where norms of diet, hygiene and social mingling changed drastically for the worse for the tribal Andamanese population. The offensive behaviour and sexual exploitation by the guards led to the disaster of syphilis epidemics. The disease spread and killed the tribal Andamanese population. What made matters worse was the lukewarm response to the brewing disaster.

The response of the British to the diminishing numbers of the tribal Andamanese was grossly inadequate and insufficient. When the first case of syphilis came to notice in 1876, confinement of the ailing tribal Andamanese was suggested as the best measure to control the spread of the disease. Therefore, the British put the ailing tribal Andamanese in the Andaman Hospital set up in Port Mouat (on the west coast of South Andaman Island) in 1869 to look after their health: 'These measures consist merely in completely isolation those affected with the disease from the rest of the community' (*Man's Annual Report* (1876–77) in Portman 1899: 609). However, the tribal Andamanese did not like confinement in the hospital and escaped to the forest: 'Great difficulty was experienced in keeping these people from going back to their jungle villages, where, of course, they would only help to spread the disease' (Portman 1899: 607).

As there was no way to check the spread of the disease, the tribal Andamanese population had declined to a dismal number, so much so that orders were issued for the closure of the Andaman Orphanage and the Homes. The Andaman Orphanage was started in 1869 and closed in 1896. This was late, as by this time there were no children left at the orphanage. The first Home was started in 1863; all Homes were closed in 1931 and the surviving tribal Andamanese were relocated to Ritchie's Archipelago where they were 'abandoned'. Their population was reduced to 90 individuals.

By 1931 the tribal Andamanese had lost skills, vigour and desire to gather forest resources and had become dependent on the British administration for food, liquor and drugs. In the 1960s they were reduced to 24 individuals and were found begging for food in Port Blair. To protect the few tribal Andamanese in 1969, the Government of India resettled them at Strait Island, a small island with an area of 6.01 km^2 located 74 km away from Port Blair. Here they live on government doles and also go for hunting and fishing around this island only.

Thus, it can clearly be seen that the vulnerability of the tribal population in terms of 'lack of immunity' to alien people, food and ways of living – be it exploitation, confinement, sedentarization or change in diet – exposed the tribal Andamanese tribes to syphilis. The disease took a toll of 7,000 people in a span of 30 years. The population of tribal Andamanese was nearly 7,000 in 1863 when the first Home was set up. It declined to 100 by 1931, with maximum killing confined to 30 years from 1865 to 1895. The killing was confined to the Andaman Islands only and to the Andamanese who 'visited' the Homes. It was the vulnerable tribal Andamanese population that succumbed to the epidemics, while the non-tribal population grew exponentially during same period. The disaster of epidemics among the tribal Andamanese clearly brings forth the case of vulnerability and disasters. The tribal Andamanese were rendered vulnerable by the British. The disaster was inevitable. The non-tribal population of the A&N Islands were 'immune' from the epidemics, but they were not spared either. They became victims of World War II.

3 World War II

A disaster of the twentieth century

The A&N Islands had barely recouped from the disaster of the epidemics of the late nineteenth century when it faced the disaster of World War II (henceforth the War). The A&N Islands were drawn into this War for three and a half years from March 1942 to September 1945. The War was a global military conflict that involved many countries and over 100 million military personnel. More than 70 million people were killed, majority of them were civilians, making it the deadliest conflict in human history. In this War, world nations were organized into two opposing military alliances called the Allies and the Axis. The Allies included the United Kingdom, France and Poland. Later they were joined by Australia, Canada, New Zealand, Newfoundland, South Africa, Soviet Union and the United States of America. The Axis consisted of Germany, Japan and Italy.

The main participants of the War were the Europeans. Many countries of Asia and Africa at this time were colonies of these European powers and hence were drawn into the War without consent. In 1939, British Viceroy Lord Linlithgow declared that India was to enter into the War, and this pulled in the A&N Islands too. During this War, the major strategy of the Axis power was to take control of the territories of the Allies. In Asia, Japan was at war with China in 1937 when the War began on 1 September 1939 after Germany attacked Poland. Japan moved east and took over French Indochina (present-day Vietnam, Cambodia and Laos) in 1940; Japan attacked Australia from 1940 onwards as it was part of the British Commonwealth. The Japanese conquered Malaysia and the Dutch East Indies on 11 January 1942. Japan invaded Burma on 15 January 1942. A month later, on 14 February 1942, they captured Indonesia, and on 15 February 1942 they took control over Singapore and Rangoon by pushing the British westward. The A&N Islands were the last territory captured by the Japanese in their westward movement in 1942.

The disaster: World War II in A&N Islands

The War actually started in the A&N Islands when the Japanese troops attacked it continuously from February 1942 (Dasgupta 2002; Roychowdhury 2004; Sareen 2001). In the A&N Islands the command of the Axis was held by the Japanese while Britain led the Allies. Witnessing the success of the westward annexation by the Japanese, the British sensed that the A&N Islands were in close range. Therefore, in fear and anticipation, the British began to evacuate the A&N Islands from January 1942. The British shifted all its officers and their families from Ross Island to mainland India, leaving behind only a handful of officers, the villagers and the tribal population. They also sent back the freedom fighters from the Cellular Jail to the jails on mainland India.

On 23 March 1942, 13,000 Japanese soldiers took control over the A&N Islands (Dasgupta 2002; Roychowdhury 2004; Sareen 2001). They released the convicts in the Cellular Jail, presuming them to be freedom fighters, which was not the case. In August 1943 the Japanese took control of Car Nicobar Island and stationed 1,000 soldiers there. Although the Japanese had possession of the entire A&N Islands in 1943, they were concentrated in Port Blair in South Andaman Island and Car Nicobar Island. This is because even in 1942, except for the settlement of Port Blair, the Andaman Islands were covered with dense forests. Among the 12 Nicobar Islands, the population was concentrated in Car Nicobar Island. The period of Japanese regime was the time of the War and was a period of disaster. The disaster was characterized by the fortification of the A&N Islands, continuous air raids, famine, and injuries and killings.

Fortification of A&N Islands

By the beginning of the War in 1942, the British had spent 80 years in the A&N Islands, but did not develop it as a military outpost. The A&N Islands were vulnerable to air raids and submarine attacks. During any war, a common practice is a 'scorched earth policy', which is a military strategy that involves destroying anything that might be useful to the enemy while advancing through or withdrawing from an area. In it, all of the assets that are used or can be used by the enemy, such as food sources, transportation, communications and industrial resources, are targeted and destroyed. In some cases, even the people in the area are killed. The 'scorched earth' policy of the British left the Japanese with scanty infrastructure. Therefore the Japanese invested heavily in building air raid shelters and anti-aircraft guns on the coast along with machine guns, cannons, search lights, military posts and bunkers along the seashore.

To facilitate the movement of troops and artillery, the Japanese focused on building transport networks and thus roads, jetties and airstrips were constructed. A circular road was built both in Port Blair and in Car Nicobar Island. The Japanese strengthened and upgraded the jetties of South Andaman Island and constructed the jetty at Mus village in Car Nicobar Island. The British had begun construction of air strip in Port Blair in 1937, but the former left it unpaved. The Japanese paved and lighted the airstrip so that aircraft could land at night too. This airstrip functions as the present-day Port Blair airport. The second airstrip was constructed at Malacca in Car Nicobar Island in 1943 and is under the Indian Air Force. In 2016 this airport was opened for civilians, though the aircraft, the Dornier 228 and the crew belong to the Indian Air Force. These two airports were the only airports in the A&N Islands till 1998. These infrastructures did provide protection but also became the target of the air raids which characterized the period of World War II in the A&N Islands.

Air raids

Constant air raids characterized the War in the A&N Islands. For the three and a half years of the War, the A&N Islands were never free from air raids. Spatially, these attacks were concentrated in Port Blair and Car Nicobar Island (Table 3.1a). Port Blair and Car Nicobar Island each had a nucleus of population and had a garrison, armouries, stores of arms and artilleries. The target of these attacks, both Japanese and British, were military institutions, infrastructure, cargo vessels and war ships. Each of the forces wanted to paralyze and maroon the other. Therefore, in the air raids the little infrastructure that existed and which the Japanese constructed were destroyed. Some of the important infrastructures damaged in the War were the wireless station, Chatham Saw Mill, the marine dockyard, two wings of the Cellular Jail, roads, jetties, government buildings and houses. Aircraft were not spared either. Another target of the air raids was war ships and cargo vessels. These ships were attacked at sea both by air bombers and by submarines. This led to shortages of food, medicine, clothes, oil, petrol, arms and ammunitions for the Japanese forces and the local population, leading to famine in the A&N Islands.

Famine conditions

As supply ships were the targets of the air raids, food imports stopped and famine gripped the A&N Islands. Though the British did initiate agriculture with convicts in the A&N Islands, food was always imported for the penal colony from mainland India and Burma (Sareen 2001). Supplies of food and other necessities was in short supply in 1942. After the British air raids of supply ships, 'people of Port Blair had to subsist on roots, leaves, wild corns and pulses' (Roychowdhury 2004). Malnourishment gripped the people of the A&N Islands and so people suffered diseases like anaemia, beriberi and scabies. Tribal Nicobarese were not spared either. The diet of the tribal Nicobarese consisted of food gathered from the forest such as keveri (traditional bread-fruit), yams, banana, pigs, coconuts and variety of seafood and was thus not dependent on food from mainland India. However, due to shortage and stoppage of supply of food from the mainland, the Japanese soldiers looted the food off the tribal Nicobarese (Roychowdhury 2004), thus leading to famine conditions even in the Nicobar Islands. During the War the population of the entire A&N Islands suffered hunger and malnourishment. The air raids destroyed infrastructure, the garrison and supplies of the defence forces, and injured and killed the civilians.

Injured and the killed

The three main modes of killing used during the War were air raids, spy cases and mass execution by the Japanese. From a sample of the killed in air raids during the War, compiled from various documented sources, it is evident that the highest killing took place at Chatham Saw Mill on 15 August 1943, in which 200 people were killed (Table 3.1).

Table 3.1 Estimated population killed in air raids of World War II in A&N Islands

Air raids (date)	*Target Sites*	*Killed*	*Injured*
1943, 15 August	Chatham Saw Mill bombed	200	150
1944, 14 April	12 planes raided Port Blair	10	14
1944, 2 October	Air raids in Car Nicobar Island	26	Unavailable
1944, 17 October	20 planes raided Car Nicobar Island	Several	Unavailable
	Total	**236+**	

Source: Based on Dasgupta (2002); Lall (1992); Sareen (2001); and Roychowdhury (2004)

In addition, the Japanese tortured and killed local people who knew English, suspecting them to be spies of the British. The Census of India (1961) notes that as a result of these tortures an unknown number of men, women and children were killed for being suspected spies. Nearly 1,200 local people were thus killed during the War.

The third mode of killing was mass execution of the 'non-working' population by the Japanese. The blockage of ships curtailed the supplies of food, medicines and other resources and there was population pressure on the limited resources. So the Japanese ruthlessly mass-executed people incapable of providing hard labour such as women, children, the elderly and the disabled. In the Havelock massacre, in July 1945, a total of 700 people from Aberdeen village, consisting mainly of women, children and aged men were taken to Havelock Island in a ship and were thrown into the sea. Only two among the 700 survived (Roychowdhury 2004). Similarly, in Car Nicobar Island, the Japanese killed several elderly people and children and their bodies were dumped into the forest near the present-day Arong village (Fieldwork 2010). An estimated number of total persons killed in the A&N Islands in the War is about 4,000 (Table 3.2).

In three and a half years, nearly 4,000 people were killed and an unknown number were injured. Added to this was the famine and a crumpled and collapsed infrastructure. Such was the disaster that when the British reoccupied the A&N Islands in 1945, Port Blair was described as a 'ghost town'. The roads were in a state of total disrepair,

Table 3.2 Population killed under Japanese during the War in A&N Islands

Population before Japanese occupation*	*Population after Japanese occupation*	*Population killed (approx.)*	*Based on*
21,316	14,000	7,317	Dhingra (2005)
		3,000	Sareen (2001)
		3,000	Roychowdhury (2004)
20,000	14,000	7,000	Mackenzie, in Dasgupta (2002)
21,000	18,000	3,000	Census of India (1961)
20,000	18,000	2,000	Dasgupta (2002)

*Only non-tribal population.

Source: Compiled by the author

the population exhibited symptoms of severe malnutrition and suffered anaemia, beriberi and scabies. Most buildings were damaged and most houses were either without walls or roof (Roychowdhury 2004).

The disaster lasted for three and a half years in the A&N Islands and came to an end after the War ended in August 1945 and the British reoccupied it. But why were the A&N Islands a target of the Japanese attacks? What made the A&N Islands vulnerable to the attacks and the famine?

Vulnerability of A&N Islands to the War

The A&N Islands were vulnerable before the War. The War accentuated and brought the vulnerability to the forefront and it turned into a disaster. The A&N Islands were rendered vulnerable for different but associated reasons. One of the reasons was the small size of the A&N Islands. For the Japanese it was easier to capture a small outlying island territory of the British than to capture mainland India. The 8,249 km^2 of the A&N Islands was easy prey and the Japanese took control of the islands in less than 12 hours.

Second, the A&N Islands have a strategic location. Situated at the head of the Indian Ocean, the A&N Islands provide access to both Southeast Asia and the east coast of mainland India. Capturing the A&N Islands was advantageous to the Japanese because they could easily target the British aircraft operating out of Burmese bases as well as the British fleet in the Indian Ocean and the submarines in the Strait of Malacca (Sareen 2001). Important cities of British India which were exposed to the missiles launched from the A&N Islands were Kolkata, Chennai, Dhaka and Vishakhapatnam. The Japanese bombing flights attacked Trincomale (on the east coast of Sri Lanka, which was a British colony) and Vishakhapatnam (on mainland India) from the A&N Islands in 1942.

Lack of support to the Japanese by the local population became another reason for vulnerability. In the early twentieth century, the freedom fighters from the mainland India were transported to the Cellular Jail of the A&N Islands. The Japanese believed that the revolutionaries engaged in the freedom struggle of India would support them in overthrowing the British. But pre-empting this, the British had stopped transporting freedom fighters to the Andaman Islands since 1938, and the ones already in the Cellular Jail were transported back to mainland India before the War broke out. When the Japanese occupied the A&N Islands, the Cellular Jail only housed the criminals sentenced for life for heinous offences such as murder, dacoity and rape. However, the Japanese were unaware of this, and immediately after the capture of the A&N Islands they released all prisoners in the Cellular Jail, thinking them to be freedom fighters. But after realizing their mistake, the Japanese captured the convicts again and put them back in the jail (Roychowdhury 2004). Thus the Japanese got no support from either the local people or the prisoners.

One of the most important reasons of vulnerability of the A&N Islands to the War was its lack of fortification. The British had not developed the A&N Islands as a military base. For the British it was only a penal colony. The resources and expenses were spent on managing convicts and using their labour for exploiting resources. The A&N Islands before the War did not offer any military advantages to the British and hence its defences were not looked at. The defence on the A&N Islands during the War was under four heads. The administration of the A&N Islands was under the Government of India. Its military was under the commander in chief of the Indian Army. But its naval and air defence was under the commander in chief, China Station and British Far East Command, respectively. The A&N Islands were transferred to the control of the South-West Pacific Command which delegated command to the general officer commanding Burma, who described the defence situation of the A&N Islands as 'deplorable'. In this power structure, the defence of the A&N Islands suffered.

The British not only neglected the defence, they also ignored the infrastructure and economic development of the A&N Islands. In 1942, before the Japanese took over the A&N Islands, it had spatially unbalanced regional development. Though the British had come to the A&N Islands in 1858, their area of interest was confined to Port Blair. They set up offices, government residences, roads, transport lines, communication networks, ports, factories, hotels, stores of all items like food, clothes, coal and other such items in Port Blair, and hence the population was concentrated there. As per the 1941 Census Report, out of the total population of 21,316 in the Andaman Islands, 19,489 (91 per cent) lived in 79 villages of the Port Blair Area known as Port Blair settlement. The remaining 32 villages of the Andaman Islands were scattered in Middle Andaman Island which was chiefly forest camps and could not be called a proper village. The situation was similar in the Nicobar Islands.

The British took over the Nicobar Islands in 1869 and set up another penal colony in Nancowry Island (one of the islands of the Nicobar archipelago). However, this colony was closed in 1888 because of sickness and the exorbitant cost of maintaining it, after which there was no British presence in the Nicobar Islands. There was no wireless or cable communication and no roads. The only means of communication for the Nicobar Islands with headquarters at Port Blair and the outside world was a steamer that visited four times a year (Imperial Gazetteer of India 1909). The communication links and infrastructure were poorly established even in Port Blair. Dass (1937), who visited the islands, notes that there is neither regular nor frequent communication to the seaports of India and other places. The A&N Islands remained cut off for most part of the year from mainland India, yet it remained over-dependent on the latter. This contributed to its vulnerability.

The A&N Islands were not self-sufficient. The penal colony was completely dependent on the mainland for all its supplies such as food, clothes, medicine, equipment, tools and other household items. More than 80 per cent of the expenditures were on purchase of items of domestic needs (Tables 3.2a and 3.3a). When this link from the mainland was severed during the War, the A&N Islands faced famine. Though the A&N Islands exported timber, cane and other minor forest products like edible birds' nests and trepan, the revenue generated did not even cover half the expenditure costs (Imperial Gazetteer of India 1909). Sherman (2009) calculates that the average cost per prisoner in Port Blair was three times more than in the jails of mainland India.

The British were the only ones that provided the link between the A&N Islands and mainland India. But in time of need, the British deserted the A&N Islands, thus contributing to its vulnerability. Lack of defence of the A&N Islands and Japanese advancement towards the west prompted the British to evacuate the A&N Islands in January 1942. Preference was to be given to the removal of the families of British officers and government servants, which was followed by the evacuation of other convicts if space in the ships was available (Sareen 2001). The focus of evacuation was the safety of the British officers followed by Indian officers and the military. The poor communication with mainland India hampered quick and efficient evacuation of the A&N Islands. In 1942 there was only one steamship, the S.S. *Maharaja*, which made approximately one trip to Port Blair in three weeks (Dass 1937). Evacuation took place from 2 January to 13 March 1942, which is a period of 10 weeks, i.e. the steamship could have made only four trips. The capacity of the ship was not more than 500. Thus, 2,000 people could be evacuated in the given period. One more ship was put in for the process of evacuation that had much lesser capacity than S.S. *Maharaja*. This ship could not have evacuated more than 1,000 people. Therefore approximately only 3,000 persons were evacuated, leaving behind the majority who were left 'unprotected' and were vulnerable to attacks and air raids. The population that was

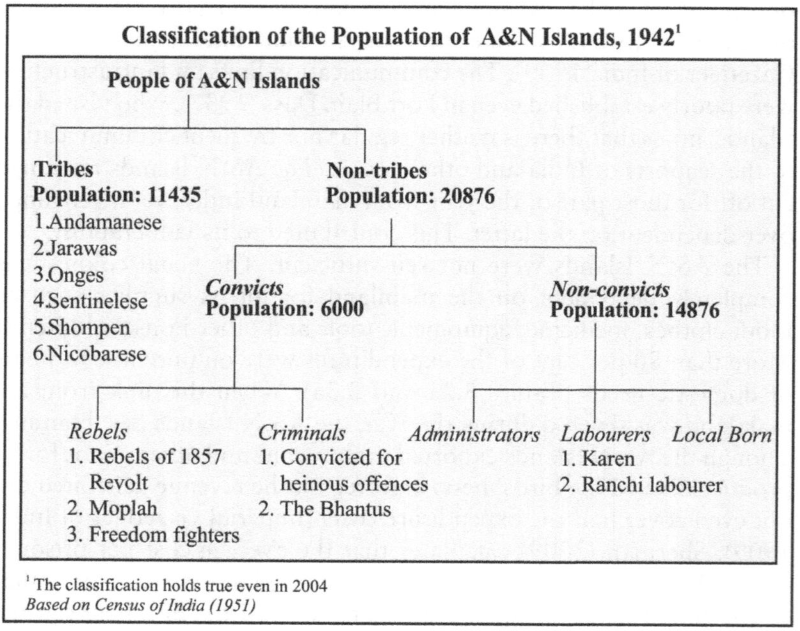

Figure 3.1 Classification of the population of A&N Islands, 1942

left behind was already vulnerable, and after the evacuation of the British and the government officers they became more vulnerable.

At the time of the War, the A&N Islands were inhabited by different groups of vulnerable people. These were the tribes and the non-tribes which can be further classified (Figure 3.1).

There were six different tribes, namely, Andamanese, Jarawas, Onges, Sentinelese, Shompens and Nicobarese; non-tribes included the convict population and the non-convicts. The total population including the tribes in 1941 was 32,311 (Census of India 1941). Of this, total tribal population was 11,435 (Nicobarese: 10,745; other tribes: 690) and non-tribal population was 20,876. There were nearly 6,000 convicts and 14,876 non-convicts. Nearly 3,000 people, all non-tribal, were evacuated before the War. Therefore, the total population immediately before the Japanese took over the A&N Islands was about 29,311. Different groups were vulnerable in different ways.

The tribes: The tribal population of the A&N Islands was 35 per cent of the total population in 1941. The Tribal Nicobarese formed 94 per cent of this group as by the early twentieth century the Andamanese tribal population had dwindled to less than 20 individuals because of disaster of epidemics. Other aboriginal tribes like Jarawas, Onges, Sentinelese and Shompens were there but few. The Jarawas occupied the western margin of the South Andaman and Middle Andaman Island, while the Onges, Sentinelese and Shompens were confined to Little Andaman Island, North Sentinel Island and Great Nicobar Island, respectively. The tribal Nicobarese were spread in 12 of the Nicobar Islands. The vulnerability of the tribes to the War was determined by their interaction with the British. The tribal Nicobarese of Car Nicobar Island were under the British and Christianity had begun to spread, therefore, they became the target of the Japanese attacks during the War. The tribal Nicobarese, subsisting on a hunting, gathering and collecting economy, could not defend themselves and succumbed to the onslaught of the Japanese.

The non-tribes: The non-tribal population constituted 65 per cent of the total population of the A&N Islands before the War. The population was concentrated in and around Port Blair. The non-tribal population was not a homogenous group. The vulnerability of the non-tribal population could be scaled from most vulnerable to least vulnerable. The most vulnerable were the convicts.

The pre-War non-tribal population of the A&N Islands can be divided into two groups – the convicts and the non-convicts. The A&N Islands were set up as a penal colony, thus, the convicts formed the major section of the non-tribal population. The first group of convicts to arrive in the Andaman Islands were from the United Province and the Central Province. Later convicts were brought from different parts of India. The convict population was dominated by convicts from Bengal and Punjab (Table 3.3).

Table 3.3 Source of origin of convicts

Name of State/Province	*1910–20*	*1932–38*
Bengal	38	331
Punjab	81	3
Uttar Pradesh	11	10
Bihar	0	18
Maharashtra	3	0
Madras Province	0	3
Delhi City	0	1
Total	133	366

Source: Based on Majumdar (1975)

The convict population was also of two types – rebels and criminals. The rebels were convicted for revolting against the British Raj. As a form of severe punishment, they were transported to Andaman Penal Colony either for a very long time or for lifetime imprisonment.

The rebels could be classified into three types on the basis of the time and nature of their rebellions. First were the rebels of the 1857 Revolt. The genesis of the Andaman Penal Colony owes its origin to this revolt on mainland India. After the 1857 Revolt, 200 convicted rebels were sent to the Andaman Islands in March 1858 as the first batch of convicts. By June 1858, the number of rebels rose to 773 and in December 1858 it swelled to 1949.

After rebels of the 1857 Revolt, the next batch was the rebels of the Moplah Rebellion of 1921–22. The Moplahs are the Muslim peasants of the Malabar region of Kerala. They revolted against the oppressive Hindu landlords. As the British officers were protecting these landlords, the rebellion was seen as an act against the British Raj. The revolt spread through the late nineteenth and twentieth centuries and finally in 1921 the Moplahs were defeated. In the 1921 uprising about 2,337 Moplahs were killed, 1,652 were injured and 5,995 were captured, while another 39,348 surrendered (Ansari 2008). With the arrests of this rebellion, numbers swelled in the Madras jail. To ease the burden of the Madras jail and to discourage such revolts in future, the Moplahs were sentenced to life imprisonment and sent to the Andaman Penal Colony. In 1922, therefore, 1,885 Moplahs including 714 women went to the Andaman Islands and were settled around Port Blair.

A third kind of rebels were the freedom fighters. With the rise of nationalistic sentiments in the late nineteenth and early twentieth centuries, groups of revolutionaries began attacking the colonial regime, and with the partition of Bengal in 1905, these attacks increased. The British labelled these revolutionaries as 'terrorists' and transported them to the Cellular Jail. The later became infamous as 'Kalapani', the most dreaded form of punishment. Freedom fighters of the Maniktola Bomb Case (1909), Delhi Conspiracy Case (1915), Lahore Conspiracy Case (1915) and Manipur Conspiracy Case (1918) were transported to the Andaman Islands.

A second type of convicts were the criminals. Two types of criminals were sent to the A&N Islands: first, those who committed heinous offences like dacoity, causing grievous hurt, burglary, theft and rape; second, those convicted under the Criminal Tribe Acts. The Government of British India passed Criminal Tribe Acts in 1871, 1910 and 1920. As per the act, nomadic communities were classified as being 'addicted to the systematic commission of non-bail able offenses', and under this act 150 tribes and castes were notified as criminal tribes (Coomar 1997). The Bhantu of United Province and Central Province was one of the notified 'criminal tribes'. The colonial government decided to transport 285 Bhantus, including 135 women, to the Andaman Islands under the care of the Salvation Army during 1924–26 (Coomar 1997). They too were settled around Port Blair in South Andaman Island.

In the War, the convicts were one of the most vulnerable non-tribal groups, because they lived in confinement and had no freedom to build protection against the attacks. They were like bonded labourers who were dependent on the British for food, clothing and shelter. When the British evacuated, they were the worst sufferers.

The non-convicts too were vulnerable. The non-convicts consisted of three groups: the administrators, the labourers and the 'local borns'. They came to the A&N Islands along with the convicts in 1858. Initially they were minuscule in number but later grew, and by 1891 they accounted for nearly one-fourth of the total non-tribal population. Their numbers further increased and in the 1941 census they constituted 71 per cent of the total non-tribal population (Table 3.4).

Among the three groups of non-convicts, the administrators were the least vulnerable. The administrators were commissioned to the A&N Islands to manage the convicts, the 'local borns' and the tribal population. The British and the Indian government officials and their families were housed in Ross Island, the administrative headquarters of the Andaman Penal Colony. This group was least vulnerable to the War as they were in charge and knew about the Japanese intention to capture the A&N Islands, and thus were the first to evacuate.

Table 3.4 Composition of non-tribal population in the Andaman Penal Colony (1874–1901)

	Convict population	*Non-convict population*	*Total non-tribal population*
1874	7,569 (82)[1]	1,663 (18)	9,232 (100)
1881	11,452 (81)	2,746 (19)	14,198 (100)
1891	11,738 (75)	3,822 (25)	15,560 (100)
1901	11,947 (74)	4,159 (26)	16,106 (100)
1941	6,000 (29)	14,876 (71)	20,876 (100)

1 Figures in parentheses represent per cent of total.

Source: Based on Imperial Gazetteer of India (1909)

The second of the non-convicts were the labourers brought to the A&N Islands by the British to extract the timber from the dense forest. The estimated yield of high-quality timber was 10,000–12,000 tons per year (Imperial Gazetteer of India 1909). The convicts could not be used as forest labourers because they were agriculturalists and thus not adept at clearing the jungle. Moreover, convicts had a tendency to flee when they were sent into the forest for work. Therefore, to exploit forest resources, the British 'imported' labourers from Burma (present-day Myanmar) and Ranchi (present-day state of Jharkhand in mainland India). A group of 166 Karens, an ethnic group inhabiting southern part of Burma, were brought to the Andaman Islands as labourers in 1925 and they settled in Middle Andaman Island. They were employed with the forest department to extract Andaman hardwoods. As the terrain and climate of the Andaman Islands was similar to Burma, the Karens adapted and their number swelled to 263 by the 1931 census. Labourers from Ranchi were 'hardy and capable workers'. They too were employed by the forest department and became a part of the Forest Labour Force in 1925 (Dass 1937). The labourers were vulnerable to the attacks during the War as the Andaman Islands were a new territory for them. They lived in the forest camps and were unaware of the impending Japanese attacks, nor did they have resources to secure themselves and thus were left alone to face the Japanese onslaught after the British evacuated.

The most vulnerable among the non-convicts were the 'local borns' who dominated the non-tribal population before the War. In the A&N Islands, 'local born' or 'local' is the term used for the population born out of convict marriages. The British wanted to create a stable society in Andaman Islands to provide cheap labour for exploiting the resources. Therefore in 1871 the British introduced a system of classification of convicts into six classes on the basis of their behaviour and conduct in the Andaman Penal Colony (Table 3.4a). The rank of a 'self-supporter' was the highest rank a convict could attain with good behaviour for ten years. They were given 'ticket-of-leave', which meant that they could live as free individuals in villages, could call wives from mainland India or could marry a female convict, raise a family and live life like normal citizens, but within the boundary of the penal colony only (Dhingra 2005). In haste to create a stable society soon, the prison terms to obtain 'ticket-of-leave' were shortened for the Moplahs and Bhantus. British also supported the growth of the non-tribal population and built houses for the self-supporters, relaxed rules of visiting Port Blair, opened schools, hospitals, and primary health centres, and granted permission to build religious places. The conducive environment led to the growth of the population of the progeny of the convicts who were known as 'local born'. Describing the 'local born' population, the Imperial Gazetteer of India (1909) notes: 'Like every other population the "local-born" comprise every kind of personal character . . . On the whole, considering their parentage, the "local born" population is of a much higher type than might be expected'. The 'local born' were the most vulnerable non-convict group. They were born and brought up in the Andaman Islands and had no connection with mainland India or place of origin of their parents. On mainland India, they were looked down upon as descendants of criminals transported for life (Dass 1937). Dass notes, 'they were socially boycotted by most of the Indian officials mainly for three reasons: first, for being born to convict parents, second, for not strictly adhering to orthodox customs of caste and religion, and third, for their moral weakness' (Dass 1937: 68). Therefore, when the War broke out and the capture of the A&N Islands was imminent, they had no place to go. The British left them to fend for themselves. At the time of the War, thus, the A&N Islands were inhabited by the vulnerable people who were deserted by their 'masters', the British. But one of the most significant reasons of the vulnerability of the A&N Islands to the War was that it was a British colony.

The A&N Islands – a British Colony

One of the crucial reasons for the disaster of War in the A&N Islands was that it was a colony of the British. Being part of the Allies, they were prone to attack by the Axis and vice versa. Had the A&N Islands not been a territory of British, the Japanese would not have attacked them. The Japanese attacked only Port Blair in South Andaman Island and Car Nicobar Island, while the remaining 570 islands remained unattacked and uncaptured. This proves that the Japanese were only interested in capturing the British territory. Thus, the A&N Islands could not escape from the disaster of War as it was a British Colony.

The A&N Islands neither had a role in the cause of the War nor were the islanders a willing participant in it. But because of their vulnerability, they were drawn into it. The islanders were mute participants who were attacked again and again, first by the Japanese and later by the British. Both the British and the Japanese troops had support from their 'administrations', but the people of the A&N Islands had no support. All through the history of the A&N Islands, neither were the people allowed to build their defences nor was it built by the powers controlling it. The end result was the disaster. Even after the disaster, the response provided by the British was 'lukewarm'.

Response of the British to war

The Japanese were defeated in the War in August 1945 and the British took formal possession of the A&N Islands on 26 September 1945. On this day, the S.S. *Sandra*, a British steamship, reached Port Blair with civil supplies and medicine. The first post-War response was to abolish the penal settlement in October 1945. The 6,000 convicts whose terms were still left were given remissions and were allowed to go back to their home in India. Of this number, 4,200 convicts were repatriated to India; the others preferred to stay behind. Second, the British administration repaired and reconstructed roads, hospitals, drains and houses. Third, several schemes were launched for relief, such as free distribution of food and fuel for 30 days, free distribution of clothes, pension for every family that suffered due to the War and an interest-free loan, up to a maximum of Rs. 2,000, repayable over 15 years. All schemes included almost everyone in Port Blair and Car Nicobar Island. Thus, post-disaster response of the colonial government was concentrated in only two worst affected places, Port Blair and Car Nicobar Island, and these two places became the hub of growth and development.

The A&N Islands came under British India in 1945 and remained so till India's Independence in 1947. Barely were the A&N Islands recouping from the impact of the War when another disaster hit mainland India and its impact was felt in the A&N Islands. In 1947, India got Independence, but the nation was partitioned. With Independence came in refugees or the partition migrants. Some of these migrants were settled in the A&N Islands, initiating another chain of events which rendered the A&N Islands vulnerable to disasters. What was the impact of the Partition of India on A&N Islands? How did the Partition of India make A&N Islands vulnerable to disasters? These questions are explored in the next chapter.

Appendix

Table 3A.1 Significant air raids on A&N Islands during the War

Attack by	*Date of bombing*	*Target*	*Disaster*
Japanese	1 February 1942	Aberdeen Jetty	Missed the target but bombed Port Blair
Japanese	5 February 1942	Four merchant vessels including HMIS *Ramdas*, the main cargo carrier of the A&N Islands	Loss of food and other cargo that affected the supply; an important link between mainland India and A&N Islands severed
Japanese	16 February 1942	Port Blair raided for 65 minutes by three bombers	Affected the settlement area of Port Blair
Japanese	February 1942	Four merchant vessels attacked	The supply of food and medicine was affected
Japanese	February 1942	Port Blair attacked by air raids as well as submarines	Settlement area of Port Blair affected
Japanese	13 March 1942	Ship *Maharaja* attacked	An important link between mainland India and A&N Islands severed; no more evacuation possible
British	22 March 1942	Air raid; all of Port Blair attacked	Damaged many buildings
British	July 1943	Japanese cargo vessel attacked by near Ross Island	Vessel sunk with supply of food, medicine, ammunition and other items
British	15 August 1943	Chatham Saw Mill attacked	200 died and 150 wounded; the southern end of the mill destroyed by direct hit

(*Continued*)

Table 3A.1 (Continued)

Attack by	*Date of bombing*	*Target*	*Disaster*
British	3 September 1943	An anchored Japanese supply ship in Car Nicobar Island attacked	Scarcity of food and medicine
British	September 1943	Hospital and other buildings of Car Nicobar Island attacked	Many people were killed
British	14 April 1944	12 planes raided Port Blair	10 people killed and 14 severely injured; shipyard blasted, one ship destroyed and telephone cables cut at several places
British	2 October 1944	Air raid in Car Nicobar Island	26 people were killed
British	17 October 1944	20 planes raided Car Nicobar Island	Several killed
British	July 1945	Frequent air raids in Port Blair and Car Nicobar Island	Several killed

Source: Based on Dasgupta (2002); Lall (1992); Sareen (2001); and Roychowdhury (2004)

Table 3A.2 Expenditure of Andaman Penal Colony (1890–1905)

Nature of expenditure	*Heads of expenditure*	*Expenditure (Rupees in thousands)*		
		1890–1	*1900–1*	*1905–6*
Supplies	Clothing for convicts and police	33	38	58
	Purchase of stores	36	58	120
	Subsistence money to convicts	92	71	88
	Salaries, establishment and contingencies	120	138	123
	Passage money and freight on stores	163	173	175
	Forest establishment and supplies	175	285	500

Nature of expenditure	Heads of expenditure	Expenditure (Rupees in thousands)		
		1890–1	1900–1	1905–6
	Commissariat establishment and supplies	414	610	718
Sub-total		*1,033*	*1,373*	*1,782*
Service	Education	5	5	6
	Ecclesiastical (Church)	8	8	10
	Jail	10	24	28
	Marine	12	22	25
	Medical	37	40	44
	Police	137	146	140
Sub-total		*209*	*245*	*253*
Building infrastructure	Public works	26	71	109
Sub-total		*26*	*71*	*109*
Other	Other charges	13	15	17
	Tea cultivation	16	30	25
Sub-total		*29*	*45*	*42*
	Total expenditure	**1,297**	**1,734**	**2,186**

Source: Imperial Gazetteer of India (1909)

Table 3A.3 Expenditure heads of Andaman Penal Colony

Nature of expenditure	Per cent of total expenditure		
	1890–1	1900–1	1905–6
Supplies	80	79	82
Service	16	14	12
Building infrastructure	2	4	5
Other	2	3	2
Total expenditure	100	100	100

Source: Based on Table 3.2a

Table 3A.4 Classification of convicts in Andaman Penal Colony

Class	*Composition*	*Degree of freedom*	*Kind of labour*	*Payment*	*Duration*
VI	Invalids Superannuated convicts Females	–	Did not work	–	–
V	Convicts degraded from higher classes for serious offences or for running away	Work in heavy iron	Severe	No payment	–
IV	Newly arrived convicts Degraded from higher class Promoted from lower class	Work in light iron	Severe	No payment	Six months for new convicts
III	Passed through probationary course	Free movement within sub-division during the day and confinement at night	Hard	No payment	Four and half years
II	Spent five years in penal colony with good conduct	Free movement within sub-division during the day and confinement at night	Employment in petty official positions in hospitals, public office, servants	Paid labour	Five years
I	Spent 10 years in penal colony – were the trustworthy convicts and were given 'ticket-of-leave' or became 'self-supporter'	Free movement within the villages – free living within the penal colony	Given land to cultivate, engage in trade, or craftsmanship	Earn his own living	Till end of sentence

Source: Compiled from Imperial Gazetteer of India (1909); Sen (2000); and Dhingra (2005)

4 The 1947 Partition of India and vulnerability of the Andaman and Nicobar Islands

The A&N Islands were barely recouping from the impact of the World War II when India became independent in 1947. With Independence also came Partition. On the basis of religion India was divided into two sovereign states: India and Pakistan. In this 'Partition of India', Punjab on the west and Bengal on the east were divided. The Partition of India was a disaster in terms of sheer numbers killed and displaced. In the riots that broke out across the country, close to 200,000 people were killed and more than 15 million people were displaced from Punjab and Bengal (Talbot and Singh 2009). Of these 15 million, 11 million were from Punjab and four million people were from Bengal. Both the Indian side of Punjab and Bengal faced an acute problem of accommodating refugees. However, the problems of refugees in Bengal was distinctly different and perhaps much more difficult when compared with Punjab. This was because refugees of Punjab came in one large tide, but in Bengal they came intermittently, every year from 1947 to 1956 (Table 4.1).

This added four million people to the already overpopulated West Bengal. The refugee problem did not abate till 1971, when Bangladesh was created. In 1981, refugees constituted one-sixth of the population of West Bengal (Census of India 1981). The problem of illegal immigrants from Bangladesh still plagues West Bengal, Assam, Meghalaya and Tripura.

Second, in Punjab almost equal numbers moved across borders while in Bengal large numbers of refugees came from across borders into India while only small numbers moved out from the Indian part of Bengal. Thus, in the case of Punjab, land and resources were left behind by the outgoing migrants and space was available to accommodate the incoming refugees, which was not the case for Bengal. There was no space available to accommodate the streams of refugees in Bengal (Chatterji 2007).

Table 4.1 Estimated number of refugees in West Bengal (1947–56)

Year	*Number of refugees (in millions)*
1947	0.344
1948	0.786
1949	0.213
1950	1.575
1951	0.187
1952	0.200
1953	0.076
1954	0.118
1955	0.24
1956	0.32
Total	**4.059**

Source: Based on Kudiasya (1996)

Third, large sections of refugees from Punjab were affluent industrialists and landowners. They also had political connections in Delhi, the capital of the India. This meant the refugees of Punjab were given compensation of funds and land. The refugees of Bengal were landless and poor, thus they hardly received any compensation. The majority of the refugees, who came to Bengal after 1947, were primarily the 'Namasudras' (Chatterji 2007). The 'Namasudras' is the name of the Hindu community originally from Bengal. They were traditionally engaged in cultivation and as boatmen. They served the rich landlords who had migrated before and immediately after Independence. These poor refugees neither possessed land nor did they receive any compensation when they arrived in India.

Prior density of population also made resettlement of Bengali migrants difficult. In Punjab the population density was 111 persons per square kilometre, while in Bengal it was 311 persons per square kilometre (Census of India 1941). Therefore, Punjab could 'accommodate' surplus population, while West Bengal had to disperse refugees. Space was a serious constraint where Bengal was concerned. The refugees had to be displaced to be settled.

The Government of India set up a committee and it was decided that the exodus of four million refugees in Bengal would be dispersed to 'uninhabited' parts of India. Thus, 88,000 families were resettled in Tripura; 25,209 families in the Dandakarnya region of east central

India; 4,000 families in present-day Uttarakhand; and 3,695 families to the A&N Islands (Kudiasya 1996). There were five significant reasons for selecting the A&N Islands, an archipelago 1,200 km away from mainland India, for resettlement.

Reasons for selecting A&N Islands for resettlement

The first reason was that the population of the Andaman Islands was 'depleted'. During the colonial period, the epidemics had killed 99 per cent of the population of the Great Andamanese, the original inhabitants of the Andaman Islands. Approximately 10 per cent of the population of Port Blair was killed during World War II. Again, as the penal colony of the Andaman Islands was abolished in 1945, the convicts were deported back to mainland India. As for other tribes, they were confined to few islands only – Sentinelese in North Sentinel Island, Onges in Little Andaman Island and Jarawas on the western margin of South Andaman and Middle Andaman Island, Shompens in Great Nicobar Island. Nicobarese were confined to Nicobar Islands only. The population in the 1940s was concentrated in Port Blair, Car Nicobar Island and Chowra Island. The 'depopulated islands' 'appeared' suitable with a potential to absorb surplus population of the refugees of West Bengal, which was already overpopulated.

The second reason was the low population density of the A&N Islands. In 1947 the A&N Islands had a population density of 3.64 persons per square kilometre, a density 100 times lower than West Bengal and India which was 311 and 103 persons per square kilometre, respectively (Census of India 1941). So the committee set up for the resettlement of population were sure that the A&N Islands could accommodate the surplus population of refugees in West Bengal.

The third reason was the similarity of climate of the Andaman Islands and that of West Bengal. The administrators and policymakers were convinced that families could adjust well because the climate of the Andaman Islands was similar to that of West Bengal, with hot summer followed by three months of heavy monsoon rainfall. The only difference was that the Andaman Islands did not have a winter season as experienced in West Bengal. The similarity of climate would support cultivation of paddy and thus made the A&N Islands suitable for the Bengali refugees.

The fourth reason was the geostrategic location of the A&N Islands in the Bay of Bengal. The Cocos Island of Burma is 30 km north of Landfall Island of the A&N Islands, while Indira Point, the southernmost point of the A&N Islands, is 150 km north of the northern tip of Sumatra Island, Indonesia. The Government of India was keen to populate the islands in the fear that any other neighbouring countries might occupy them. Therefore, they hurriedly 'populated' them with the Indian population to maintain the claim over the A&N Islands.

The fifth reason was that the A&N Islands were 'perceived' to be rich in natural resources. The Andaman Islands have 29 commercially exploitable species of trees, and more than 95 per cent of them were under forest cover in 1947. As the A&N Islands are fringed by coral reefs, both the Andaman Sea and the Bay of Bengal are rich in fish. Thus the A&N Islands' strategic location, potential resources, low population density and 'availability' of land made them viable for resettlement of the Bengalis after the Partition of India. The resettlement was carried out under various schemes.

Schemes of resettlement in A&N Islands

The resettlement in A&N Islands took place under four schemes – (1) resettlement of partition refugees from West Bengal (1949); (2) 'colonization scheme' (1952); (3) development of plantation and resettlement of Sri Lankan Tamils (1965); and (4) was resettlement of ex-servicemen (1969).

Refugee families from West Bengal were resettled under the first scheme. The scheme was in operation for three years, from 1949 to 1952 and was replaced by the 'colonization scheme' in 1952. The colonization scheme was drawn up by the interdepartmental committee for relief and rehabilitation under the Ministry of Rehabilitation in the report 'Possibilities of Colonisation and Development of the Andaman and the Nicobar Islands'. The report is also known as the Shivdasani Report and was passed by the Cabinet in 1952. This scheme proposed for the resettlement of 4,000 agriculturalist families (especially, the Bengali partition refugees) over 20,000 acres of land in five years. The scheme was sponsored in the First Five-Year Plan under the Ministry of Rehabilitation, and was to be carried out by the A&N Administration. This scheme continued till 1961; by that time a total number of only 2,882 families were resettled in the A&N Islands.

Among these families, 197 families belonged to the *Ranchiwalas* who were 'imported' to the A&N Islands as skilled labour. The resettled families (henceforth resettlers) were not familiar with the dense forest and could not clear the forests because originally they were agriculturalists and were thus not adept at working through the jungle. Hence, the A&N Administration brought labourers from the tribal population of the Chotanagpur plateau. They were known as

Ranchiwalas. It was a similar method as adopted by the British during the colonial period. The other non-Bengali families were brought from Tamil Nadu and Kerala, where climatic conditions and cropping patterns were similar to the A&N Islands.

The resettlement of the Sri Lankan Tamils repatriated from Sri Lanka under the Indo-Ceylon Agreement of 1964 initiated the third resettlement scheme in the A&N Islands. The Indo-Ceylon Agreement, also known as the Sirima-Shastri Pact, was an agreement deciding the fate of stateless 'plantation' Tamils in Sri Lanka. During the nineteenth and early twentieth centuries, the British recruited a large number of Tamils of Indian origin to work in tea, coffee, rubber and coconut plantations in Sri Lanka. By 1946 their numbers had grown to 780,000, forming 11.7 per cent of the population. After the Independence of Sri Lanka in February 1948, the government of Sri Lanka passed the 'Ceylon Citizenship Act' in August 1948 and declared the Tamil plantation workers as 'stateless' and their citizenship has been a subject of continuous dispute between Sri Lanka and India. After 11 rounds of discussions between India and Sri Lanka, the Indo-Ceylon Agreement (1964) was signed under which (1) 525,000 persons would be granted Indian citizenship and progressively repatriated to India over a period of 15 years; (2) 300,000 persons would be granted Sri Lankan citizenship during the same 15-year period; (3) both repatriation and granting of Sri Lankan citizenship phased in over 15 years would, as far as possible, keep pace with each other in proportion to relative numbers; and (4) the status and the future of the balance of 150,000 persons were to be the subject of separate agreement between the two governments (Kanapathipillai 1995).

A Special Area Development Plan was initiated by the Ministry of Rehabilitation in 1964 for the development of 'such special areas as may be indicated by the Prime Minister from time to time'. The 'special areas' are those areas where for climatic, geographical or other reasons, economic and social development has been retarded and which though sparsely populated is richly endowed with natural resources (Ministry of Rehabilitation 1965). In 1965, the Special Area Development Plan and the resettlement scheme were clubbed together. The fourth and fifth Five-Year Plans put forward resettlement of 10,000 families on 77,000 acres of land. The A&N Islands are on the same latitude as Tamil Nadu and Kerala and share a similar climate, and therefore initiating rubber plantation in Katchal Island was proposed and the repatriated Sri Lankan Tamils were sent there to develop rubber plantations (Ministry of Rehabilitation 1965). It was believed that this would solve two problems – first, the repatriated Sri Lankan

Table 4.2 Years of resettlement and number of families resettled (1949–80)

Types	*Years of resettlement*	*No. of families resettled*
Bengali refugees	1949–75	3,695
Ex-servicemen	1970–80	330
Ranchiwala	1959–61	197
Keralite	1954–61	183
Tamil – Sri Lankan	1965	72
Burmese	1965	50
Tamil – Indian	1955–60	43
Mahe, Pondicherry	1954	4
Total	**1949–80**	**4,574**

Source: Based on Directorate of Economics and Statistics (1965) and Biswas (2009)

Tamils could be resettled and second, the rubber plantation would lead to economic development for the A&N Islands.

The India-China War of 1962 brought the need for populating the border areas, and again in 1969 the retired servicemen from Indian Army known as 'ex-servicemen' along with their families were settled in Great Nicobar Island under the fourth scheme of resettlement which lasted till 1975. Therefore the resettlement in the A&N Islands proceeded at various phases from 1949 to 1980 under the four major schemes. The total number of families resettled under all these schemes from 1949–80 was 4,574 (Table 4.2).

The refugees from East Bengal account for 81 per cent of the resettlers. The families from other parts of India together accounted for 19 per cent of the total. Different areas of A&N Islands were selected and cleared to resettle the families under various schemes.

Location of resettlement

The families were resettled in different locations of the A&N Islands but the core was Port Blair (Figure 4.1). The 1949–52 resettlers were settled around Port Blair. Under the colonization scheme, settlers of the first period (1953–55) were settled in Ferrargunj (in South Andaman Island) and Rangat (the southern part of Middle Andaman Island). In the second period (1956–61), families were settled in northern part of Middle Andaman Island and Diglipur (in North Andaman Island). The major factor in the relocation was the presence of a natural harbour, as shipping was the only mode of transportation in that period.

Under the third scheme, resettlement started in the remaining parts of the Andaman Islands and in Katchal Island (part of the Nicobar Islands). The fourth scheme opened the Great Nicobar Island (Dhingra 2005). The resettlement led to drastic changes in the socio-economic situation of the A&N Islands.

Changes in A&N Islands as a result of resettlement

The most significant change brought by resettlement was the increase in population, which also led to increase in the number of villages, infrastructure, agricultural land, livestock population, administration, imports and expenditures. In 1947, the population of the A&N Islands was only 30,000, which increased to more than 63,548 in 1961, an increase of 105 per cent. The percentage of population growth was very high all throughout the resettlement period (Table 4.3).

By 1981 the population had increased to 188,741, and in 2001 (the last census before the Tsunami of 26 December 2004) it was 356,152. In these 54 years, from 1947 to 2001, the population of the A&N Islands increased on three accounts: (1) resettlement; (2) natural growth and (3) in-migration of the opportunists.

Resettlement was the first reason of this increase. A total of

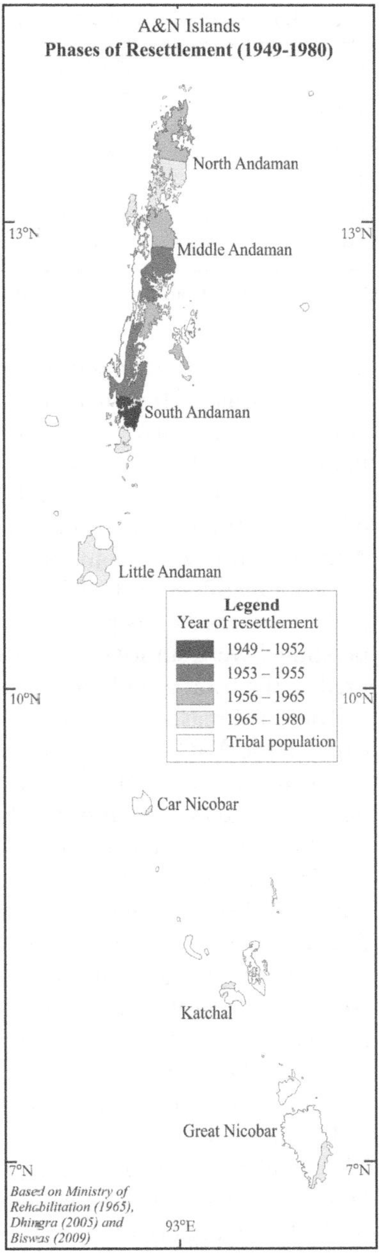

Figure 4.1 A&N Islands: phases of resettlement (1949–80)

Table 4.3 Percentage population growth in A&N Islands (1951–2001)

Years	*Andaman*	*Nicobar*	*A&N Islands*
1951–61	158.33	21.27	105.19
1961–71	90.81	48.77	81.17
1971–81	69.35	40.57	63.93
1981–91	52.54	28.74	48.70
1991–2001	30.14	7.19	26.94

Source: Based on Census of India (2001a)

4,574 families were resettled in the period between 1949 and 1980. At an average of five members per family, nearly 22,870 people were added to the population from 1951 to 1981. During 1951–81 the population of the A&N Islands increased by 509 per cent.

The second cause of the swell in population was the natural increase of this resettled population together with the 'local born' population. Adding to this was the mobile population of the government employees from mainland India who preferred to relocate to the A&N Islands and contributed to its population growth. Another 5,400 armed forces personnel were stationed in the A&N Islands in 1981 which contributed to the total population.

The third reason of increase was in-migration of opportunists who came to the A&N Islands. The educated people were attracted to 'government jobs' with higher pay (island special duty allowances), residential quarters, free ship travel to the mainland once a year and other benefits. The uneducated people were required for the rising labour requirement for clearing forest lands as well as for the construction works. Fishermen got opportunity and trade was profitable. Thus, the A&N Islands became a lucrative destination for both educated and the uneducated people. With supportive government policies, the A&N Islands became the prospective destination for the opportunity seekers. Therefore, along with the officially resettled families, almost an equal number of families went to the A&N Islands in search of better jobs and employment (Dhingra 2005).

The population increase became the trigger factor for other changes. The number of villages increased from 201 in 1951 to 491 in 1981. During 1951–61, 198 new villages were created in the A&N Islands. In the following decade, 1961–71, only nine new villages were added. In the next two decades, 1971–81 and 1981–91, 101 and 13 new villages were added, respectively. Thus a relation exists among resettlement, growth of population and increase in the number of villages.

Table 4.4 Infrastructure in A&N Islands (1951–81)

Infrastructure type	*Number*	
	1951	*1981*
Metalled road (in km)	190	648
Wharfs and jetties	5	40
Power generation capacity (in MW)	1.1	7.5
Airport/airstrip	2	2
Schools/colleges/training institutes	40	252
Hospitals/health centres	18	80

Source: Based on Directorate of Economics and Statistics (1965) and Census of India (1981)

Resettlement also led to increase and expansion of infrastructure (Table 4.4). Investments in transport and other infrastructure have remained high in all 11 Five-Year Plans. It has fluctuated from 54 per cent in the Second Plan to 83 per cent in the Fourth Plan. With resettlement and better infrastructure, land under agriculture also increased from 2,000 ha in 1941 to 50,000 ha in 1981. As the A&N Islands do not have any large-scale industry, they have to import goods such as medicine, clothes, footwear, food items, milk powder, electronic items, vehicles, construction material and other items. There has been an immense increase in the import of not only food grains but also of all manufactured goods. The planners envisioned to make the A&N Islands an economically rich and self-dependent territory. However, over the last 70 years, the A&N Islands have become completely dependent on mainland India for its sustenance.

The arrival of refugees to the A&N Islands happened in different time periods and they occupied different islands. In the first and second phases, the Andaman Islands were resettled and the Nicobar Islands in the third phase. While the Government of India had drawn the plan for the resettlement scheme and transportation to the A&N Islands, they saw the empty islands, resources and geopolitical significance, but what they failed to see and realize was the tribal population, its special needs and its way of life. What was introduced was a pattern of development which imitated that of mainland India. Moreover, even this development was spatially unbalanced. The resettlement led to the vulnerability of the A&N Islands.

Impact of resettlement on vulnerability of A&N Islands

The resettlement process contributed to the vulnerability in three significant ways – (1) the population exceeded the carrying capacity of the A&N Islands, (2) resettlement of vulnerable population and (3) by marginalization of tribal population.

Population exceeded carrying capacity

In the A&N Islands, the population growth during the decade of 1951–61 was 105.19 per cent and for the Andaman Islands, it was 158.33 per cent. In the Nicobar Islands, the growth of population doubled in the decade of 1961–71 compared to the decade of 1951–61. This immense population growth was much higher than the carrying capacity of the A&N Islands. The carrying capacity of the A&N Islands was calculated by Murthy (2005) that took into account land, climate, water resources and levels of agricultural technology. As per this study, the A&N Islands can support a population of 160,000 at a density of 3 persons per hectare of agricultural land with intermediate levels of agricultural technology. The maximum, with highly advanced levels, could go to 5 persons per hectare, i.e. a population of 250,000 (Murthy 2005). As per the Census of India (2001a), the population of the A&N Islands was 356,265, which is more than twice the carrying capacity at intermediate level of agricultural technology. The increase in population was not the only one factor that led to the vulnerability to disasters; the other significant factor was that the people who were resettled were already vulnerable and carried their vulnerabilities to the A&N Islands.

Resettlement of vulnerable population

Refugees from the Partition of Bengal, repatriated Sri Lankan Tamils and the ex-servicemen formed the bulk of the resettlers. The Bengalis formed 80.78 per cent of the total families resettled, and included in the 'depressed class' list prepared under the Government of India (Schedule Castes) Order of 1936. They were poor, marginal farmers or landless labourers who served the rich landlords. But after Partition 'they abandoned the little they possessed in the east only when they were driven out by extreme violence or by intolerable hardship . . . they fled for their lives, taking nothing with them, temporary refuge on the other side of the border [Kolkata]' (Chatterji 2007). They were already vulnerable and thus suffered immensely in the Partition disaster. They had not even recouped from the trauma of the Partition and their losses when they were resettled in the A&N Islands. Thus, their resettlement was not out of choice, but as they had no other option they agreed to be resettled in the A&N Islands.

Similar was the situation of the Sri Lankan Tamils who were repatriated as they became 'stateless' in Sri Lanka. They were born and brought up in the plantations of Sri Lanka and then suddenly it was declared that they did not belong to the place and thus were repatriated. Though they worked in the tea, coffee and rubber plantations, they were labourers with no rights over the land. They were 'thrown out' of their country of birth. They too were living in this trauma of being 'stateless' when they were resettled in the A&N Islands. Here too, they were not given all the 'promised' benefits, but were left on their own in remote islands with no basic amenities.

The successive batches of resettlers were given fewer benefits than earlier ones. The batch of 1949 was given 10 acres of cultivable land. The 1950 batch was given '5 acres of plain land' and the 1951–52 batches were offered only '5 acres of jungle covered land'. The ex-servicemen resettled in Great Nicobar Island were given 10 acres of land for cultivation and the families employed by the Rubber Board were given a half acre of land for homestead. As resettlers initially were 'supported' by the A&N Administration, they became dependent and lacked the initiative to develop the A&N Islands as they had come under force and not by choice. But, they did increase in numbers and also with the support of the government they encroached on the land and resources of tribal population, the original inhabitants of the A&N Islands, and marginalized them in their own island.

Marginalization of tribal population

The A&N Islands are home to a tribal population. Resettlement did not take place over 'vacant land'. It was 'home' to the six aboriginal tribes, four in Andaman Islands and two in Nicobar Islands. The tribes of the Andaman Islands were Sentinelese, Great Andamanese, Onges and Jarawas, while Shompens and Nicobarese occupied Nicobar Islands. Among these tribes, the Sentinelese tribe are confined to North Sentinel Island (a small island with an area of 59.67 km^2 lying on the west of South Andaman Island); they attack anyone who venture near their island and thus have been left alone. Their numbers are still not precisely known. The Great Andamanese population had dwindled to less than 30 by 1947 and were found begging in Port Blair. To protect them, in 1969 the Government of India resettled them at Strait Island, an island with an area of 6.01 km^2 located 74 km away from Port Blair. Presently the A&N Administration provides food, clothing, education and medical facility to the Great Andamanese. Their population was already on the verge of extinction after the epidemics of the nineteenth century, and therefore the resettlement post-Independence did not affect them. Among the tribes of the Andaman Islands, it was the Onges and the Jarawas who were severely affected by the resettlement.

Before resettlement, the Onges were spread across the Little Andaman Island. The core area of Onges was Hut Bay, as this was a safe harbour with an abundance of fresh water. These very characteristics drew the administrators at Hut Bay and it became the centre of resettlement in Little Andaman Island. But 'the areas selected were in the middle of the Onge lands, or that the Bay on which the Andaman Labour Force built their jetty wharf, and breakwater was Onge territory' (Dhingra 2005). In the mid-1960s, resettlers comprising 858 families, 399 from the mainland, 165 from Car Nicobar Island, and 119 from South Andaman Island and Middle Andaman Island, were moved to Little Andaman Island (Directorate of Economics and Statistics 1965). In 1961 the non-tribal population of Little Andaman Island was only 129, but swelled to 7,214 by 1981.

After usurping the prime land and displacing the Onges, the administration demarcated a reserve area for them and made settlements which were similar to refugee settlements, with a school, dispensary, well, pigpen and coconut plantation. This was completely in contrast to the way of life of the tribal Onges and has pushed the Onges to the verge of extinction: 'shattered by the swift destruction of their habitat, though they are in greater numbers than the [Great] Andamanese, the Onges, in a similar sense, are a threatened tribe' (Dhingra 2005).

The largest among the tribes of the Andaman Islands (following the disaster of the epidemics of the nineteenth century), the Jarawas were also marginalized. The core territory of the Jarawas was on the western margin of South Andaman Island and Middle Andaman Island but they roamed the entire forests to collect food and shifted location as and when required. Jarawas did not accept the invasion of their territory mutely. They fiercely attacked the settlers when non-tribal villages were set and forests were being cleared. Between 1948 and 1973, there were 66 attacks by Jarawas in which humans, pets, cattle, dogs and elephants were injured or killed (Krebs 2007). The Jarawas retaliated to the resettlement in their 'home land'. However, instead of retreating from Jarawa territory, the Inter-Departmental Team which was constituted for the development of the A&N Islands in 1965 termed the area as 'Jarawa-infested' and stated that,

> even after years, we have not been able to establish any contact with them. The Jarawas have been uniformly hostile to all outsiders with the result that half of Middle Andaman is treated as a *Jarawa infested* area where it is difficult for any outsider to venture.
> (Ministry of Rehabilitation 1965)

And the resettlement continued unabated.

In the Nicobar Islands too, the Nicobarese and the Shompens were affected by resettlement. The Shompens occupied the eastern margin of the Great Nicobar Island. When the resettlement of ex-servicemen was planned, the Shompens were forced to retreat to the forest. A Shompen hut-complex was constructed at Campbell Bay in Great Nicobar Island where Shompens come as and when required to collect ration and clothes.

The Nicobarese tribe accounts for 97 per cent of the total tribal population of the A&N Islands. Where the number of other tribes does not exceed 400, the Nicobarese are close to 30,000. The strong presence of tribal council and the Andaman and Nicobar Islands (Protection of Aboriginal Tribe) Regulation of 1956 has given the Nicobarese some protection. However, with the third and fourth schemes, a non-tribal population was permitted into the Nicobar Islands. In the colonial period in Car Nicobar Island, the Japanese had constructed the airstrip which was taken over by the British Royal Air Force after World War II. It was transferred to the Indian Air Force in 1956. This allowed non-tribal population in the Car Nicobar Island. Resettlement in the A&N Islands became a critical factor which contributed to the vulnerability of the tribal Nicobarese population later during the Tsunami disaster.

The impact of the resettlement was such that the area occupied by tribes has shrunk by more than 90 per cent in the Andaman Islands from 1947 to 2001 (Figure 4.2). In 1947, tribes occupied large part of the island, while the non-tribes were confined to areas around Port Blair. But by 2001, the area of the tribes had shrunk while the area for the non-tribes had expanded. The change is more prominent in the case of the Andaman Islands. In the Nicobar Islands, the area of Nicobarese has reduced, only slightly. It was not that the planners were unaware of the plight of the tribal population of the A&N Islands, but with the massive pressure to resettle the refugees and the 'excess' population, they ignored the needs of the tribal population.

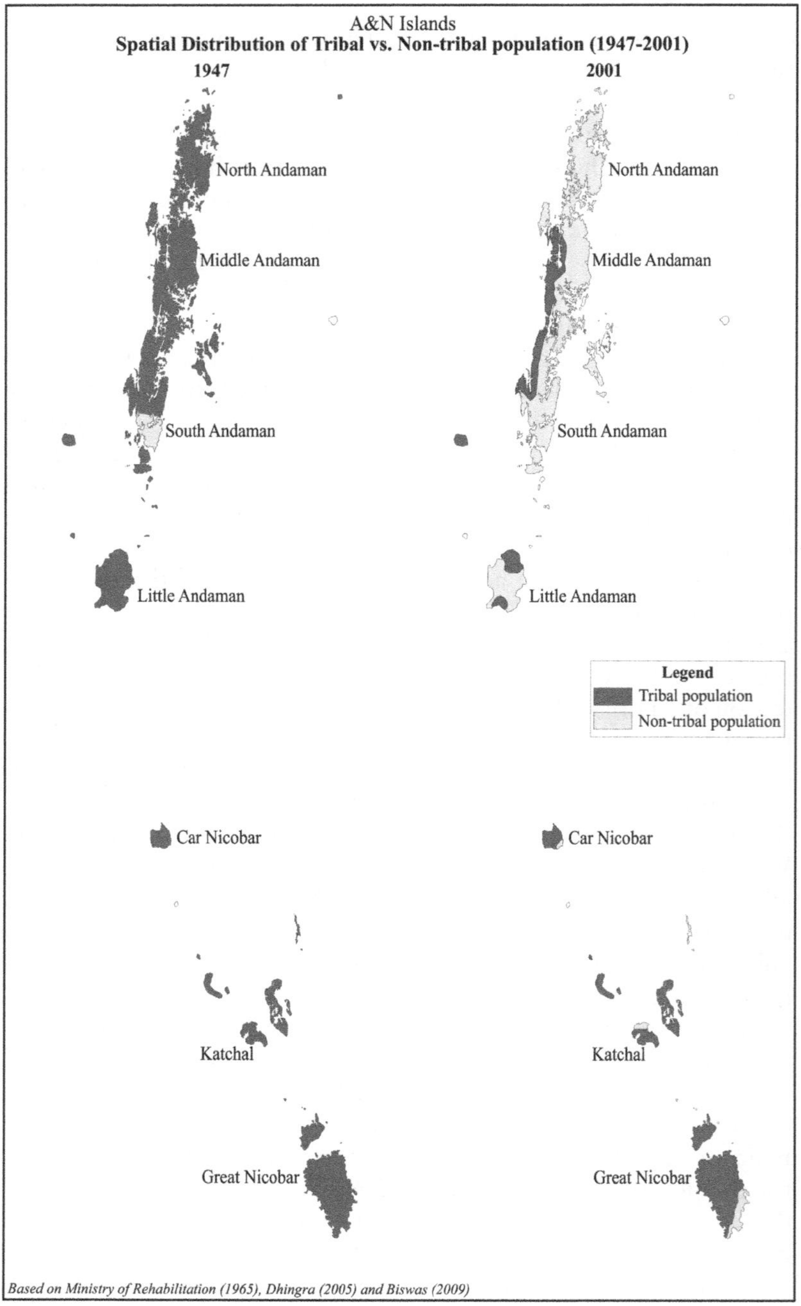

Figure 4.2 A&N Islands: spatial distribution of tribal vs. non-tribal population (1947–2001)

In order to protect the tribal population, the Andaman and Nicobar Islands (Protection of Aboriginal Tribe) Regulation (henceforth 'Regulation') was passed and implemented by the Government of India in 1956. The Regulation conferred power to the chief commissioner to designate areas 'predominantly inhabited by aboriginal tribes as reserved areas'. The tribes were confined to few areas and large part was opened up for the non-tribal population. The Sentinelese and Great Andamanese were confined to North Sentinel Island and Strait Island, respectively. The Onges were allotted two pockets in Little Andaman Island while the Jarawas were restricted to the western margin of South Andaman Island and Middle Andaman Island and the Shompens got confined to the central part of Great Nicobar Island. As the Nicobarese lived in small islands, where large-scale resettlement was not possible, the entire Nicobar Islands were declared a 'protected area', except for the airfield and port of Car Nicobar Island, port of Kamorta Island and eastern coast of Great Nicobar Island.

In the 'protected' area, entry of non-tribes, purchase of land by non-tribes and trading by non-tribes was prohibited. Penalties were fixed in case the regulation was violated. With the exception of the government officials on duty, entry into the reserved area was prohibited. Power was conferred to officers (all police above rank of constable and all officers of the forest department) to arrest without warrant persons who commit offences punishable under the said regulation.

In principle, the Regulation aimed at protecting the aboriginal tribes of the A&N Islands; however, in reality, several amendments soon followed the Regulation which made the tribal population vulnerable. These amendments can be grouped into three categories: (1) those pertaining to the change in the boundaries of the reserve areas; (2) those that deal with the relaxation for government officials and defence personnel to enter the reserve area; and (3) those that regulate licences for business and trading in the reserved areas. These amendments diluted the importance of the original Regulation of 1956. For example, the amendment that families of government officers do not need a 'tribal pass' to enter the reserve area was misused, and many unauthorized individuals took up residence in the reserved area.

A gross violation of the Regulation was the construction of the Andaman Trunk Road (ATR) through the Jarawas Reserve Area (Figure 4.3). The construction of the road started around 1969 and was completed in 1989. The stretch of 129 km out of the total 360 km runs north to south and cuts across Jarawas reserve forest. The road brought vulnerability to the Jarawas in the way similar to the Andaman

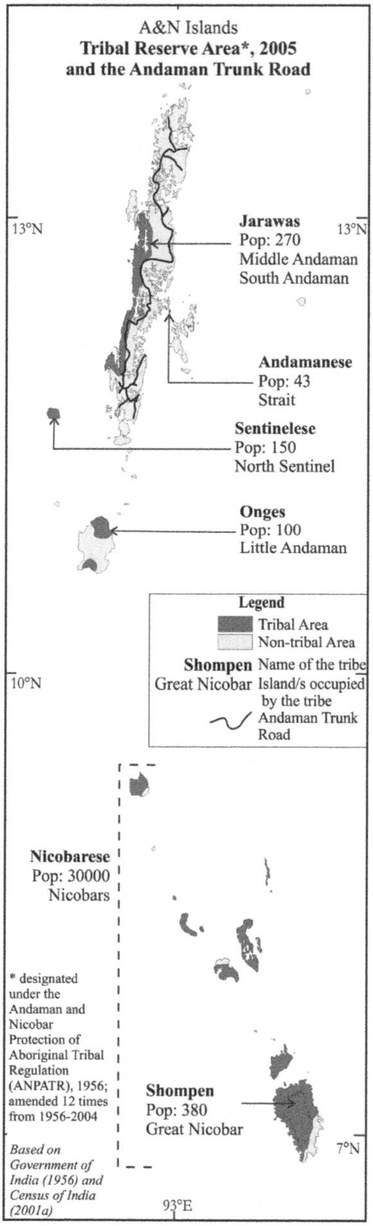

Figure 4.3 A&N Islands: tribal reserve area, 2005 and the Andaman Trunk Road

Homes had created vulnerability for the Andamanese. The ATR brought the Jarawas in close contact with the non-tribal population and exposed them to disease and other forms of exploitation. In the 2000s the common sight on ATR was Jarawas begging to cars for tobacco and betel nuts. Some Jarawas also bartered with honey and forest products. This inter-mixing for Jarawas and the non-tribal was banned, and a 'convoy system' was introduced in 2004 to pass through the Jarawa reserve on the ATR. However, to date 'Jarawa sightseeing' forms a part of the itinerary of the tourists bound for Baratang Island and beyond.

Because of the ATR, anthropologists fear that

> this might be the beginning of the induction of a culture of consumerism and materialism that is so contradictory to their [Jarawas'] traditional means of livelihood and also very challenging in relation to their indigenous behavioural patterns towards each other and the equal distribution of resources around them.
>
> (Dhanuja 2003)

Most of the Jarawas have become addicted to chewing betel nut (*paan*), and they began to barter forest produce for the intoxicants. Prior to 1997, Jarawas were 'hostile' to any outsider entering 'their' forest. In 1997 Jarawa boy 'Enmey'

Table 4.5 Type of diseases introduced among the tribal Jarawas (1998–2001)

Year	*Disease*	*Reported cases (number)*	*Deaths (number)*
1998	Pneumonia	49	0
1999	Measles	95	0
1999	Mumps	27	0
2000–01	Malaria	73	0

Total Jarawas population – 266 as per Census of India 2001.

Source: Based on Ministry of Home Affairs (2003)

was found wounded in Kadamtala in Middle Andaman Island. He was taken to the hospital and treated for the wounds and was later released with gifts, after which the Jarawas began frequenting the ATR. Since 1997 the Jarawas keep a friendly relationship with outsiders. Since 2001 there have been no incidences of 'Jarawa attacks'. Immediately after this contact of 1997, a series of diseases broke out and most of the Jarawas were affected (Table 4.5). Within three years after the 'contact', a total of 244 cases of disease were reported, but timely medical interventions allowed cure and no one died from these diseases.

A survey conducted among the Jarawas in 2002 showed that 43.4 per cent of them suffered skin diseases which originated from the use of dirty clothes. About 80 per cent of the population suffered from respiratory tract infection, mortality due to which was reported to be high (Ministry of Home Affairs 2003). The cause of respiratory tract infection could be traced to the exposure of Jarawas to toxic fumes of vehicles plying on the ATR. If the contact of the Jarawas continues unabated along the ATR, possibility of them facing the same fate as that of the Great Andamanese, cannot be ruled out.

The A&N Islands have been treated by mainland India as a 'dump yard' where refugee population was resettled. The underdevelopment of A&N Islands even after 70 years of Independence is proof of this neglect. The utter disregard for developing and harnessing its natural resources or caring for tribal population created vulnerability against the backdrop of which needs to be seen the Tsunami of 2004. The making of the Regulation itself is testimony to the vulnerability of the tribes of A&N Islands. Otherwise, why would a community need a regulation to be protected in their own lands?

5 The Tsunami

A disaster of the twenty-first century

On 26 December 2004, the day after Christmas, at 6:28 a.m. (IST), an earthquake measuring Mw 9.3 on the Richter scale occurred at 3°19′12″ N latitude and 95°51′0″ E longitude off the west coast of northern Sumatra at a depth of 30 km (Figure 5.1).

The epicentre of the earthquake was 300 km south-south-east of Great Nicobar Island. This was second largest recorded earthquake in history (after the Chilean earthquake of 1960, which had a magnitude of 9.6). In the A&N Islands, the maximum intensity of shaking was VII (very strong) on the MSK Scale, where the ground shook for four to five minutes.

The earthquake occurred when the Indian-Australian plate slid under the Burma micro-plate, thereby raising the Burma micro-plate by about 10 m. The earthquake was followed by rupturing of the western border of the Burma micro-plate for 1,200 km northwards, which lasted for eight minutes (Bilham 2005). The rupture took place in two phases. In the first phase, the rupture began at the epicentre and moved 400 km northward at a speed of 2 km per second. After a pause of 100 seconds, the rupture gradually continued northwards towards the A&N Islands. A total of 18,000 aftershocks of $M \geq 3$ from 6 January to 16 March 2005 were recorded (Mishra et al. 2007). In this earthquake, the energy released equalled 23,000 Hiroshima-type atomic bombs. The energy released was such that the North Pole shifted by 2.5 centimetres in the direction of 145° east longitude. The earthquake changed the earth's shape by one part in 10 billion; as a result, the length of the day has been reduced by three microseconds (Correspondent 2004). The earthquake disturbed $100 \times 70 \times 1.5\ km^3$ of water and thus set in motion a tsunami (henceforth Tsunami) which moved in all directions with an initial speed of 700 km per hour, which is equal to the speed of a jet plane. Within 30 minutes of its generation, the Tsunami engulfed the entire A&N Islands (Figure 5.2).

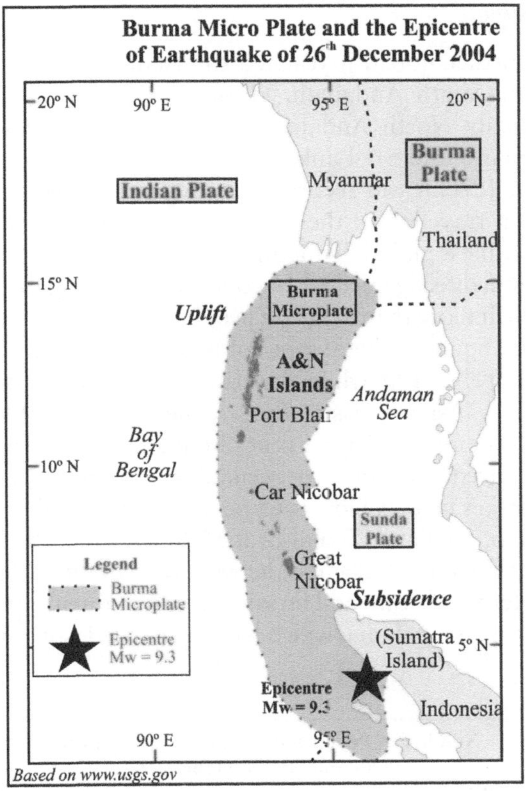

Figure 5.1 Burma micro-plate and the epicentre of earthquake of 26 December 2004

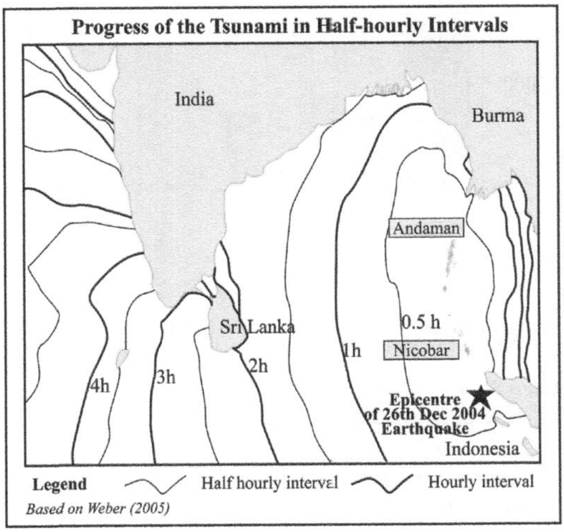

Figure 5.2 Progress of the Tsunami in half-hourly intervals

Moving further north, it reached the islands of Car Nicobar, Little Andaman, South Andaman, Havelock, Baratang, Middle Andaman and finally North Andaman. The Tsunami waves reached the Indian mainland and Sri Lanka at 9:30 a.m. (IST). By seven hours, i.e. 1:30 p.m. (IST) it was recorded on the eastern coast of Africa. The Tsunami then travelled to the Atlantic Ocean and the Pacific Ocean and thus became a global Tsunami.

The Tsunami directly affected 14 countries bordering the North Indian Ocean in which more than 100,000 people were killed (Figure 5.3).

The official figure of the number of people killed in the Tsunami has been placed at 184,189 while the financial loss was estimated to be $9,930 million (Asian Disaster Preparedness Centre 2006). Indonesia accounted for more than 71 per cent of the total killed. Sri Lanka, India and Thailand accounted for 19 per cent, 7 per cent and 3 per cent, respectively. Indonesia's financial loss is 43 per cent of the total financial loss of $9,930 million. Thailand accounted for 22 per cent of the financial loss, followed by Sri Lanka, India and the Maldives. In the audit India's rank was third in terms of death and fourth in terms of financial loss. This was not the first tsunami in India, which has experienced tsunamis in the past.

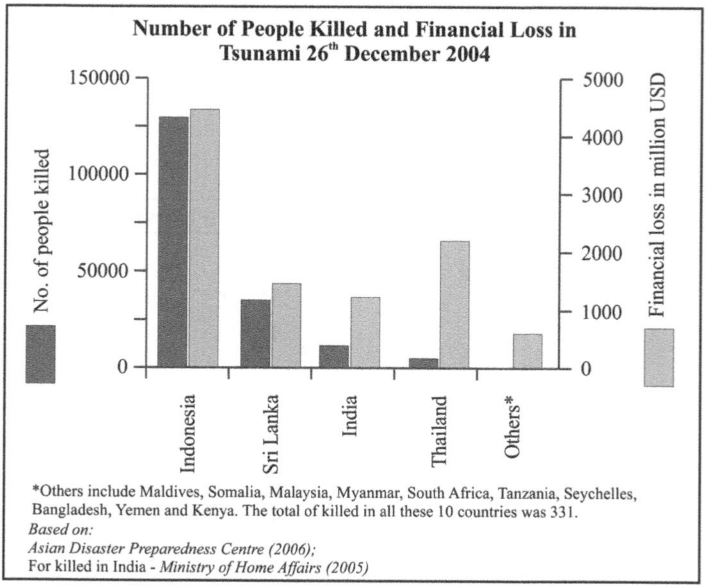

Figure 5.3 Number of people killed and financial loss in the Tsunami of 26 December 2004

Tsunamis in India

The Indian Ocean experiences one tsunami roughly every three years, while in the Pacific Ocean tsunamis occur at least eight times every year (Rastogi and Jaiswal 2006). There are 21 tsunamis recorded on the Indian coast (includes the tsunami on the Makran coast, Iran in 1008 and the tsunami in Sri Lanka in 1882). Of these 21, seven were recorded from the west coast, eight from the east coast and six arose in the cluster from the A&N Islands (Figure 5.4).

Tsunamis are thus not new for the A&N Islands. Traditional beliefs of the aboriginal tribes of the A&N Islands also provide proof of their existence. The Onges, one of the six aboriginal tribes of the A&N Islands, call tsunami *giyangejebey*, which means 'solid earth becoming fluid, like the sea-water' and thus explain a tsunami – the angry spirits

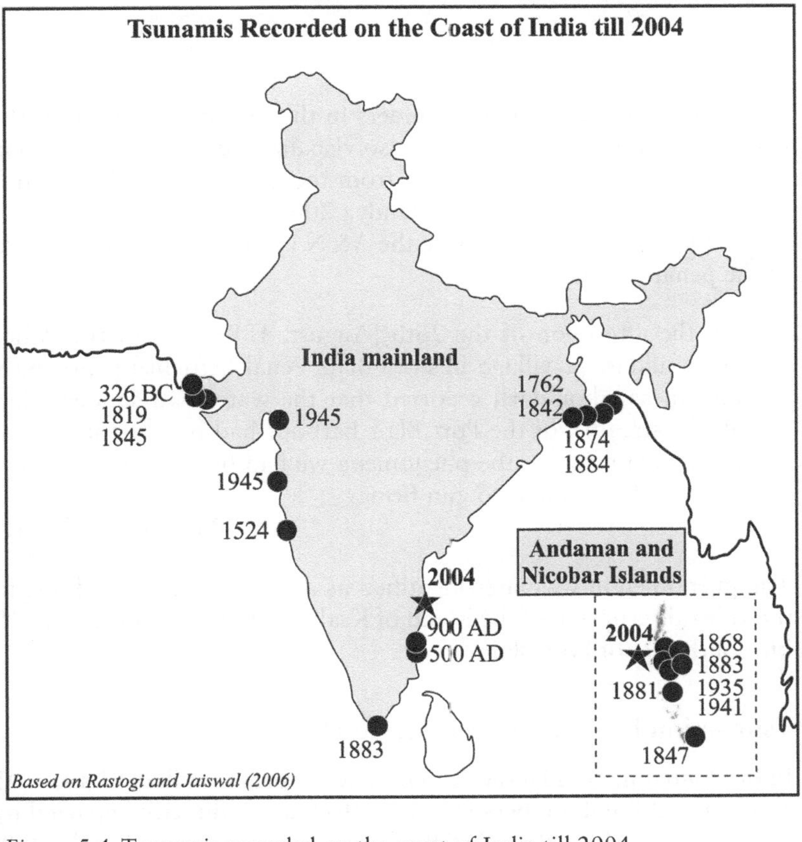

Figure 5.4 Tsunamis recorded on the coast of India till 2004

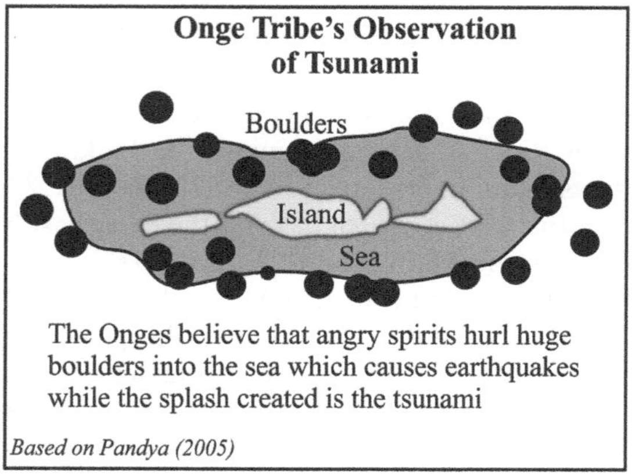

Figure 5.5 Onge tribe's observation of Tsunami

of the ancestors throw huge boulders in the sea and the splash is the tsunami (Figure 5.5). The Onges also visualize tsunami as a breathing process, wherein if water recedes from the land very quickly, then it would come back very rapidly (Pandya 2005).

The third proof of tsunamis in the A&N Islands is the British record of the penal colony:

> On the afternoon of the 26th [August, 1883] one of the Police at Shouldari [a village in Port Blair Penal Settlement, presently known as Chouldari] reported that the water had suddenly left all the streams, of the Port Blair harbour had behaved in a most erratic manner, . . . the phenomena we had been witnessing were explained, as similar to gun firing.
>
> (Portman 1899: 661)

The phenomenon was later identified as a tsunami, generated due to the volcanic eruption of the island of Krakatau, which is located 2,419 km south-east of Port Blair.

Tsunami in India, 26 December 2004

In the Tsunami, in India the total number of persons killed was 12,405 and nearly 2.8 million persons were affected, i.e. directly impacted by the Tsunami. More than 235,000 houses were damaged and 1,089

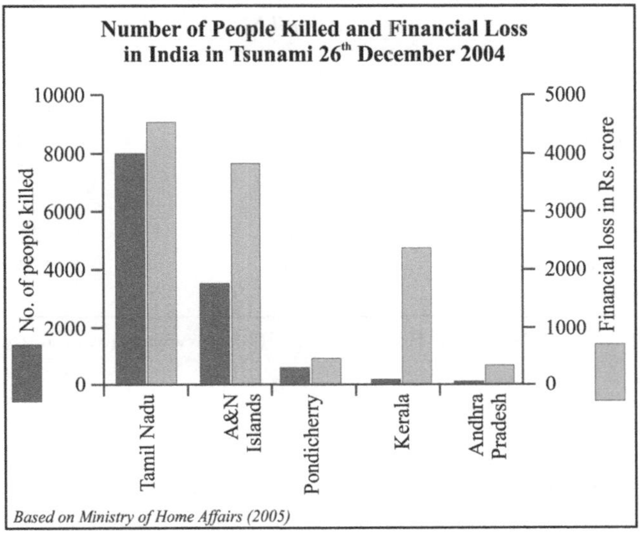

Figure 5.6 Number of people killed and financial loss in India in the Tsunami of 26 December 2004

villages were impacted. The total cost of damage in India was estimated to be at Rs. 115.45 billion. The Tsunami-affected Tamil Nadu, Andhra Pradesh, Kerala and the union territories of Pondicherry and the A&N Islands (Figure 5.6).

Tamil Nadu accounted for maximum numbers killed and maximum financial loss. The A&N Islands ranked second both in terms of number of people killed and financial loss (Table 5.1a). However, totals mask the reality. It is not Tamil Nadu but the A&N Islands which bore a higher and serious brunt. When per cent of pre-Tsunami totals was calculated, it was evident that the impact on A&N Islands was severe than in Tamil Nadu (Table 5.1).

In the A&N Islands 1 per cent of the population was killed and 100 per cent of the population was directly affected while for Tamil Nadu the figures are 0.01 per cent and 1 per cent, respectively. On other parameters, be it villages affected, crop area damaged, house damage or livestock loss, the per cent of total damage was high for the A&N Islands when compared to per cent of pre-Tsunami total damage for Tamil Nadu. In assessment of the damage in financial terms, loss for the A&N Islands was 234 per cent of its gross state domestic product (GSDP) whereas for Tamil Nadu it was 2.07 per cent.

Table 5.1 Tsunami: Tamil Nadu vs. A&N Islands

Per cent of total	*Tamil Nadu*	*A&N Islands*
Population killed	0.01	1
Population affected	1	100
Villages affected	2	38
Cropped area affected	0.29	67
Houses damaged	1	20
Livestock killed	0.01	21
Damage as per cent of total GSDP	2	234

Source: Based on Ministry of Home Affairs (2005a) www.tn.gov.in and www.and.nic.in

In the Tsunami the A&N Islands were devastated. The smaller islands were marooned and no contact could be established with the 'outside world' for a day. During the fieldwork in the A&N Islands in 2010 (five and a half years after the event), people recalled that four waves hit the coast on 26 December 2004, the second being the largest and the highest. In Car Nicobar Island, the first wave arrived 20 minutes after the earthquake. Here the waves approached the coast inwards as a 'huge wall of black water', but at Port Blair in South Andaman Island, the water rose consistently and inundated low-lying areas. This difference occurred because Car Nicobar Island lies in open sea, whereas Port Blair has an indented coastline that provides a sheltered harbour. The intensity of the Tsunami waves decreased in Middle Andaman Island and North Andaman Island as the waves hit these long and indented coasts obliquely.

Two main reasons – the speed of rupture of the plate and the shape of the islands – have been identified that affected the wave intensity and its height (Bilham 2005). The Burma micro-plate ruptured on its western margin for about 1,200 km in two phases (Bilham et al. 2005). In the first phase the rupture began at the epicentre and moved 400 km northward at a speed of 2 km per second. After a pause of 100 seconds, in the second phase the rupture continued northwards towards the A&N Islands gradually. Thus, the first phase of rupture was more rapid than the second phase. This affected the speed of water displacement and hence the speed of the Tsunami. In the first phase of rapid rupture, the speed of water displacement was more than in the second phase of slow rupture.

Table 5.2 Tsunami wave height and inundation across A&N Islands

Island	*Average wave height/ run up*[1] *(in m)*	*Average inundation*[2] *(in m)*
North Andaman	< 2	22.5
Middle Andaman	< 1	35
Baratang	< 1	< 50
South Andaman	5.7	143
Little Andaman	10	> 500
Average for Andaman Island	*2.05*	*61.34*
Car Nicobar	10.29	835.3
Nancowry group	> 15	1,500
Great Nicobar	3.31	440
Average for Nicobar Island	*10.53*	*925.1*

1 Vertical height of the waves above the normal sea level.
2 Horizontal extent covered by the waves.
Source: Based on Ramasamy et al. (2006) and Ghosh et al. (2007b)

The waves that hit the southern part of the A&N Islands, especially the Nicobar Islands, were larger in size and energy than the waves which hit the northern part consisting of the Andaman Islands (Table 5.2).

On average, the height of the Tsunami in the Nicobar Islands was five times higher than in the Andaman Islands, except Little Andaman Island. The average wave height recorded from the Andaman Islands was about 2 m, while in the Nicobar Islands the recorded average wave height was above 10 m. In Car Nicobar Island the wave height was 20 m at Kimius village, 19 m at Kakana village and 15 m in Mus Village. Similar was the case with inundation. The average inundation in the Andaman Islands was limited to a thin belt of coastal area, whereas in the Nicobar Islands the average inundation was close to 1 km (Ghosh et al. 2007b).

The second factor determining the intensity and height of the waves was the shape of the islands, especially the presence of bays. A tsunami moves at a speed relative to the water depth; with decrease in depth of water, the speed of the tsunami decreases and its height increases. As the tsunami moves away from the point of its origin, its speed is reduced due to friction with the sea floor. When it encounters the continental shelf, the wave length decreases to less than 20 km, but at the same time, the wave height increases proportionately. In shallow waters, height increases from 1 m to more than 30 m. In the narrow channels, waves set close and become steep. The A&N Islands have several bays as well as narrow channels. Katchal Island, one of the most affected islands, is surrounded by four bays – West, East, North and South – and Car Nicobar Island has the bays at Kimius and Sawai. These bays led to steep wave heights that were most destructive. When the Tsunami approached the A&N Islands, it refracted, turned and twisted around the islands and passed through the channels between the islands. Therefore, instead of approaching from one side of the island, the Tsunami surrounded the islands from all sides, creating a wraparound effect. With this wraparound effect, the islands were inundated from all sides. In larger islands, the waves moved along the estuaries and backwaters and inundated more than a kilometre of large islands even though there was no wraparound effect. The average size of the Andaman Islands is 245.83 km^2, while the Nicobar Islands is 149.44 km^2. Out of 12 Nicobar Islands, three islands are less than 10 km^2 in size, three are in the range of 10–70 km^2, sizes of the other five are 100–200 km^2 and only one island, Great Nicobar Island, has an area of more than 1,000 km^2. The small area of the Nicobar Islands intensified the wraparound effect of the Tsunami. The Andaman Islands are elongated except Little Andaman Island, hence the waves moved parallel to the coast and hit the coast obliquely, which reduced the wave height and its area of inundation. Large-scale death, damage, destruction and collapse of the basic infrastructure in the A&N Islands was attributed to the steep wave height of the Tsunami.

The killed and the affected in Tsunami

Officially the number of people killed in the A&N Islands was put at 3,513 by the A&N Administration and the Ministry of Home Affairs in June 2005. This list is grossly erroneous because it was created by questioning the survivors about the missing family members. No one counted the number of dead bodies recovered or found. The bodies were mass cremated (eyewitness account, fieldwork, 2010). Thus there was no way to verify the number of killed, especially in cases where all members of the family were killed or were missing. Some of the households included names of family members in the killed list who had died prior to the Tsunami, to claim compensation. Thus, the exact number of persons killed is not known. However, as the figure 3,513 was released by the Administration and the Ministry of Home Affairs, it is this which has been taken into consideration.

There were many people who were carried away into the sea by the Tsunami, and whose bodies were not recovered. Such cases were declared as 'missing'. As per provisions of Indian Evidence Act, 1872, if a person is not heard for seven years or more he or she is presumed to be dead. Therefore, in this research, both the 'missing' and the 'killed' have been termed as 'killed'. This list was far from complete, as the details of the 3,513 individuals were not available. Therefore, to analyze the spatial pattern of the killed and the missing, an alternative that was sourced was the list of people given the ex-gratia payments by the Government of India for the killed and the missing. This list had total of 2,596 cases. This figure, 2,596, had a shortfall of 917 people as the total numbers killed was placed at 3,513 by the Government of India. The cause for the shortfall could not be ascertained. The most probable reason is that all family members were killed in the Tsunami and there was no one to claim compensation. Therefore, in the absence of any other authentic document regarding the number of people killed, 2,596 was taken as the number killed for all purposes of this research.

Of this total 2,596, 432 were 'killed' and 2,164 were declared 'missing'. A sharp spatial variation exists for the killed. The Andaman Islands account for 2 per cent of the total killed while the bulk, 98 per cent, were from the Nicobar Islands (Figure 5.7). In the Andaman Islands, 85 per cent of the killed were from Little Andaman Island and the remaining 15 per cent from South

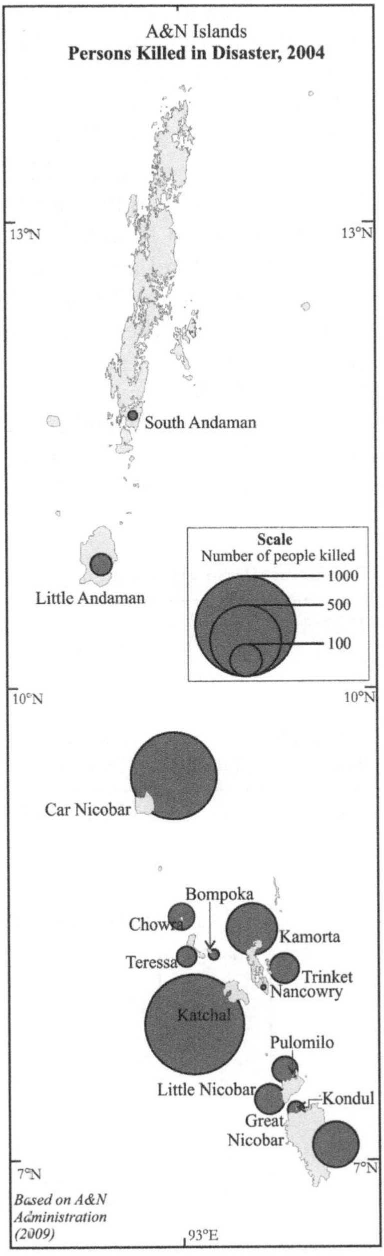

Figure 5.7 A&N Islands: persons killed in disaster, 2004

Andaman Island. There was no deaths reported from any other Andaman Islands. In contrast, all of the Nicobar Islands reported deaths. Katchal Island took the lead and accounted for maximum deaths, 38 per cent of the total.

Katchal Island was followed by Car Nicobar Island, which accounted for 28 per cent of the total killed. Together, Katchal Island and Car Nicobar Island accounted for two-thirds of the total killed in the A&N Islands. Kamorta Island, Great Nicobar Island, Little Nicobar Island and Trinket Island accounted for 25 per cent of the total. The other six islands (Chowra, Teressa, Bompoka, Nancowry, Kondul and Pulomilo) account for the remaining 8 per cent of the total killed. Nearly 1 per cent of the total population of the A&N Islands was killed in the Tsunami (Table 5.3).

In the Andaman Islands 0.02 per cent of its population was killed, while in the Nicobar Islands 6 per cent were killed; the Census of India 2011 records the toll (Figure 5.8). In the Nicobar Islands the

Table 5.3 Population killed in the disaster, 2004

District	*Total population (2001)*	*Number killed*	*Per cent of total killed*
Nicobar	42,068	2,542	6.04
Andaman	314,082	54	0.02
Total A&N Islands	**356,150**	**2,596**	**0.73**

Source: Based on Census of India (2001a) and A&N Administration (2009)

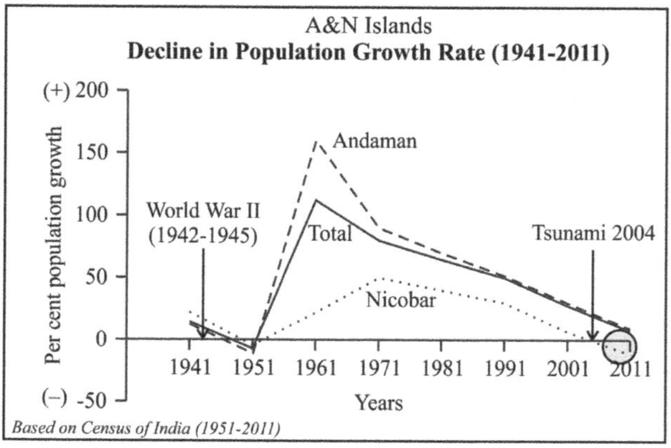

Figure 5.8 A&N Islands: decline in population growth rate (1941–2011)

population declined by 5,249 between 2001 and 2011. Again, it was the Nicobar Islands where per cent killed to total pre-Tsunami population was highest (Figure 5.9). The populations of islands of Pulomilo, Little Nicobar and Trinket were reduced by 44 per cent, 25 per cent and 20 per cent, respectively. In each of the islands of Katchal, Bompoka and Kondul, 18 per cent of the population was killed. Along with the killed were the injured, the homeless and the dislocated households.

Besides those killed, nearly 50,000 people were affected directly in the Tsunami in the A&N Islands. This figure has been derived from the figures of total number of damaged houses. It is based on the assumption that one house has five members, and because 10,000 houses were damaged, hence the total number of people affected directly was 50,000.

But these people were not physically injured. While in other 'natural disasters', the number of injured is three to five times more than the number of dead, in a tsunami, the number of injured is less than the persons killed (Goodman 2005). The Centre for Research on the Epidemiology of Disasters (CRED) database shows that the number of injuries in 'natural' disasters is approximately twice the number of disaster-related deaths over the past 20 years (Paul 2011). As a case, in the Bhuj earthquake (Gujarat State) on 26 January 2001, about

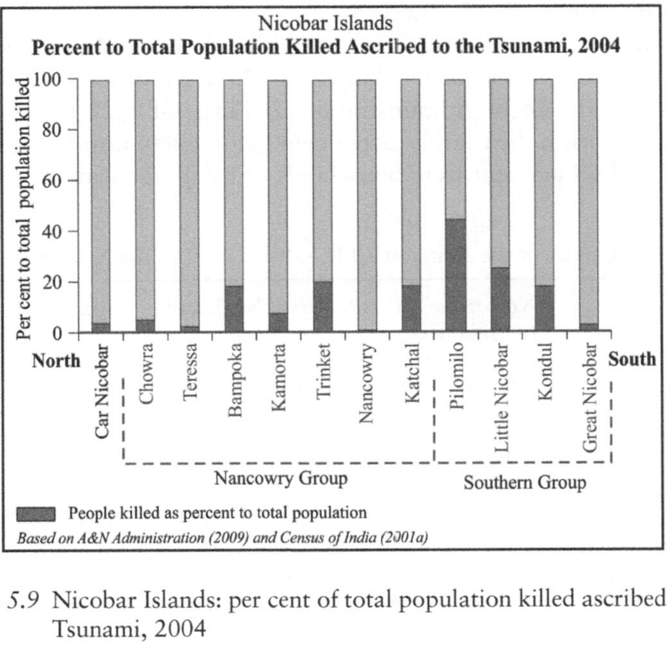

Figure 5.9 Nicobar Islands: per cent of total population killed ascribed to the Tsunami, 2004

20,000 persons were killed, while 167,000 persons were injured, the ratio of killed to injured was 1:8. However, in the Tsunami 2004, in the A&N Islands 3,513 were killed, while 1,514 were injured.

Of the total injured, 2 per cent were from the Andaman Islands while 98 per cent were from the Nicobar Islands. In the Andaman Islands, the injured were only from Little Andaman. Great Nicobar accounted for 55 per cent of the total injured, followed by Car Nicobar (17 per cent), Trinket (13 per cent) and the remaining 14 per cent from Kamorta (10 per cent), Bompoka (2 per cent) and Nancowry (2 per cent). The type of injury was not given in any government report, but the fieldwork in 2010 confirmed that common injuries were fractures and wounds due to hitting a tree or a pole. In fact, more than physical injury, a more serious secondary impact was the displacement of the survivors, their loss of livelihood and the emotional trauma.

In the Tsunami, the people living on the coasts of the Nicobar Islands were worst affected. Their houses were swept away in the incoming and the returning waves. There was no trace of habitation left on the coasts. With no place to take shelter, the people became completely dependent on the A&N Administration. The A&N Administration in turn decided to move people to a safer place. Thus in the evacuation and relief process, the households with damaged houses or houses that were swept away were dislocated. The distance of dislocation varied for different households (Table 5.4).

Of the total 9,797 directly affected households, 9,280 households were relocated in the same island and 517 were dislocated to a different island (A&N Administration 2011a). Of 9,280 households relocated within the island, 32 per cent belonged to non-tribal households and 68 per cent were households of tribal Nicobarese. The new

Table 5.4 Impact of the Tsunami 2004 – dislocation in A&N Islands

Dislocation	*Non-tribal households*		*Tribal Nicobarese households*		*Total households*	
	Number	*Per cent of total*	*Number*	*Per cent of total*	*Number*	*Per cent of total*
Dislocated within the island	2,980	32	6,300	68	9,280	100
Dislocated to a different island	305	59	212	41	517	100
Total households	**3,285**	**34**	**6,512**	**66**	**9,797**	**100**

Source: Based on A&N Administration (2011a)

site of houses for these households was 4–6 km inland from the sea. As the households remained on their own island, the movement was more like relocation rather than dislocation. Relocation means to move (a building, company etc.) to a different location, while dislocation means to put out of place, that is, to put out of proper relative position or to displace.

Of the 517 households dislocated to different islands, 212 belonged to tribal Nicobarese and 305 were non-tribal households. The 212 households of the tribal Nicobarese belonged to the islands of Bompoka, Trinket, Pulomilo and Kondul. These four islands were declared unfit for habitation post-Tsunami, hence these households were accommodated to other islands: 84 households from Bompoka to Teressa; 52 households from Trinket to Kamorta; 41 households from Pulomilo to Little Nicobar; and 35 households from Kondul were dislocated to Great Nicobar. These islands did not have any non-tribal households.

But the 305 non-tribal households from the Nicobar Islands were dislocated because they were staying 'illegally' in these islands. The Nicobar Islands are protected by the Andaman and Nicobar Islands (Protection of Aboriginal Tribe) Regulation of 1956, which prohibits the entry of non-tribal households in the Nicobar Islands (with few exceptions). These households comprised of the descendants of the employees of the A&N Administration who after retirement settled in the Nicobar Islands. These non-tribal households were dislocated from the islands of Car Nicobar, Katchal, Nancowry and Teressa and were relocated to Port Blair and Bambooflat in South Andaman Island. Of these 305 households, 178 were from Car Nicobar, 111 from Katchal, 13 from Nancowry and three from Teressa.

Though the impact of dislocation emerges strongly couple of years after the disaster, the signs of trauma are evident instantaneously. Immediately after the Tsunami, in January–February 2005, to analyze the psychological impact of the Tsunami, a medical team from the National Institute of Mental Health and Neuro Sciences (NIMHANS), Bangalore visited 74 relief camps in the A&N Islands. Of the total 475 cases treated, 37 per cent showed symptoms of adjustment disorder, 21 per cent had depression symptoms, 12 per cent had panic disorder while another 11 per cent suffered from post-traumatic stress disorder (PTSD) (Table 5.5).

Symptoms of PTSD were slightly higher for the displaced (12 per cent) than the non-displaced (10 per cent). Incidences of psychiatric morbidity (psychiatric morbidity refers to incidence of both physical and psychological deterioration as a result of a mental or psychological condition) was high among the displaced survivors in the form of panic disorder, anxiety disorder and somatic complaints while non-displaced suffered high adjustment disorders which are a lesser form of psychopathology (Math et al. 2008).

The impact of the Tsunami was not confined to the moment of its occurrence. In 2007, two and a half years after the Tsunami, a team of doctors from Regional Medical Research Centre (RMRC), Port

Table 5.5 Mental health problems among the Tsunami survivors in A&N Islands

Diagnosis	*Survivors*		*Total (%)*
	Non-displaced (%)	*Displaced (%)*	
Adjustment disorder[1]	120 (51.9)	58 (23.8)	178 (37.5)
Depression[2]	53 (22.9)	49 (20.1)	102 (21.5)
Panic disorder[3]	14 (6.1)	43 (17.6)	57 (12.0)
Post-traumatic stress disorder[4]	24 (10.4)	29 (11.9)	53 (11.2)
Anxiety disorder[5] (not specified)	4 (1.7)	22 (9.0)	26 (5.5)
Somatic complaints	1 (0.4)	14 (5.7)	15 (3.2)
Alcohol dependence syndrome*	2 (0.9)	5 (2.0)	7 (1.5)
Schizophrenia*	3 (1.3)	3 (1.2)	6 (1.3)
Phobic disorder	1 (0.4)	3 (1.2)	4 (0.8)
Acute stress reaction	1 (0.4)	2 (0.8)	3 (0.6)
Dissociative disorder	2 (0.9)	0 (0.0)	2 (0.4)
Recurrent depressive disorder	2 (0.9)	0 (0.0)	2 (0.4)
No symptoms of any disorder	4 (1.7)	16 (6.6)	20 (4.2)
Total	**231 (100)**	**244 (100)**	**475 (100)**

* Relapse of pre-existing mental disorder.

1 Adjustment disorder is a clinical term for a group of symptoms such as stress, feeling sad or hopelessness and physical symptoms that can occur after one goes through a stressful life event.

2 Depression is a mood disorder in which feelings of sadness, loss, anger or frustration interfere with everyday life for weeks or longer.

3 Panic disorder is a type of anxiety disorder where there are repeated attacks of intense fear that something bad will happen.

4 Post-traumatic stress disorder is a type of anxiety disorder. It can occur after experiencing of seeing a traumatic event that involved the threat of injury or death.

5 Anxiety disaster (not specified) refers to a pattern of constant worry and anxiety over many different activities and events.

Source: Math et al. (2008)

Blair and NIMHANS, Bangalore together with members of Action-Aid International conducted a survey to assess the prevalence and pattern of alcohol consumption in the post-Tsunami period. The study included 18,018 individuals from 5,161 households across the A&N Islands.

This study reported an increase in the sale of alcohol during 2005–06 in the A&N Islands (Regional Medical Research Centre, National Institute of Mental Health and Neuro Science and Action Aid 2008). In the A&N Islands 35 per cent of the male population and 6 per cent of the female population are alcohol consumers, meanwhile the national average for the same is 21 per cent and less than 5 per cent, respectively (Regional Medical Research Centre, National Institute of Mental Health and Neuro Science and Action Aid 2008). The second finding was that drinking was most prevalent among the tribal Nicobarese. Also, the study emphasized an increase of alcohol consumption among the tribal Nicobarese – 11.4 per cent for male and 10 per cent for female. Thus, one-tenth of the population of Nicobar Islands has been declared as alcohol dependent in the post-Tsunami study. The research documented a high prevalence – more than 50 per cent – of psychological problems across the population of the A&N Islands. The PTSD symptoms were prevalent among the population two and a half years after the event. Ten per cent of the male population and 20 per cent of the female population showed PTSD symptoms (Regional Medical Research Centre, National Institute of Mental Health and Neuro Science and Action Aid 2008). These symptoms can be attributed to the Tsunami, and interactions with the survivors in 2010 confirmed that many are still grappling with the memories of the Tsunami. However, the impact of the Tsunami stretched beyond the people to their home, school, work and the entire place.

Appendix

Table 5A.1 Rank of A&N Islands in terms of per cent of total damage of Tsunami in India

	Tamil Nadu	*A&N Islands*	*Pondicherry*	*Kerala*	*Andhra Pradesh*	*Total*
Population killed	8,009	3,513	599	177	107	12,405
Per cent of total	*64.56*	*28.32*	*4.83*	*1.43*	*0.86*	*100*
RANK	1	2	3	4	5	
Population affected (in millions)	0.9	0.36	0.04	1.3	0.2	2.8
Per cent of total	*32.13*	*12.75*	*1.54*	*46.56*	*7.02*	*100*
RANK	2	3	5	1	4	
No. of women who became widows	562	148	28	31	18	787
Per cent of total	*71.41*	*18.81*	*3.56*	*3.94*	*2.29*	*100*
RANK	1	2	4	3	5	
No. of children orphaned	289	143	39	3	6	480
Per cent of total	*60.21*	*29.79*	*8.13*	*0.63*	*1.25*	*100*
RANK	1	2	3	5	4	
No. of houses damaged	190,000	21,100	10,061	13,735	481	235,377
Per cent of total	*80.72*	*8.96*	*4.27*	*5.84*	*0.20*	*100*
RANK	1	2	4	3	5	
No. of villages affected	376	192	33	187	301	1,089
Per cent of total	*34.53*	*17.63*	*3.03*	*17.17*	*27.64*	*100*
RANK	1	3	5	4	2	

	Tamil Nadu	*A&N Islands*	*Pondicherry*	*Kerala*	*Andhra Pradesh*	*Total*
Cropped area affected (ha)	19,168	11,010	792	7,763	302	39,035
Per cent of total	*49.10*	*28.21*	*2.03*	*19.89*	*0.77*	*100*
RANK	1	2	4	3	5	
Boats damaged	52,638	1,401	6,678	10,882	12,189	83,788
Per cent of total	*62.82*	*1.67*	*7.97*	*12.99*	*14.55*	*100*
RANK	1	5	4	3	2	
Livestock lost	1,653	27,331	2,685	0	86	31,755
Per cent of total	*5.21*	*86.07*	*8.46*	*0.00*	*0.27*	*100*
RANK	3	1	2	5	4	
Damages in monetary terms (Rs. in billions)	45.29	38.27	4.66	23.71	3.43	115.45
Per cent of total	*39.23*	*33.14*	*4.04*	*20.54*	*2.97*	*100*
RANK	1	2	4	3	5	

Source: Based on Ministry of Home Affairs (2005a)

6 Damage and the Tsunami

The 26 December 2004 Mw 9.3 earthquake and the resultant Tsunami left behind a trail of damage and destruction. This chapter assesses the damage in the A&N Islands that has been ascribed to the earthquake and the Tsunami. In this massive earthquake, the Burma microplate was uplifted by about 10 m while a subsidence was noticed at its south-eastern end near the epicentre of the earthquake (Das et al. 2007). What contributed to the damage were the morphological changes that followed the earthquake and the Tsunami. These morphological changes were very marked and distinctly clear all along the coasts of the A&N Islands.

The bathymetry of near-shore waters played a critical role in the height of the Tsunami. Wide continental shelves of shallow waters resulted in higher waves as compared to narrow shelves with deep water levels. Coasts adjacent to wide continental shelves of 30 m or more suffered the worst damages (Sankaran 2005). The western side of the Nicobar Islands has a broader continental shelf than the eastern margin, and hence, the western boundary of Car Nicobar Island, the Nancowry group of islands and Great Nicobar Island were worst affected.

The A&N Islands were uplifted on the north-western part while its south-eastern part subsided. The exposed coral reefs and new beaches in the north-western part and the submerged mangrove stands in South Andaman Island and Nicobar Island are the stark evidence of this movement of tectonic plates (Figure 6.1). The shoreline of the A&N Islands changed with the uplift, subsidence, drowning of beaches and formation of new beaches. This was most evident in the Nancowry group of islands (Table 6.1).

The Nancowry group of islands includes the six islands that lie close together – namely, Chowra, Teressa, Bompoka, Trinket, Kamorta, Katchal and Nancowry – and are also known as the central group of Nicobar Islands.

Of all these seven islands, the change was most pronounced in Trinket. It was a narrow island of 36.30 km^2, the shape resembling that of a seahorse, but post-Tsunami, Trinket was divided into three smaller islands as the sea encroached on the indented coast of the island as a result of subsidence.

The margins of Kamorta, to the west of Trinket Island, were deeply indented on all its sides and thus the Tsunami waves reached the interior of the islands. The Nancowry group of islands had several creeks flowing into the sea. The Tsunami waves moved inward and eroded as well as widened the low-lying areas surrounding the creeks. The total land submerged in the Nancowry group was 38.29 km^2. The relative location of the islands with respect to each other impacted the extent of morphological change. Nancowry Island is sheltered on three sides by the islands of Katchal, Trinket and Kamorta on the west, east and north, respectively. Therefore it was only the southern part of Nancowry Island that was exposed to the waves. Again, Nancowry Island tapers towards the south. This deflected the waves to the east and the west of the island, thereby compounding the impact on Katchal and Trinket, leaving Nancowry Island relatively unaffected. The dense forest and the rich coral

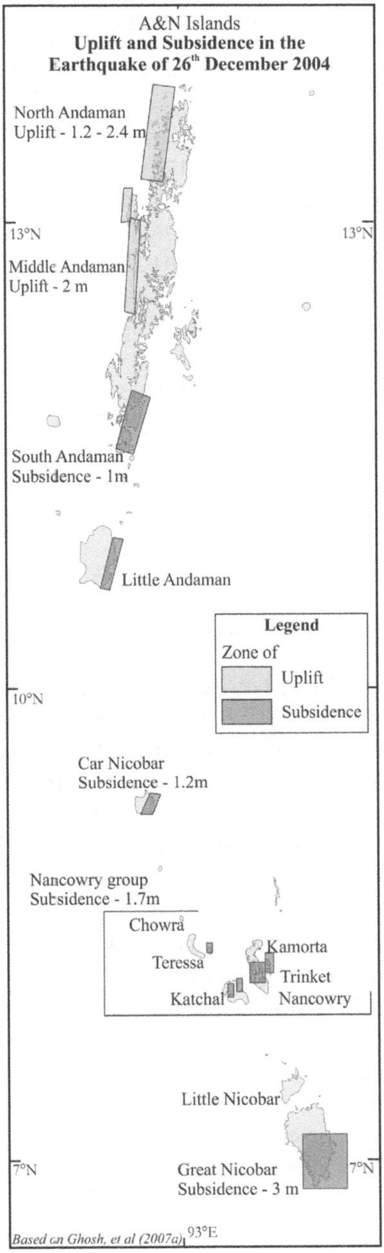

Figure 6.1 A&N Islands: uplift and subsidence in the earthquake of 26 December 2004

Table 6.1 Land submerged in Nancowry group of islands

Islands	*Area (in square kilometres)*		
	Pre-Tsunami	*Post-Tsunami*	*Land submerged*
Katchal	174.3	160.73	13.57
Kamorta	188.03	176.38	11.65
Trinket	36.26	29.25	7.01
Teressa	101.26	97.64	3.62
Nancowry	66.82	65.6	1.22
Bompoka	13.46	12.82	0.64
Chowra	8.2	7.62	0.58
Total	**588.33**	**550.04**	**38.29**

Source: Based on Das (2005)

reefs of the A&N Islands too were severely impacted in the earthquake and the Tsunami.

Forest destruction

In 2003, before the Tsunami (henceforth pre-Tsunami), 84 per cent of the A&N Islands were covered with forests (Forest Survey of India 2003). In aftermath of the Tsunami, 335 km^2 of forest area was lost and the percentage declined to 80 per cent (Table 6.2). The destruction of forest was uneven. In the Andaman Islands, 84 per cent was under forest pre-Tsunami, of which 122 km^2 was lost accounting for 2 per cent of its pre-Tsunami total. In the pre-Tsunami period, 86 per cent of the geographical area of the Nicobar Islands was under forest. Of this, 213 km^2 was lost which was 14 per cent of its pre-Tsunami total (Forest Survey of India 2003, 2005).

The Forest Survey of India classifies forests into very dense, moderately dense and open on the basis of their crown densities. The very dense forests have crown density of 50 per cent, meanwhile the crown density of moderately dense forest and open forest are 40 per cent and 10 per cent, respectively. In the A&N Islands, all three forest types, very dense, moderately dense and open, have registered a decline in 2005 when compared to 2003 figures (Table 6.3).

The open forest registered a decline of 8 per cent, and moderately dense forest and very dense forest recorded a decline of 6 per cent and 3 per cent, respectively. The loss was minimum for the category

Table 6.2 Tsunami and the decline in forest cover: a comparison, 2003 vs. 2005

(areas in square kilometres)	*Andaman*	*Nicobar*	*Total*
Total geographical area	6,408	1,841	8,249
Pre-Tsunami forest cover (2003)	5,387	1,577	6,964
Per cent of total geographical area	*84.07*	*85.66*	*84.42*
Post-Tsunami forest cover (2005)	5,265	1,364	6,629
Per cent of total geographical area, 2005	*82.16*	*74.09*	*80.36*
Forest cover lost (2003–05)	−122	−213	−335
Loss as per cent of total pre-Tsunami forest cover	*−2.26*	*−13.51*	*−4.81*

Source: Based on Forest Survey of India Report (2003, 2005)

of 'very dense forests' because these forests cover the slopes of hills of higher altitudes and are away from the coast.

The forests in the A&N Islands include the equatorial evergreen, tropical evergreen, deciduous and the coastal forest consisting of littoral forests and the mangroves. In the interiors there are deciduous, tropical evergreen and equatorial evergreen forests. Mangroves and littoral forest hug the coast of the A&N Islands. The littoral forest is the coastal forest which is found along the beaches. Mangroves are located on the coasts and are in contact with the brackish water. Both littoral forest and mangroves are part of the coastal ecosystem along with sea grass and the coral reefs. Mangroves dominate the Andaman Islands while littoral forests are extensive in the Nicobar Islands. In the Tsunami it was the coastal forest that was most affected.

Of the littoral forest, the palms of nypa and pandanus (commonly known as nipa palm and screw pine, respectively) were either uprooted or died due to ingress of saline water. Both these palms were virtually wiped out from the Nancowry group of islands and Great Nicobar Island post-Tsunami (A&N Administration 2006). The second most affected forest was the mangroves. Mangroves extend over 966 km^2 in the A&N Islands and account for 20 per cent of the total mangroves in India (Dam Roy and Krishnan 2005). Of this total 966 km^2, 96 per cent is in the Andaman Islands and 4 per cent in the Nicobar Islands. Therefore damage to mangroves was confined to the Andaman Islands. In the Andaman Islands, mangroves are found along creeks, marshes and backwaters such as Shoal Bay (in South Andaman), Yerrata (Middle Andaman), Austin Creek, Kalighat Creek and

Table 6.3 Decline in forest crown density: a comparison, 2003 vs. 2005

	Andaman Islands	*Nicobar Islands*	*A&N Islands*
Very dense forest **(crown density > 50%)**			
Pre-Tsunami (2003)	2,613	862	3,475
Post-Tsunami (2005)	2,572	787	3,359
Change	−41	−75	−116
Per cent change	−1.57	−8.70	−3.34
Moderate dense forest **(crown density between 40% and 50%)**			
Pre-Tsunami (2003)	2,202	607	2,809
Post-Tsunami (2005)	2,185	461	2,646
Change	−17	−146	−163
Per cent change	−0.77	−24.05	−5.80
Open forest **(crown density < 10%)**			
Pre-Tsunami (2003)	572	108	680
Post-Tsunami (2005)	508	116	624
Change	−64	8	−56
Per cent change	−11.19	7.41	−8.24

Source: Based on Forest Survey of India Report (2003, 2005)

Cadell Bay (North Andaman). The uplift and subsidence had damaged mangroves in three different ways (Sankaran 2005).

First, due to the uplift of the Andaman Islands, the high tide water level receded. This exposed the roots of the mangroves, which caused the death of mangrove stands along North Andaman Island, especially in Deshbandhugram, Laxmipur, Milangram and Swarajgram. Second, the physical force of the Tsunami was so intense that mangroves were uprooted, as reported from Mahatma Gandhi National Marine Park at Wandoor in South Andaman Island (Sankaran 2005). Third, in South Andaman Island and all of the Nicobar Islands, the subsidence had a drowning impact on the mangroves. Mangroves thrive in the intertidal zone, however most varieties are not salt-resistant. As the roots of the mangroves were permanently submerged under the sea water, the mangroves wilted and withered. This situation was reported at Wandoor and in Mahatma Gandhi National Marine Park and along Sippighat

Creek and Chouldari Creek. At present, the brown band fringing the islands instead of mangroves confirms the destruction. The destruction did not end here. The mangroves and the littoral palms that were uprooted, unlocked and released mud and pollutants that were later deposited on the coral reefs by the returning waves, thereby impacting the rich coral reefs of the A&N Islands.

Impact on coral reefs

Of India's total 2,329.9 km^2 of coral reefs, 41 per cent are in the A&N Islands. The other areas of coral reefs in India are Lakshwadeep with 816.16 km^2 (35.03 per cent), Gujarat with 460.16 km^2 (19.75 per cent) and Tamil Nadu with 94.36 km^2 (4.05 per cent) (Department of Ocean Development and Space Application Centre 1997).

The A&N Islands are surrounded by shallow seas and coasts. The tropical climate and extensive mangroves support the growth of coral reefs around the A&N Islands. The coral reefs are of extensive size on the western margin of the A&N Islands. This is because the land gradually descends towards the west and provides shallow continental shelves for the growth of coral reefs (Turner et al. 2001). The A&N Islands have two types of coral reefs: fringing reefs and barrier reefs. Fringing reefs grow near the coastline around islands and continents. They are separated from the shore by narrow, shallow lagoons. The barrier reefs are also parallel to the coastline but are separated by deeper and wider lagoons. The Andaman Islands have fringing reefs, whereas the Nicobar Islands have barrier reefs (Hoon 1997).

Coral reefs are often called the rainforests of the sea, both due to the vast amount of species they harbour and to the high productivity they yield. Covering less than 1 per cent of the ocean floor, reefs support an estimated 25 per cent of all marine life, with over 4,000 species of fish alone. Not only do coral reefs provide habitat to large number of species, but their beauty is also one of the major tourist attractions. According to a United Nations estimate, the total economic value of coral reefs range from USD 100,000 to 600,000 per km^2 per year (International Coral Reef Initiative 2016).

Coral reefs are a tropical, shallow water ecosystem, largely restricted to the area between 30° north and 30° south latitudes. Corals generally have very specific environmental requirement for light, temperature, water clarity, salinity and oxygen. Even in the clearest seas, reef-building corals are restricted to depths of less than 30 m and are generally found in much shallower depths. Reef building corals, known as hermatypic corals, have micrcosopic algae with zooxanthin in their bodies that gives the colour to the corals. The coral and the algae share a symbiotic relationship, both drawing nutrients from each other for survival (Correspondent 1998).

The delicate coral reef system of the A&N Islands was affected in the Tsunami and 400 km^2 area of coral reef was brutally destroyed (Alongi 2008). Even though the impact on the coral reefs emanated from the Tsunami, it was not the waves that damaged it. There were secondary causes that impacted the reefs. In the Andaman Islands, the damage was because of uplift, whereas in the Nicobar Islands it was on account of subsidence and wave action. The western margin of the North Andaman Island was uplifted, thereby exposing the coral polyps. In the absence of sea the corals died, leaving behind bleached, lifeless reefs. The area of the reef exposed ranged from 0.5 km^2 in Snark Island to 4.5 km^2 around North Reef Island. In Smith Island and Ross Island 97 per cent of the coral reefs were killed (Alongi 2008). In contrast to the North Andaman Island, in South Andaman Island the force of the Tsunami was such that the coral reefs were dislodged and smashed. The islands of Alexandra, Bellie, Boat, Chester, Grub, Redskin, Jolly Bouy and Snub of the Mahatma Gandhi National Marine Park were damaged by the sheer force of the waves (Alongi 2008).

The reefs of the Nicobar Islands were doubly damaged: first, the Tsunami waves uprooted the corals, and second, what remained was buried under the debris unlocked by the uprooting of coastal forest (Alongi 2008). The returning Tsunami wave also dumped the debris and sediments on the remaining corals. The corals were thus buried under the load of the debris and sediments. Coral reefs of entire Nicobar Islands including Car Nicobar, Teressa, Kamorta, Nancowry, Katchal, Trinket, Little Nicobar and Great Nicobar were damaged. The absence of mangroves in the Nicobar Islands made them more vulnerable compared to the Andaman Islands. In the Andaman Islands, mangroves that were not uprooted arrested the debris flowing back into the sea and thereby protected the coral reefs. As mangroves were absent in the Nicobar Islands, the debris, sediments and muck was deposited directly on the coral reefs by the returning waves (Saxena et al. 2008).

The Tsunami uplifted submerged land, destroyed forests and submerged and killed corals. It is clear in the case of forests and corals that the exposed and the vulnerable were the ones which suffered the maximum. If this principle applies in case of the physical and natural world, what about the humans?

Affected livelihoods

The impact of the Tsunami was far and wide. The submerged coasts, inundated lands, uprooted forests and damaged coral reefs completely shattered the livelihoods of the people of the A&N Islands. Among the three sectors of economic activities – primary, secondary and tertiary – it was the primary sector, especially fishing, that was most impacted. The fishing boats and trawlers were destroyed and the damage to coral reefs led to the decline in fish catches.

Impact on fishing

The coastline of A&N Islands is 1,912 km, which is one-fourth of the total coastline of India. Its 0.6 million km^2 of the Exclusive Economic Zone (EEZ) contributes to one-third of India's total EEZ. The territorial waters of India extend 12 nautical miles from the shoreline of the A&N Islands, thereby increasing the potential area for fishing. The A&N Islands are surrounded by a narrow continental shelf of around 35,000 km^2, of which 1,000 km^2 has rich coral reefs that support a diverse variety of fish. Fisheries potential for the A&N Islands have been estimated to be 0.15 million tonnes, of which only 19 per cent is currently harnessed (Directorate of Economics and Statistics 2008b).

Fishing is the livelihood both for the Andaman Islands and the Nicobar Islands, but more so for the latter, for the tribal Nicobarese. Potential for fishing in the A&N Islands is high, but commercial fishing has not developed. In the pre-Tsunami period a total of 57 fish landing sites and two cold storage facilities (located at Port Blair and at Wandoor) were set up in the Andaman Islands. For any coastal economy, fishing is the main source of income. Despite being surrounded by tropical sea that is rich in fishery resource, the commercial fishery has not developed in the Nicobar Islands. For the tribal Nicobarese, fishing is for subsistence, as fish is the main source of protein in their diet. The tribal Nicobarese used their canoes, locally known as *hoodies*, for fishing. Thus, commercial fishing is confined to the Andaman Islands, and so the damage to fishing in the Nicobar Islands went unregistered.

In the Tsunami, 1,800 boats were damaged and 1,500 fishing gears and equipment were lost. Of the total 2,964 licensed fishermen, 2,759 were affected (93 per cent of the total). The Andaman Islands accounted for 67 per cent of the boats damaged while the Nicobar Islands accounted for the remaining 33 per cent. In the spatial picture of damage, it is the Andaman Islands that claim the larger share of the damage. Within the Andaman Islands, 56 per cent of the damaged boats were from South Andaman, 20 per cent from Middle Andaman, 11 per cent from North Andaman and 8 per cent from Little Andaman. In the Nicobar Islands, 26 per cent of the damaged boats were from Car Nicobar, 22 per cent from Nancowry, 17 per cent from Chowra, 13 per cent from Teressa, 11 per cent from Great Nicobar and others (Digal 2006). Loss of fishing equipment and gears followed the same spatial pattern of damage to boats.

In the aftermath of the Tsunami, the fish catch declined drastically in the A&N Islands (Figure 6.2). Nearly 150 tonnes of fish was exported in 2003–04, i.e. the year preceding the Tsunami. The fish catch of 2005 was 68 per cent less than the catch of 2004.

Monthwise data of the fish catch for 2004 and 2005 shows that the fish catch for January through April 2005 was 90–95 per cent less than

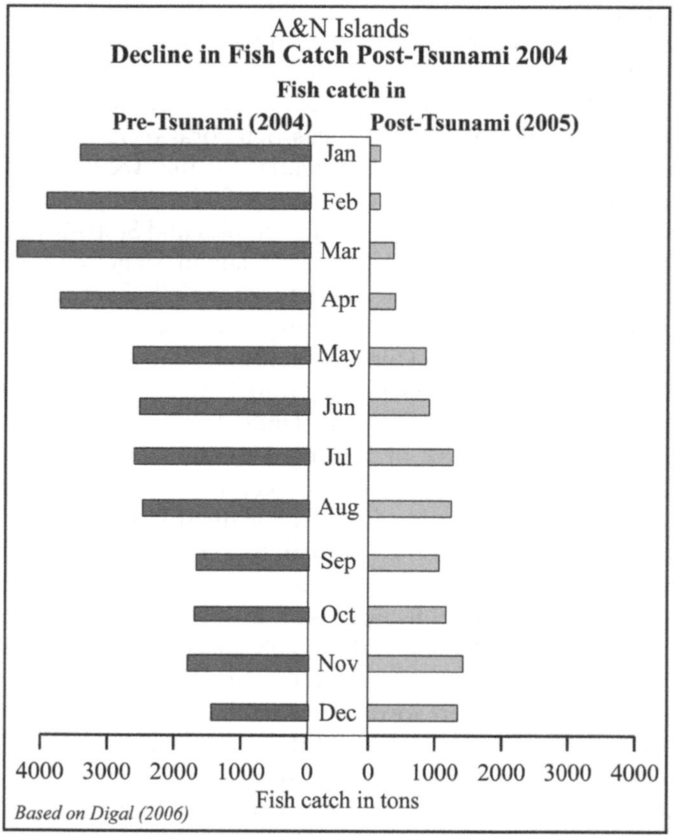

Figure 6.2 A&N Islands: decline in fish catch post-Tsunami 2004

the figures for 2004 for the same months. But fishing picked up with time, and by December 2005 the fish catch was about 8 per cent less than in 2004. This trend of the decline in fish catch and its recovery can be attributed to three reasons. First, in the Tsunami the fishing boats were damaged and the fishing gears were lost. In the absence of trawlers, boats, nets and other equipment, it was impossible for the fishermen to go out for fishing. Second, coral reefs were dead and washed out after the Tsunami, thus the bulk of the fish population was killed. The nearshore water became turbid with returning waves which carried mud, muck and sludge. A study concluded that coral reef was filled with silt and sand after the Tsunami. Due to complete washout of coral reefs, reef fish including lobster were eliminated (Sircar 2006). But more than lack of equipment or fishes, the third significant reason for the decline in the fish catch was the fear of the sea. Fieldwork in 2010 confirmed that tribal Nicobarese fishermen were so scared of the sea after the Tsunami that they did not venture into the sea to fish for three to five months. But as fishing was their only means of livelihood, the fishermen of the Andaman Islands and the tribal Nicobarese did return to the sea eventually.

Even though completely devastated in the post-Tsunami period, the fishing industry recovered within a year. Another sector of livelihood which had long-term impact but which has not been able to recover to date is the agriculture.

Impact on agriculture

In the pre-colonial A&N Islands, the tribes of the Andaman Islands did not engage in agriculture while the tribal Nicobarese maintained coconut plantations. Agriculture was introduced in the Andaman Islands by the British administration in 1858. The convicts were used as labourers for clearing the forests and carving out land for cultivation. Agriculture got a boost pre-1947, when especially the convicts and settlers from mainland India (the self-supporters, the Moplahs and the Bhantus) were transported to the A&N Islands. They were given 1–5 acres of land for cultivation and setting up villages. In the 1901 census, agriculture was a means of subsistence for 57 per cent of the population. In 1927, the colonial government set up the Department of Agriculture. The Japanese who occupied the A&N Islands during the World War II also focused on the expansion of agriculture. Under the Japanese, nearly 3,036 acres (1,228.64 ha of paddy) were brought under cultivation. In 1947, when the Indian administration took over the A&N Islands, the area under agriculture was 2,266 ha (Sareen 2001). Under the Resettlement Scheme introduced in 1949, a total of 3,281 households were resettled in the A&N Islands and each household was allotted 10 acres of land for cultivation. Thus, by the end of 1961, the area under cultivation increased to 9,633 ha (Dhingra 2005). At the time of closure of the Resettlement Scheme in 1975, land under agriculture was 50,000 ha. It has not officially increased ever since. This is because in 1980, as per the Forest Conservation Act, the area under forest was declared a reserved forest and a ban was put on conversion of forest land for any other purpose. However, unabated encroachment of forest land for expansion of agriculture still continues. In 2001 land available for cultivation was 6 per cent of the total geographical area. Of this, 60 per cent is in the Andaman Islands and 40 per cent in the Nicobar Islands.

The agricultural land in the A&N Islands is concentrated in the narrow valleys and along the coast for two reasons: first, a large part of the Andaman Islands has rugged hills, and flat land is only available in the narrow valleys between the hills and along the coast; second, the initial settlements came up near harbours, as ships were the only mode of travel among various islands before the construction of the Andaman Trunk Road started in 1969. Therefore, land for agriculture was cleared close to the settlements and thus was confined to the harbour area. It is only during the mid-1970s, after the construction of the road, that land for agriculture was cleared along the road as in the case of Middle Andaman Island.

In the Nicobar Islands agriculture includes coconut groves of tribal Nicobarese and the rubber plantation of the Rubber Board of India along with paddy and other crop cultivation of Great Nicobar Island. In the pre-colonial A&N Islands, the tribal Nicobarese bartered coconuts for rice, iron, clothes, and other goods with the ships that halted at the coast. Post-Independence, the tribal Nicobarese sold copra to the A&N Administration. Agricultural land, including coconut groves of the tribal Nicobarese, is confined to the belt of 3–4 km inside along the coast.

The rubber plantation and cultivation of different crops were introduced in the Nicobar Islands during the resettlement scheme from the mid-1960s. The rubber plantation is located in Katchal Island, away from the coast, and hence was unaffected in the Tsunami. As the agricultural fields of Great Nicobar Island and the coconut groves of the tribal Nicobarese were on the coast, they suffered great damage. The damage assessment report records that 'nature's fury has flattened as many as 0.5 million coconut trees, accounting for nearly 40 per cent of the copra industry on which the livelihoods of thousands of families were dependent' (Bhagat 2005).

Agriculture in the A&N Islands was affected in theTsunami in three ways: (1) standing crops were damaged, (2) fertile coastal agricultural land turned saline and (3) agricultural land was submerged. The standing crops were damaged in two ways. The northern part of the A&N Islands was uplifted and thus the water table was lowered, creating moisture stress for the standing crops in Diglipur and Ariel Bay, both in North Andaman Island. The crops wilted in the absence of moisture. In the southern part the agricultural land was inundated by the Tsunami waves and were thus destroyed. In terms of crop loss in A&N Islands, paddy was most affected, and registered a loss of 44 per cent in 2005–06 when compared to the figures for 2003–04. The traditional variety of paddy is planted in September and harvested in January, thus the paddy was near to its harvest when the Tsunami came. Other crops including oilseeds, cashew nuts and sugarcane registered a decline by 39 per cent in the same period. Production of fruits and vegetables recorded a decline by 28 per cent and 17 per cent, respectively. Coconut production fell from 95 million nuts in 2003–04 to 78 million nuts in 2005–06, which is a decline of 17 per cent (Digal 2006).

Salinization of cultivation land also affected the agriculture. The Tsunami waves inundated the coastal areas and covered them in salt water. With the heat and ensuing evaporation, a salt crust turned the land infertile. In spite of the fact that monsoons since 2005 have lowered the salinity, large tracts have become unusable. Submergence is another cause that led to a decline in agriculture. The land under agriculture registered a decline after the Tsunami. Of the total 50,000 ha land under agriculture, a total of 8,068.71 ha (16 per cent) was lost due to submergence. Of this only 3,862.07 ha (48 per cent) was reclaimable and the remaining 4,206.64 ha (52 per cent) became unusable. Out of this total land affected, 24 per cent was in the Andaman Islands, while 76 per cent was in the Nicobar Islands. The average distance of inundation in the Tsunami was more in flat lands mostly confined to the Nicobar Islands. Not only was agriculture more severely affected in the Nicobar Islands, but it also registered large-scale death of the livestock.

Perished livestock

Livestock rearing is an integral component of the livelihoods of both non-tribal agriculturalists and the tribal Nicobarese. Large numbers of livestock perished in the Tsunami. As per the A&N Administration's

Table 6.4 Perished livestock in the Tsunami

	Cattle	*Goat*	*Pigs*	*Other*	*Total*
Andaman Islands					
Pre-Tsunami[1]	70,789	63,460	8,517	714	143,480
Killed in Tsunami[2]	565	2,045	606	0	3,216
Killed as per cent of pre-Tsunami total	0.80	3.22	7.12	0	2.24
Nicobar Islands					
Pre-Tsunami[1]	8,976	15,760	43,684	131	68,551
Killed in Tsunami[2]	3,221	14,578	37,840	NA	55,639
Killed as per cent of pre-Tsunami total	35.88	92.50	86.62	0	81.16
A&N Islands					
Pre-Tsunami[1]	79,765	79,220	52,201	845	212,031
Killed in Tsunami[2]	3,786	16,623	38,446	0	58,855
Killed as per cent of pre-Tsunami total	4.75	20.98	73.65	0	27.76

Source: Based on [1]Directorate of Economics and Statistics (2006); [2]Digal (2006)

damage reports, about 58,855 livestock were killed, which accounted for one-third of the pre-Tsunami livestock population (Digal 2006). Of the total livestock killed in the A&N Islands, 5 per cent were from the Andaman Islands while the Nicobar Islands accounted for the remaining 95 per cent (Table 6.4).

In the pre-Tsunami (2003) period the total number of livestock in the A&N Islands was 0.21 million (Directorate of Economics and Statistics 2008a). It included cattle (consisting of buffaloes, cows, oxen and bull), goats, pigs, horses, sheep and rabbit. Cattle, goat and other livestock accounted for 75 per cent of the total livestock population of the A&N Islands, the remaining being a population of pigs, accounted for 25 per cent.

The composition of livestock in the Andaman Islands and the Nicobar Islands differed. In the Andaman Islands, cattle and goats take the lead, while in the Nicobar Islands, in complete contrast, pigs dominate. Since cattle are reared by the non-tribal population, 89 per cent of the cattle were in the Andaman Islands and 11 per cent in the Nicobar Islands that too were confined to the non-tribal enclaves. The goat population was also concentrated in the Andaman Islands (80 per cent) as well as in the non-tribal-dominated Nicobar Islands (20 per cent). In the Tsunami, 2 per cent of the livestock of the Andaman Islands and 81 per cent of the livestock of the Nicobar Islands perished. When compared to pre-Tsunami totals, 5 per cent cattle, 21 per cent goats and 74 per cent of the pigs were killed.

In the Andaman Islands, a total of 3,216 livestock were killed. Goats accounted for 64 per cent of the total killed livestock. The 606 pigs and 565 cattle that were killed accounted for the remaining 36 per cent. When compared as per cent of total pre-Tsunami livestock population, 0.8 per cent of the cattle, 3 per cent of the goats and 7 per cent of the pigs were killed in the Andaman Islands.

The impact on the livestock was much more severe in the Nicobar Islands, where 81 per cent of its livestock perished. In the pre-Tsunami period, the Nicobar Islands led in its pig population and accounted for 84 per cent of the total pig population of the A&N Islands. The pigs of Nicobar Islands were endemic to the islands and were reared in a semi-feral system (Jeyakumar et al. 2014). Pigs are an integral part of the economy and ecology of the tribal Nicobarese. Pigs and coconuts are in fact the wealth of tribal Nicobarese. Before currency was introduced in Nicobar Islands, pigs and coconuts were used as items for barter. The tribal Nicobarese bought cloth, rice, ornaments and other goods from the ships that halted on the shores of the Nicobar Islands, trading with pigs and coconuts.

Among the tribal Nicobarese, pigs held ritualistic importance too before the spread of Christianity and Islam. The tribal Nicobarese were animistic before embracing organized religion. As per their animistic beliefs, in order to please the ruling and the guarding spirits of the sea and the forest, the tribal Nicobarese performed various ceremonies in which pigs were sacrificed, and their blood was used to sanctify the ceremony as it was believed to be pure and sacred (Census of India 1971). In every village of the Nicobar Islands, pigs were and still are given as gifts on various occasions and are the most precious possession of tribal Nicobarese. Before colonization, the tribal Nicobarese did not have many options of food available on land as the small islands are surrounded by a vast expanse of sea. As pigs were easily available, it became the main ingredient of food. Since it constituted the main diet, and the wealth of the Nicobarese, pigs became the object of reverence.

As most of the pigs were kept as pets and roamed freely in the settlement, pigs too were affected and killed in the Tsunami. Of the total livestock killed in Nicobar Islands, pigs accounted for 68 per cent followed by goats and cattle at 26 per cent and 6 per cent, respectively. Compared to pre-Tsunami livestock figures, 36 per cent of the pigs, 93 per cent of the goats and 87 per cent of the pigs were killed. During the fieldwork conducted in 2010 in Car Nicobar Island, the villagers recollected that about 5,000 pigs out of a total of about 5,100 were killed in just two villages of Arong and Kimius, which is a staggering 98 per cent. The cattle and the goats belonged to the non-tribal population, but the pigs were the wealth of the tribal Nicobarese. In the aftermath of the Tsunami, the pig population was almost wiped out from the Nicobar Islands, thereby incapacitating the livelihood of the tribal Nicobarese.

In the Tsunami, not only the primary sector of the economy was devastated, but the secondary sector, although minuscule in the A&N Islands, were destroyed.

Impact on the small-scale industrial units

Industrial output contributes 22 per cent of the gross domestic state product (GSDP) and employs 19 per cent of the total workforce of the A&N Islands (Planning Commission 2008). In the pre-Tsunami period there were 1,838 small-scale industrial units (SSIU) in the A&N Islands. Of the total 1,838 SSIUs, 1,773 (96 per cent) were located in the Andaman Islands and 65 (4 per cent) were in the Nicobar Islands. In the Tsunami, out of the total 1,838 units, 726 units were damaged. Of these, 162 units were completely damaged and another 564 were partially damaged (Digal 2006). Of the 726 damaged units, 94 per cent were in the Andaman Islands and 6 per cent were in the Nicobar Islands. The damaged units accounted for 38 per cent and 72 per cent of the pre-Tsunami totals in the Andaman Islands and Nicobar Islands, respectively. Thus, even though the number of damaged units was 15 times more in former, damage as per cent of pre-Tsunami total was more in the latter. Another emerging sector of the economy, which is the source of livelihood for many, is tourism and it was also adversely affected.

Impact on tourism

Even though no tourists were killed in the A&N Islands, a sharp dip in the number of tourists after the Tsunami was reported. The tourism industry is a recent one in the A&N Islands. The promotion of the A&N Islands as tourism destination began from 1978 (Correspondent 1978). By the late 1970s, the A&N Administration felt a need for a sector that would generate revenue without taxing it. Tourism was the best option, as the A&N Islands have beautiful beaches, dense forests, wetlands, mangroves, enchanting coral reefs and a rich biodiversity.

Tourism appeared as the best revenue generating source in which the investment was low, as it was based on exploiting the natural beauty of the place without harming it in any way. Therefore, in the sixth Five-Year Plan, tourism promotion was given a boost. The first advertisement of tourism promotion appeared in the local newspaper *Daily Telegrams* a few days later (fieldwork 2010). The sector got another boost in 1991 when the year was declared as the 'India Tourism Year' (Netobel 2004). Furthermore, the Ministry of Tourism launched the campaign 'Incredible India' in 2002 to promote tourism in India along with the National Tourism Policy that promoted 'eco-tourism' as a thrust area along with enhancing the employment potential of the sector. The A&N Administration in its vision statement of 2005 stated that the limited scope for industrial activity on the islands coupled with the decline in the wood-based industry pursuant to the Supreme Court judgment dated 7 May 2002 has led to tourism being identified as a thrust sector for economic development, revenue and employment generation on the islands (Planning Commission 2008).

With the impetus from all the sectors, A&N Islands tourism recorded a significant growth. The number of tourists to the A&N Islands has grown from 9,500 in 1980 to 90,000 in 2001 and more than 307,000 in 2015 and is still growing (Directorate of Economics and Statistics 2015). Even though the number of tourists is expanding, yet tourism as a sector contributes to about 8 per cent of the GSDP and employs less than 1.5 per cent of the total workforce of the A&N Islands (EQUATIONS 2009).

The entire A&N Islands is known for its natural beauty, but tourism is confined to some of the Andaman Islands only. Port Blair, the capital of the A&N Islands, together with Ross Island and Viper Island, are popular tourist destinations for historical tourism, as Ross Island consists of the remnants of the British headquarters of the penal colony. Viper Island has the gallows where convicts of the penal colony were hanged, while the infamous Cellular Jail is in Port Blair. Another popular tourist destination is Havelock Island. The Radhanagar Beach of Havelock Island was declared as the best beach of Asia in 2004 by the November edition of *Time* magazine in Asia (Perry 2004). Similar to the beaches of Havelock Island are the beaches of Neil Island. A further attraction of Neil Island is the natural arch created by wave erosion. The mud-volcano and limestone cave of Baratang Island adds to the itinerary of the tourists. Further north, under Diglipur *tehsil* are the Ross and Smith Islands that are connected by sandbar during low tide which disappears during the high tides. An added attraction is the turtle nesting from December to February, when the olive ridley turtles come in large numbers on the beaches of the Andaman Islands to lay eggs. The Cinque Island, Long Island and Little Andaman Island are also open for tourists, but tourism has not yet developed in these islands.

The Nicobar Islands have not been open up for tourism. The Andaman and Nicobar Protection of Aboriginal Tribes Regulation of 1956 forbids any economic activity, with few exceptions, in the Nicobar Islands as it might harm the tribal Nicobarese. Therefore, the impact of the Tsunami on tourism was limited to only the Andaman Islands.

In the Tsunami, the nascent tourism industry of the A&N Islands got the blow when the number of tourists plunged to an all-time low in 2005. The number of tourists fell from 109,582 in 2004 to 32,381 in 2005, a fall of 70 per cent in one year (Figure 6.3).

December is the peak of tourist season for the Andaman Islands. There were nearly 800 foreign tourists and around 10,000 domestic tourists in the A&N Islands at the time of the Tsunami (EQUATIONS et al. 2008). No tourists were killed during the event. This is because the small budget hotels are concentrated in and around the main market of Port Blair. This is located at least 2–5 km away from the coast on high ground. Other hotels near the coast are located on elevated land, like Hotel Sinclair and Hotel Megapode Nest, and hence were saved. In Havelock Island, though roads were damaged, the jetty was affected and

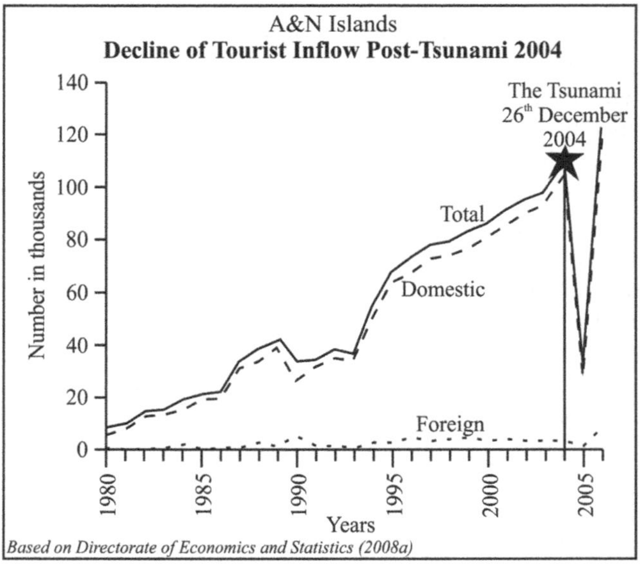

Figure 6.3 A&N Islands: decline of tourist inflow post-Tsunami 2004

hotels were inundated, yet no tourists were killed. However, the panic, the fear of the sea and the collapsed infrastructure of not only Havelock Island but the entire A&N Islands led to a decline in the number of tourists. Bookings for the hotels were cancelled for the entire year.

Initially the impact of the Tsunami was not felt severely as the inflow of media persons and the non-governmental organization (NGO) workers sustained the sector. Till about March 2005, the hotels had full occupancy. However, after this period, the tourist flow to the island dwindled. It was not only the hotel owners who were affected. Everyone associated with tourism, be it the local food vendors, the souvenir shop owners or the auto and the taxi owners – all suffered because of the decline in the number of tourists. Fieldwork in 2010 confirmed that the small traders dependent on tourism, like vendors and food stall owners of Chidiyatapu and Wandoor (the popular tourist place 24 km south of Port Blair) in South Andaman Island and vendors of Havelock Islands, were severely affected. Even though no tourists were killed, in the Tsunami the small population dependent on tourism lost their means of livelihood.

Another damage that intensified the impact of the Tsunami on the already damaged livelihood was the damage to infrastructure. A good and sturdy infrastructure is the prerequisite of any kind of development. The collapse of infrastructure is a disaster in itself. In the Tsunami, the limited infrastructure of the A&N Islands completely collapsed.

Damage to infrastructure

Large-scale damage to infrastructure was reported from the A&N Islands. Almost 70 per cent of the public and private infrastructure of the Nicobar Islands was wiped out (Bhagat 2005). In the Tsunami, the total economic loss in the A&N Islands was estimated at Rs. 38.27 billion (Ministry of Home Affairs 2005a). This was 234 per cent of the GSDP of the A&N Islands in 2004–05. This included the damages to infrastructure, its associated loss to trade and damage to houses.

Infrastructure serves many purposes and has been broadly classified into social infrastructure that includes schools and hospitals, and physical infrastructure which includes power, transport and communication networks. Social infrastructure provides for growth and development of the population, while physical infrastructure's role is in providing services and connecting places for ease of the flow of goods, services and people. In the Tsunami both the social and physical infrastructure of the A&N Islands were extensively damaged. The following section assesses the damage to the infrastructure of the A&N Islands by two different perspectives – first, island-wise damage for each of the sectors of infrastructure to assess the spatial picture of damage. However, this assessment does not bring out the intensity of the damage; therefore, a second perspective is given, where the damage has been assessed for each of the islands as per cent of pre-Tsunami totals.

Damage to schools

In the pre-Tsunami A&N Islands, there were 369 schools, of which 213 were primary, 59 middle, 49 secondary and 48 senior secondary. The A&N Islands have a four-tier schooling system. At the lowest level are the primary schools (classes I–V), followed by middle schools (classes VI–VIII), secondary schools (classes IX–X) and senior secondary schools (classes XI–XII). Among these schools, there is an overlap in the classes taught. For example, senior secondary schools have classes for primary school, middle school and secondary school as well. Since the highest class taught is 12th, it is called senior secondary school. Similarly, a secondary school has classes I–X and middle school has classes I–VIII. Primary schools have only classes I–V. Thus higher order schools have more classes and a larger number of students and teachers while primary school has fewer students and teachers.

Of the total 369 schools, 299 were in the Andaman Islands and 70 in the Nicobar Islands. In the Tsunami, out of this total of 369 schools, 67 were damaged: 32 in the Andaman Islands and 35 in the Nicobar Islands. Schools damaged as a percentage of the pre-Tsunami total were 11 per cent for the Andaman Islands but 50 per cent for the Nicobar Islands. Damage to schools was of two types: partial and complete. In partial damage, only a part of the school was damaged and it could be repaired and restored. These schools could function even with the damage. On the other hand, completely damaged schools fully collapsed and required reconstruction. Of the 67 schools damaged, 34 schools were partially damaged while 33 were completely damaged (Figure 6.4).

Out of 32 damaged schools of the Andaman Islands, 24 were partially damaged and eight (five in Little Andaman and three in South Andaman) were completely damaged. In the Nicobar Islands, of the 35 damaged schools, 10 were partially damaged and 25 were completely damaged. Here, except in Great Nicobar Island, the schools were completely damaged in all other islands. Even though the number of schools damaged in the Andaman Islands and the Nicobar Islands was near equal, as a per cent of pre-Tsunami totals, the schools damaged was more severe in the latter.

The schools became non-functional not only because of the collapse of the school building but also because many teachers were killed. In Katchal Island, all of the 47 teachers posted in the 11 schools were killed (Dhar 2005b). In Kendriya Vidyalaya of Car Nicobar Island, out of 10 teachers, four were killed (Majumdar 2005). The schools suffered further as the undamaged schools were converted into relief camps. Moreover, the fear of the Tsunami was such that even till February and March teachers refused to take posting in the Nicobar Islands (fieldwork 2010). The classes were hampered and schools resumed only after April 2005. Damage to schools affected the population of children, but it was the collapse of the medical services that severely impacted all affected islanders. In any disaster, medical facilities are most critical, as it is in medical centres that doctors, medicines and rest is provided. But in the Tsunami, medical facilities of the A&N Islands collapsed.

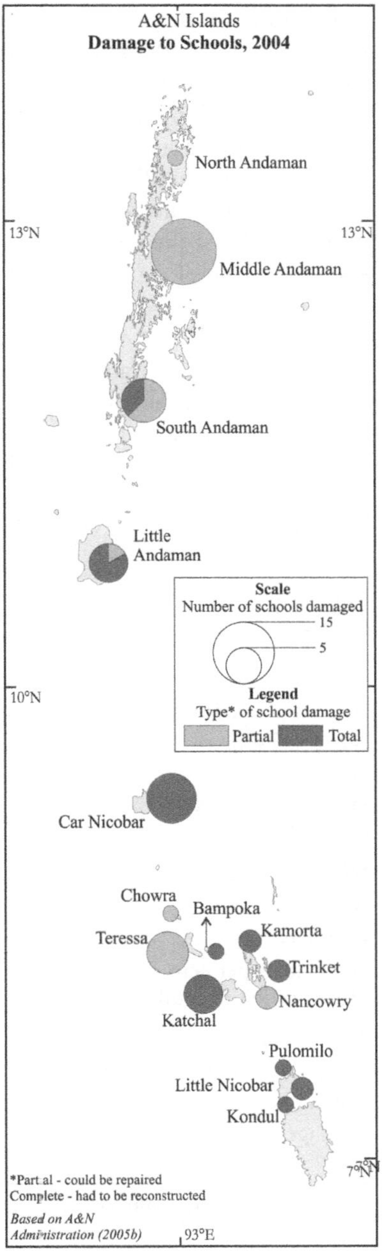

Figure 6.4 A&N Islands: damage to schools, 2004

Damage to medical services

In the pre-Tsunami period, the medical infrastructure consisted of 130 establishments, including hospitals, community health centres (CHCs), primary health centres (PHCs), urban health centres (UHCs) and primary health sub-centres (PHSCs), with a total of 1,005 beds for 350,000 population. Of these, 805 beds were in the Andaman Islands and 200 in the Nicobar Islands (Directorate of Economics and Statistics 2006). In the Tsunami, of the total 130 medical establishments, 31 were damaged, two in the Andaman Islands and 29 in the Nicobar Islands (Table 6.5).

Incidentally, none of the hospitals or CHCs were damaged, and it was the PHCs and the PHSCs which collapsed. Of the total 19 PHCs, four were damaged, and of the 103 PHSCs, 27 were damaged. In the Andaman Islands, of the 16 PHCs and 73 PHSCs, only one was damaged in each case. Therefore out of total 95 medical service infrastructures, two were damaged which was 2 per cent of the pre-Tsunami totals. In the Nicobar Islands, out of the 35 pre-Tsunami medical infrastructures, 29 were damaged accounting for a loss of nearly 83 per cent. The Nicobar Islands had one hospital, Bishop John Richardson Hospital in Car Nicobar Island, which was not damaged. All the three PHCs of Nicobar Island, namely, PHC Katchal, PHC Teressa and PHC Campbell Bay, were affected. Out of the 30 PHSCs, 26 were damaged. The damage to the medical service infrastructure incapacitated the Nicobar

Table 6.5 Tsunami and the damage to medical infrastructure

	Hospital	*CHC*[1]	*PHC*[2]	*PHSC*[3]	*Total*
Andaman Islands					
No. pre-Tsunami	3	3	16	73	95
Damaged	0	0	1	1	2
Damage as per cent of pre-Tsunami total	0	0	6.25	1.37	2.11
Nicobar Islands					
No. pre-Tsunami	1	1	3	30	35
Damaged	0	0	3	26	29
Damage as per cent of pre-Tsunami total	0	0	100	86.67	82.86
A&N Islands					
No. pre-Tsunami	4	4	19	103	130
Damaged	0	0	4	27	31
Damage as per cent of pre-Tsunami total	0	0	21.05	26.21	23.85

1 CHC: Community health centre.
2 PHC: Primary health centre.
3 PHSC: Primary health sub-centre.

Source: Based on Directorate of Economics and Statistics (2006) and A&N Administration (2005b)

Islands. It is usually said that you can survive for three weeks without food, but cannot live for three days without water. Thus potable water is the most essential requirement for sustenance. In the Tsunami the water supply was affected.

Damage to water supply

In the aftermath of the Tsunami, there were reports from the Nicobar Islands that in some of the islands the officers of the administration involved in the relief and rescue were using bottled water for bathing while in some of the islands there was a scarcity of drinking water (fieldwork 2010). The drinking water supply was hard hit. In the pre-Tsunami period the drinking water sources of A&N Islands included piped water in some islands and wells in most of the other islands. In the aftermath of the Tsunami, the wells became unusable as these were inundated and contaminated with the sea water. An increase in dysentery was reported from the relief camps of Car Nicobar Island (Shankar 2014). The water pipes were damaged, thereby hampering the water supply in places with piped water.

In the pre-Tsunami period only some parts of the A&N Islands were covered with a piped water supply. Of the total 501 inhabited villages, 241 villages had a piped water supply which was 48 per cent of total villages. Of these 241 villages, 206 villages were in the Andaman Islands and 35 villages were in the Nicobar Islands. These villages accounted for 65 per cent and 21 per cent of the total villages of the Andaman Islands and Nicobar Islands, respectively. Rainwater is the source of the piped water supply. The rainwater is collected then through pipes and is taken to the settlements.

Port Blair and the surrounding areas, including villages, have a well-developed water supply network. Rainwater is the source of water supply to Port Blair. Rainwater is collected in natural reservoirs, small check dams and wells, and is treated before being transferred to pumping stations for distribution. There are five reservoirs that collect rainwater for supply to Port Blair. These are Dilthaman Diggi which is the smallest of all and has capacity to hold 0.20 MLD (mega-litre per day) of water. The others are Lambaline Diggi (0.70 MLD), Nayagaon Diggi (1.17 MLD), Jawahar Sarovar (2.70 MLD), and the largest is the Dhanikhari Dam Reservoir (18.20 MLD). The water treatment plants are located at Lambaline, Dairy Farm, Garacharma and Dilthaman tank and the pumping stations are at Police Hill, Buniyadabad, AIR colony and Dairy Farm. Water is transferred through a network of pipes from each of these sites to the consumers.

Thus the piped water supply infrastructure consists of a water collection centre, water treatment plant, pumping station and distribution lines network (Andaman Public Works Department 2009). Each of the water supply infrastructures was adversely affected in the Tsunami.

In the Andaman Islands, the damage to water supply was because of the upliftment and subsidence rather than the Tsunami per se. Distribution pipes were worst affected due to ground shaking; the joints broke apart, thereby disrupting the water supply. The pipeline from Dhanikhari Dam to Port Blair reported breakages at 38 points. In the Lambaline area in Port Blair land subsided, therefore sea level rose and affected the water pipeline. Of all the reservoirs, the Dhanikhari Dam Reservoir and the Chouldari Dam Reservoir sustained damages due to earthquake. Cracks developed in the foundation and seepage occurred at Dhanikhari Dam Reservoir. The crest of Chouldari Dam Reservoir buckled under the stress generated due to the earthquake (Tang et al. 2006).

In the Nicobar Islands, Car Nicobar and non-tribal settlements of Great Nicobar had piped water supply. The remaining 10 islands (namely, Bompoka, Chowra, Teressa, Trinket, Katchal, Kamorta, Nancowry, Pulomilo, Kondul and Little Nicobar) depended on wells for water. In pre-Tsunami in Car Nicobar Island water was supplied through the desalinization plant installed by the Central Salt Research Institute. The reverse osmosis system treated brackish water and then supplied it to the settlements. After the Tsunami, the salt content of the water increased and the water quality fell. The plant also suffered damage as it got inundated. In Great Nicobar Island, the pipes for distribution of water were washed away, thereby incapacitating the settlement. Not only was there no water in the period after the Tsunami, there was no electricity either. The power supply infrastructure also collapsed.

Damage to power supply and communication

In the A&N Islands thermal power is the main source of generating electricity for commercial purposes. Of the total electricity generated, 95 per cent is from thermal power plants, while the remaining 5 per cent is from hydel power plants (Tang et al. 2006). The diesel for generating the thermal power is brought from mainland India in tankers. The islands are separated by the large expanse of the sea, therefore transmission over long distances is not economical. Thus, there is no single power grid in the A&N Islands. Power houses have been set up on different islands to cater to its needs.

Power generation in the A&N Islands began in 1926 at Ross Island through a steam-driven diesel generator of 100 KW capacity. Another 100 KW capacity was added in 1929 at Port Blair. In 1951 one power house with 1.1 MW electricity generating capacity was installed at Port Blair. The number of power houses was 34 in 2001 with generating capacity of 44.25 MW. In 2002 the Kalpong hydroelectric power project in North Andaman Island added 5.25 MW of capacity to the total. Another 20 MW capacity was added in 2003 with the setting up of a diesel-based power plant in Bambooflat in South Andaman Island by Suryachakra Power Corporation Limited. Thus, in 2003, before the Tsunami, in the A&N Islands there were 35 power plants with 69.5 MW power generating capacity (Tang et al. 2006). Of this total 69.5 MW, 88 per cent of the installed capacity was in the Andaman Islands and 12 per cent in the Nicobar Islands.

As per the Ministry of Power (2011), a village will be declared as electrified if (i) basic infrastructure such as distribution transformer and distribution lines are provided in the inhabited locality as well as the dalit basti hamlet where it exists; (ii) electricity is provided to public places like schools, panchayat office, health centres, dispensaries, community centres etc.; and (iii) the number of households electrified should be at least 10 per cent of the total number of households in the village. This definition came in effect from 2004–05. Prior to this, the 1997 definition was used. Therefore the Census of India (2001b) used the 1997 definition, according to which a village will be deemed to be electrified if the electricity is used in the inhabited locality, within the revenue boundary of the village for any purpose whatsoever (Deendayal Upadhyaya Gram Jyoti Yojna 2004).

In pre-Tsunami A&N Islands, out of 501 inhabited villages, 329 villages were electrified, while the remaining 172 villages were without power. In the Andaman Islands 30 per cent of inhabited villages were without power supply, while in the Nicobar Islands this percentage was 44 per cent. The power supply network consists of diesel-generating set, the building in which the set is installed, the transmission transformers and the transmission lines. Damage to any one part can halt the power supply. In the Tsunami the entire network of power supply was severely affected.

Damage to the power supply infrastructure was of three kinds – damage to the generation sets, damage to transmission lines and damage to buildings where the generation sets were installed (Tang et al. 2006). Out of the total 69.5 MW power generating capacity in the A&N Islands, 38.75 MW was damaged in the Tsunami. Of the 38.75 MW damaged, 88 per cent was in the Andaman Islands and 12 per cent in the Nicobar Islands. Maximum damage to generation capacity was in South Andaman Island followed by North Andaman Island. This was on account of the closure of the Suryachakra Power Corporation and the Kalpong Power Plant. Damage to both accounted for 65 per cent of the total power generation capacity loss. The plant of the Suryachakra Power Corporation was completely shut down since seawater inundated its control room. The boundary wall of the plant was breached and equipment was damaged. Its supply of 20 MW of power remained suspended. The only hydro-electric power plant, Kalpong Power Plant in North Andaman Island too suffered damage because of the ground shaking of the earthquake. The power generator was in working order but there was misalignment at the foundation and cracks developed on the dam (Tang et al. 2006).

Of the 12 Nicobar Islands, two islands did not have any power supply before the Tsunami. In seven islands power generating capacity, transmission lines and the power house were damaged, while in three islands, Great Nicobar, Kamorta and Nancowry, only the transmission lines were destroyed. In Car Nicobar 3.35 MW generation capacity was damaged, followed by 0.58 MW in Katchal, 0.42 MW in Teressa, 0.17 MW in Chowra and 0.07 MW in the remaining Little Nicobar, Pulomilo and Trinket (Ministry of Home Affairs 2005a).

Even though ‘quantitatively’ more damage (88 per cent of the total 38.75 MW power generating capacity damaged) occurred in the Andaman Islands, the impact of the loss was more intensive in the Nicobar Islands. In the Andaman Islands there were 36 diesel generating power houses with installed capacity of 65 MW. Of this 37 MW was dislocated and shut down due to the Tsunami. The remaining 28 MW that was not damaged was used to support the affected areas, while in the Nicobar Islands the power supply completely stopped. The impact was severe for the Nicobar Islands (Figure 6.5).

In the Andaman Islands, of the pre-Tsunami total installed power generation capacity, 56 per cent was damaged and in the Nicobar Islands 54 per cent was damaged. Among the Andaman Islands only in Little Andaman Island was power generation capacity completely damaged. In three islands, namely, North Andaman, Middle Andaman and South Andaman, the damage was 78 per cent, 74 per cent and 49 per cent, respectively, of the pre-Tsunami total installed power generation capacity. Among the Nicobar Islands, complete damage to the power generation capacity was reported from six out of 10 islands. These were Chowra, Katchal, Little Nicobar, Pulomilo, Teressa and Trinket. In Car Nicobar Island, 99 per cent per cent of the power generation capacity was damaged. In the remaining three islands, namely,

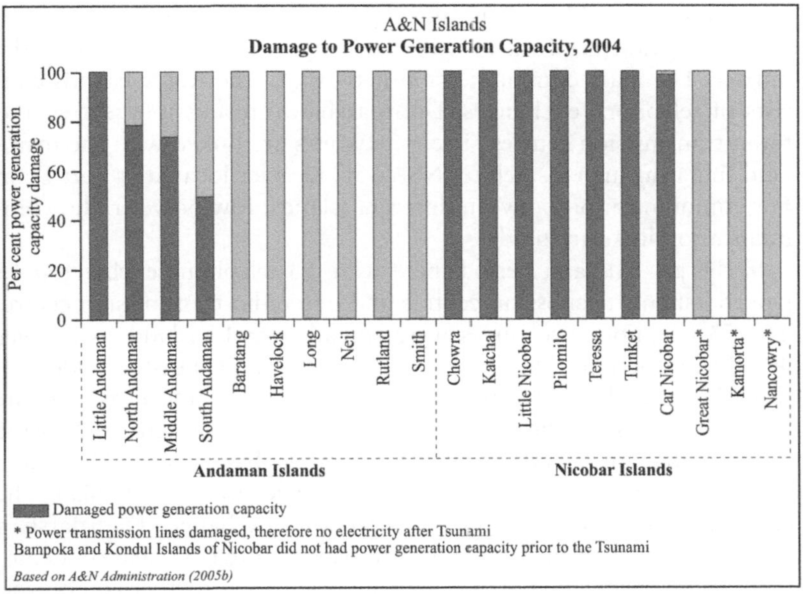

Figure 6.5 A&N Islands: damage to power generation capacity, 2004

Kamorta, Nancowry and Great Nicobar, the power generation capacity was not damaged, however the transmission network was damaged and hence power supply remained suspended until the lines were restored by February 2005.

Thus, it can be concluded that the Nicobar Islands suffered severely even though quantitatively the Andaman Islands accounted for a larger share of the damage of the installed power generation capacity. This is because in the pre-Tsunami period, the Nicobar Islands had negligible power supply infrastructure when compared to the Andaman Islands and when this minimum was also damaged, its impact was acute. Moreover in the Andaman Islands raw materials, labour and engineers were available to fix the damage. On the other hand, the Nicobar Islands became inaccessible after the Tsunami as the transport and communication networks had collapsed.

A sector completely dependent on power supply is the communication infrastructure. All equipment requires electricity to operate. In the Tsunami, along with the power supply the communication network also collapsed. Communication infrastructure is imperative in disaster response. The ability to communicate over long distances in short time periods is essential for rescue and supply of relief material. Damage to communication network in a disaster aggravates and intensifies the impact. The main communication network of the A&N Islands consists of telephone exchanges, Doordarshan transmission centres and radio transmission centres. These networks are linked with the mainland India through a satellite INSAT-3C receiver located at Port Blair. For communication between different islands, low power radio communication links are used.

In the pre-Tsunami period there were 53 telephone exchanges, 15 Doordarshan transmission centres and one radio transmission centre. The 53 telephone exchanges were spread over 14 islands with over 40,000 connections (Directorate of Economics and Statistics 2006). Of these 53, 42 exchanges were in the Andaman Islands and 11 were in the Nicobar Islands. Of the 15 Doordarshan transmission centres, 10 were in the Andaman Islands and five in the Nicobar Islands. There was only one radio transmission centre that was located in Port Blair. The communication infrastructure of telephone exchange, Doordarshan transmission centre and radio transmission centre consisted of a communication exchange building, satellite receivers and other equipment.

The communication network completely collapsed after the earthquake. In North Andaman Island, Middle Andaman Island and northern part of South Andaman Island, damage was due to the ground shaking in which wires snapped and transmission poles were razed to the ground. In the southern part of South Andaman Island, Little Andaman Island and in the Nicobar Islands, the exchange was wiped off or submerged in the Tsunami. In Port Blair, the satellite antenna for INSAT-3C lost alignment and connection with mainland India was severed. Telephone and Doordarshan buildings located on or close to the coast like in Bambooflat in South Andaman Island, Hut Bay in Little Andaman Island, Car Nicobar Island and Campbell Bay in Great Nicobar Island were washed off. In the islands of Katchal, Kamorta and Teressa, equipment was damaged (Tang et al. 2006). In the islands of Nancowry, Katchal, Teressa, Kamorta and Great Nicobar, the communication links were suspended for over two days. As a result of the breakdown of the communication network, all connection between the Nicobar Islands and the Andaman Islands was severed. Connection could be established with Car Nicobar Island and Great Nicobar Island only by afternoon on the day that Tsunami hit the coast, while it took more than a week to restore the communication network in the remote islands of Nicobar. The reason for the delay in re-establishing the communication network was the damaged transport infrastructure.

Damage to transport infrastructure

In the A&N Islands, all three modes of transportation – airways, roadways and waterways – are present. Dependence of the A&N Islands for transport of goods and services on the airways is less, even though it is the fastest mode of travel between the islands and between the mainland and the island. Airways are mostly used for passenger movements; and as the cost of air transportation is very high, not many islanders use it. Two modes of transport are the most important: sea transport and road transport. Sea transport is used for connecting mainland India with the A&N Islands as well as connecting the different islands. Road transport is used for transportation within the islands. Therefore each of the modes is important for connecting the A&N Islands with the rest of the world not only for transfer of people but also for all goods and services. In the Tsunami, all three modes were severely damaged.

Damage to roads: There are two types of roads present in A&N Islands: the metalled and the non-metalled roads. Metalled roads are motorable while the latter become unusable during the rains. (In the following section, the term 'road' refers only to the metalled roads.) Roads of the A&N Islands can be broadly grouped in two kinds – trunk roads and settlement roads. Trunk roads are major roads built for travelling long distances, while settlement roads link places within a settlement. The latter are urban and rural roads. The urban roads are wider and have higher traffic density than the rural roads. Pre-Tsunami, the A&N Islands had three trunk roads – the Great Nicobar Trunk Road (GNTR), the Little Andaman Trunk Road (LATR) and the Andaman Trunk Road (ATR). The total length of trunk roads pre-Tsunami was 450 km. Besides the trunk roads there are settlement roads, both urban and rural. Urban roads are concentrated in and around Port Blair, and their total length pre-Tsunami was 117 km. The total length of rural roads was 425.29 km in the A&N Islands. Thus, the total road length of the A&N Islands pre-Tsunami was 992.29 km (trunk road: 450 km; urban roads: 117 km; rural roads: 425.29 km).

Out of the total 992.29 km, 207.3 km were damaged accounting for 20 per cent of the total road length (Digal 2006). Of the total road damaged 17 per cent was in the Andaman Islands and 83 per cent was in the Nicobar Islands. It was damage to the trunk roads that severed the connection between the settlements in these islands. All three of the trunk roads suffered damage, but it was the GNTR that suffered the maximum damage (Table 6.6).

The north-south stretch of 51 km was completely wiped out. The east-west segment was already in dismal condition before the Tsunami.

Table 6.6 Damage to trunk roads in A&N Islands, 2004

Trunk road	*Length (km)*	*Width (m)*	*Islands covered*	*Connecting points*	*Damage post-Tsunami*
Andaman Trunk Road	333	3.6	South Andaman, Baratang, Middle Andaman and North Andaman	Chidiyatapu to Ariel Bay	Stretch in South Andaman along coast severely damaged
Great Nicobar Trunk Road	93	3.6	Great Nicobar Island	Kopen Heat to Campbell Bay and Campbell Bay to Indira Point[1]	> 90 per cent damaged
Little Andaman Trunk Road	24	3.6	Little Andaman Island	Dugong Creek to Harminder Bay	8 km along Hut Bay damaged
Total	450				

[1] Indira Point is the southernmost tip of India located in Great Nicobar Island at 6°44′ north latitude and 93°50′ east longitude.

Source: Based on A&N Administration (2005b)

The remaining segments also collapsed after the earthquake of magnitude Mw 9.1. The second most affected trunk road was the LATR. The stretch from Ramkrishanpur settlement till Hut Bay was washed off, as this part was very close to the coast while remaining parts were unaffected (Andaman Public Works Division 2006).

Segments of the ATR in the South Andaman Island lying close to the coast were washed away. Due to ground shaking and lateral spreading, cracks developed on the roads. Embankments, culverts and sluice gates in the roads were breached and the roads were washed away. The Chengappa Bridge connecting the Middle Andaman Island and North Andaman Island was damaged. Its middle section fell from its bearing making the bridge unusable. In the earthquake of 26 December 2004, the spans dislodged from their piers. The bridge remained out of operation for more than six months (Rai et al. 2006). The urban roads lying close or on the coast in the South Andaman Island were severely damaged such as the Junglighat jetty road, Bambooflat jetty road, and roads along Marina Park, Haddo and Corbyn's Cove.

In 2001, out of the total 501 inhabited villages, 252 were connected by roads while the remaining 249 villages were not (Census of India 2001b). Of 252 villages connected with roads, 80 per cent were in the Andaman Islands and the remaining 20 per cent were in the Nicobar Islands. There were no roads in seven out of 12 of the Nicobar Islands. These are Bompoka, Chowra, Kondul, Little Nicobar, Nancowry, Pulomilo and Trinket, and all roads of the Nicobar Islands were confined to five islands only. Roads in the Nicobar Islands connected their villages, and all villages were located on the coasts. In the pre-Tsunami period, only 20 per cent of the villages of the Nicobar Islands had roads. The remaining 80 per cent of the villages did not have any roads. There were only pathways or mud-roads. Even the little roads that Nicobar Islands had were damaged in the Tsunami.

Not only did Nicobar Islands account for 83 per cent of the total road length damaged, but road length damage as a percentage of the pre-Tsunami total was also high (Table 6.7).

In the Andaman Islands, 4 per cent of the pre-Tsunami roads were damaged, while in the Nicobar Islands 80 per cent of the pre-Tsunami roads were damaged. Therefore, in terms of road damage it was the Nicobar Islands which were more affected. The damages to roads led to the delay in rescue of stranded villagers. Without the roads, the reconstruction of houses got delayed as the house construction materials could not be transported to the site. The relief items could not be supplied on time either. Thus the 207.3 km of the damaged stretch of the road compounded the impact of the Tsunami. The damage to the airports and airstrips aggravated the delay, as relief supplies could not be airdropped.

Table 6.7 Road damage in A&N Islands, 2004

Islands	*Road length (in km)*		*Damage as per cent of total pre-Tsunami*
	Pre-Tsunami (2001)	*Damaged in the Tsunami*	
Great Nicobar	102.4	92.5	90.33
Car Nicobar	72.8	64	87.91
Teressa	4.5	3	66.67
Katchal	26.66	13	48.76
Little Andaman	58.33	8	13.72
South Andaman	426	26.8	6.29
Roads in other islands	301.6	0	0
Total	**992.29***	**207.3**	**20.89**

* Urban road length: 117 km; rural road length: 875.29 km; total road length: 992.29 km.

Source: Based on Directorate of Economics and Statistics (2006) and A&N Administration (2005b)

Damage to airports/airstrips: The fastest mode of travel between the A&N Islands and mainland India is air transport. In case of the remote Nicobar Islands spread over 250 km, air transport becomes critical as it aids in quick rescue and evacuation and also for the delivery of relief to the affected areas. But there are prerequisites for air transport to operate. The infrastructure required for air transport comprises aircraft, landing strips and the air traffic control tower. In the pre-Tsunami period the A&N Islands had three airstrips and one airport. There are a couple of crucial differences between an airstrip and an airport. Airstrips are for the use of armed forces, while airports are for the use of commercial flights. However, in case of emergency airstrips are used for emergency landing of commercial flights. Another crucial difference is that airstrips are mostly for defence purposes, hence they do not have a passenger cargo handling facility and a passenger terminal, which are essential components of an airport.

The three airstrips of the A&N Islands are at Diglipur (North Andaman), Malacca (Car Nicobar) and Campbell Bay (Great Nicobar), while the A&N Islands' lone airport is located at Port Blair (Figure 6.6). Among the four airstrips and airports of A&N Islands, the longest and the largest air strip is that of Port Blair airport, followed by Car Nicobar, Campbell Bay and Diglipur. It was the British who constructed first airstrip of the A&N Islands at Port Blair in 1937. It was a *kutcha* strip for landing and taking off of fighter planes. In 1942, during the World War II, when the Japanese took over the A&N Islands, this airstrip at Port Blair was paved, lighted and made operational for regular take-off and landing of aircraft. Bombings during the World War II made this strip non-functional. But in 1958, the strip was revived after the director general of civil aviation took over the airstrip, and an air link with mainland India was established in 1960. For security concerns the air strip was transferred to the Indian Navy in 1984. In 2004, the civilian flights operated only from this airport in Port Blair. It is known as Veer Savarkar International Airport, though no international flights operate to date.

In 1943, the second airstrip was also constructed by the Japanese during the occupation of the Nicobar Islands at Malacca in Car Nicobar Island. The British reoccupation of Car Nicobar Island in 1945 brought this airstrip under the command of the Royal Air Force. In 1956 the command of the base was transferred to the Indian Air Force. To build defences on the A&N Islands, two other airstrips were added: one at Campbell Bay in Great Nicobar Island in 1998 and a second at Diglipur in North Andaman Island in 2000. In the Tsunami, only one out of the four airstrips was severely damaged:

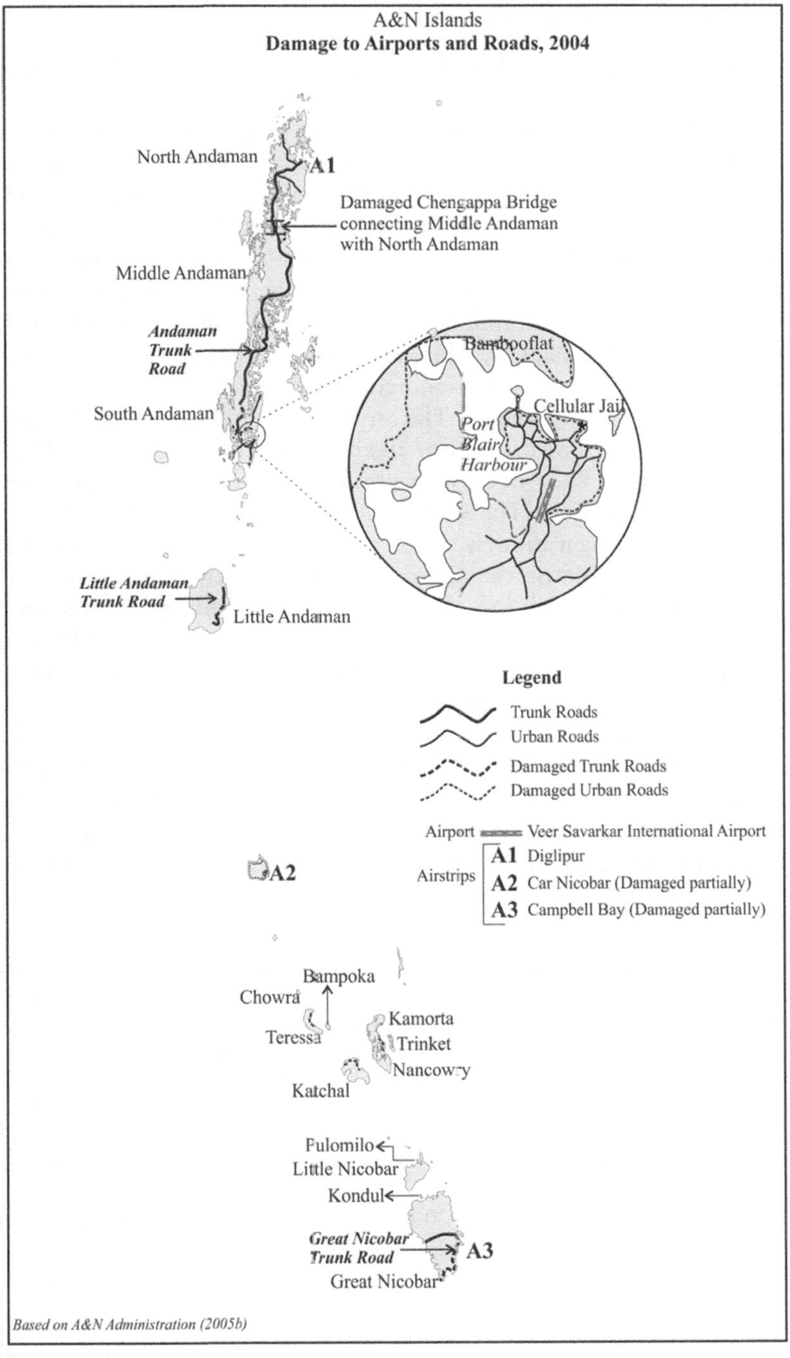

Figure 6.6 A&N Islands: damage to airports and roads, 2004

the airstrip at Car Nicobar Island under the command of the Indian Air Force (Table 6.8).

Ground shaking due to the earthquake severely damaged the Car Nicobar airstrip. Though this airstrip was made of sturdy material, it was made of rectangular panels 6 m wide, 3.75 m long and 550 mm thick which were joined together. It was at these junctions where the damage occurred (Rai et al. 2006). The damage was also accentuated after the Tsunami, as numerous large aircraft made frequent landings to evacuate stranded people and to supply relief items for the people in Nicobar Islands. Along with the airstrip, the air traffic control tower of Car Nicobar Island collapsed due to the shaking. None of the aircraft was damaged. The airport at Port Blair and the airstrip at Diglipur were constructed of flexible material and hence suffered minor damage. They developed cracks on the runway which were repaired immediately. The Campbell Bay airstrip did not suffer any major damage even though it was closest to the epicentre (Rai et al. 2006). This is because the airstrip was 1.5 km away from the coast but was at a height of 5 m above sea level, and thus was saved from the Tsunami inundation. Even after immense damage, the only point of supply of relief in Car Nicobar Island was the airstrip. Air dropping of food items was resorted to for small islands, but it failed as the site of earlier settlement was deserted and the location of the surviving people in the dense forest was difficult to gauge. Moreover, the smaller islands of Nicobar did not even have helipads for landing. Thus, the only option to connect with other islands was the sea transport, the lifeline of the A&N Islands. However, the seaports, wharfs and jetties

Table 6.8 Damage to airstrips and airports of A&N Islands, 2004

Airport/ airstrip	*Island*	*Year of construction of airstrip*	*Dimension of runway (m length/ width)*	*Status post-Tsunami*
Malacca	Car Nicobar	1943	2,717/43	Severe damage
Port Blair	South Andaman	1937	3,290/45	Minor damage
Diglipur	North Andaman	2000	975.36/30	Minor damage
Campbell Bay	Great Nicobar	1998	1,050/30	No damage

Source: Based on A&N Administration (2005b), www.ourairports.com and fieldwork (2010)

were also damaged. This emerged as the most severe impediment in rescue, evacuation and supply of relief materials.

Damage to wharfs/jetties: Sea transport is the lifeline of the islands. The A&N Islands are dependent on mainland India for all its requirements and therefore all goods and services are 'imported'. 'Import' means to bring in (merchandise, commodities, workers etc.) from a foreign country for use, sale, processing, re-export, or services. However, in the case of the A&N Islands, the term 'import' refers to goods, services and labourers brought from mainland India and not from a foreign country. Food grains are the dominant item of imports. The minuscule agricultural produce of the A&N Islands is not able to meet the food grain requirement. Therefore, food items like rice, sugar, wheat, cereals, edible oils, processed foodstuffs such as tea, coffee, baby food, health drinks, milk powder and butter, to name a few, are imported. Other essential commodities imported are petroleum and petroleum products, liquefied petroleum (LP) gas; agricultural goods such as fertilizers, pesticides, seeds, water pumps and agricultural implements; construction material such as cement, steel, asbestos sheets, steel pipes, ply boards and paint; durable consumer goods like radios, refrigerators, television sets, and bicycles, motorcycles, cars and other vehicles; books, paper, stationery items, bags; personal use items such as soap, shampoo, clothes, footwear, medicine and other items. In short, every requirement of the A&N Islands from a needle to a car is fulfilled by mainland India. The bulk of all the aforesaid goods and many other goods and services are transported to the A&N Islands by the sea route. Thus, when sea transport infrastructure was damaged in the Tsunami, the A&N Islands came to a standstill.

An important component of sea transport is the large ships that carry the cargo from mainland India to Port Blair, and the smaller vessels that transfer materials to the remote islands. Immediately after the earthquake, the ships were put out in the sea and thus the Tsunami waves passed under them without damaging them. As a result, not many ships were damaged in the Tsunami. Out of 65 inter-island vessels, only seven were damaged (Ministry of Home Affairs 2005a). Even though there was no significant damage to the vessels, sea transport was worst affected because the wharfs and jetties of the A&N Islands collapsed. The ships and boats cannot ply without wharfs and jetties. Wharfs and jetties are structures where ships berth for embarkation and disembarkation of passengers and cargos. Wharfs are raised platforms constructed parallel to the coast for the berthing of large ships. They are constructed of concrete and have cargo-handling equipment like cranes and forklifts. Jetties are raised platforms but are constructed perpendicular to the coast. The construction material of jetties varies from rubble, stone, wood and concrete or a mix of all. Jetties are for the berthing of smaller ships that are manually loaded and unloaded. In the Tsunami, the wharfs and jetties of the A&N Islands were severely affected.

The A&N Islands had seven wharfs and 47 jetties in the pre-Tsunami period. Of the seven wharfs, five were located in the Andaman Islands and two were in the Nicobar Islands. Of the five wharfs of the Andaman Islands, three are around Port Blair harbour at Haddo, Chatham and Hope Town, and these three together handled 97 per cent of the cargo exchange between the A&N Islands and mainland India, which was about 800,000 million tonnes of cargo and 275,000 passengers in 2005–06 (Andaman Lakshadweep Harbour Work 2011). The cargo from mainland is first unloaded at the three wharfs in Port Blair. Sometimes the ships from Kolkata berth at Mayabunder wharf and part of the cargo is unloaded there. The cargo unloaded at Port Blair is then distributed to the remaining four wharfs (Mayabunder in Middle Andaman Island, Hut Bay in Little Andaman Island, Mus in Car Nicobar Island and Campbell Bay in Great Nicobar Island) for further disbursement to other small islands. From these wharfs the goods are carried in smaller vessels to the different islands.

On the basis of the berthing, jetties of the A&N Islands can be classified into three types: inter-island jetties, ferry vehicle jetties and local ferry jetties. The inter-island jetties are constructed of concrete and can handle large ships that transport cargo and passengers. The ferry vehicle jetties have a ramp where vehicles can be driven on and off the ship. These are used by buses, trucks and passengers alike to cross the narrow stretch of the sea, such as between Port Blair and Bambooflat and between South Andaman Island and Baratang Island and others. The third and most numerous in numbers are the local ferry jetties. These are smaller in size and support berthing of small boats chiefly used for fishing and for passenger transport. In the pre-Tsunami period there were 12 inter-island jetties, eight ferry vehicle jetties and 27 local ferry jetties in the A&N Islands. Of these totals, Andaman Islands had eight inter-island, eight ferry vehicle and 22 local ferry jetties. The remaining four inter-island and five local ferry jetties were in the Nicobar Islands. All of the 10 inhabited islands of the Andaman Islands had either a wharf or a jetty or both, but, four out of 12 of the Nicobar Islands, namely, Bompoka, Little Nicobar, Kondul and Pulomilo, did not have either a wharf or a jetty. The residents of these islands used canoes to travel and connect with the other islands. Prior to the Tsunami, the wharfs and the jetties were concentrated in the Andaman Islands.

In the Tsunami, out of the total 54 wharfs and jetties, 30 were damaged, of which 19 were from the Andaman Islands and 11 from the Nicobar Islands. Among the Andaman Islands, South Andaman accounted for maximum where 13 wharfs and jetties were damaged, followed by Little Andaman (2), Middle Andaman (2) and North Andaman (1). Out of the 11 damaged wharfs and jetties of the Nicobar Islands, Car Nicobar accounted for three and Great Nicobar accounted for two, while the lone jetties in each of the islands of Chowra, Kamorta, Katchal, Nancowry, Teressa and Trinket were damaged. Numerically more jetties were damaged in the Andaman Islands yet the consequence of the damage was more severe in the Nicobar Islands. In the Andaman Islands, 44 per cent of the pre-Tsunami total wharfs and jetties were damaged while in the Nicobar Islands the damage was 100 per cent.

The damaged wharfs and jetties sustained varied forms of damage. Some were partially damaged while others were completely damaged. The former had repairable damages and could be made functional, while the latter could not be repaired and had to be reconstructed. In the A&N Islands, of the total 30 damaged wharfs and jetties, 15 were partially damaged and 15 were completely damaged (Figure 6.7). In the Andaman Islands, out of 19 damaged wharfs and jetties, 12 were partially damaged while seven were completely damaged. The impact of jetties damaged was more severe in the Nicobar Islands, as most of its wharfs and jetties had completely collapsed. This delayed the response: evacuation was slow and supply of relief materials was delayed. Even if ships reached the shore, it was impossible to unload relief items in the absence of the jetties.

Other foreshore structures of sea transport such as breakwaters, spillways and passenger hall buildings were also impacted. The worst affected foreshore structures were the ferry terminal at Haddo wharf and the Phoenix Bay complex at Port Blair. The control tower of Little Andaman Island was destroyed and the one at Chatham Island developed major cracks. All the foreshore structures of the Nicobar Islands suffered intense damage. Damage to infrastructure affects all people dependent on it. But more than infrastructure damage, what causes more pain and grief is the loss of home. In the Tsunami nearly 10,000 houses were destroyed, thereby making about 50,000 people homeless.

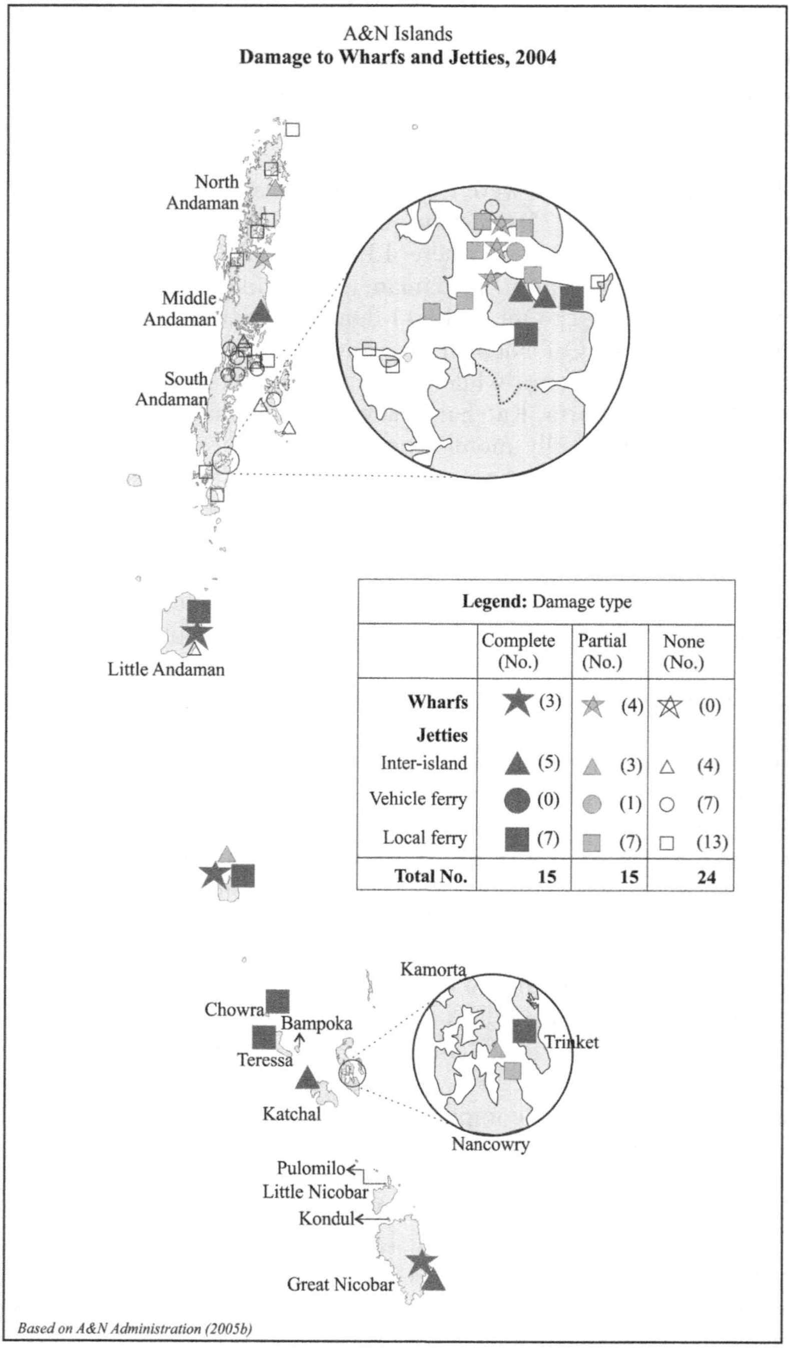

Figure 6.7 A&N Islands: damage to wharfs and jetties, 2004

Damage to houses

The Census of India defines a 'census house' as

> a building or part of a building used or recognized as a separate unit because of having a separate main entrance from the road or common courtyard or staircase, etc. It may be occupied or vacant. It may be used for a residential or non-residential purpose or both.
> (Census of India 2001e)

But a house is not merely a 'building' or a 'dwelling place'. It is a home, a place that provides sense of belongingness and sense of security to the household inhabiting it. It is a shelter, one of the basic necessities for human survival. In the Tsunami, more than 50,000 people were rendered homeless.

In the pre-Tsunami period, the A&N Islands had 104,238 census houses (Census of India 2001b). Of this total, 89 per cent were in the Andaman Islands and 11 per cent in the Nicobar Islands. In the Tsunami, 17,857 houses were damaged, that is 17 per cent of the pre-Tsunami total houses. Of the total houses damaged, 10,551 (59 per cent) were in the Andaman Islands and 7,306 (41 per cent) in the Nicobar Islands (Figure 6.8). In the Andaman Islands, South Andaman accounted for 76 per cent followed by Little Andaman (20 per cent), Middle

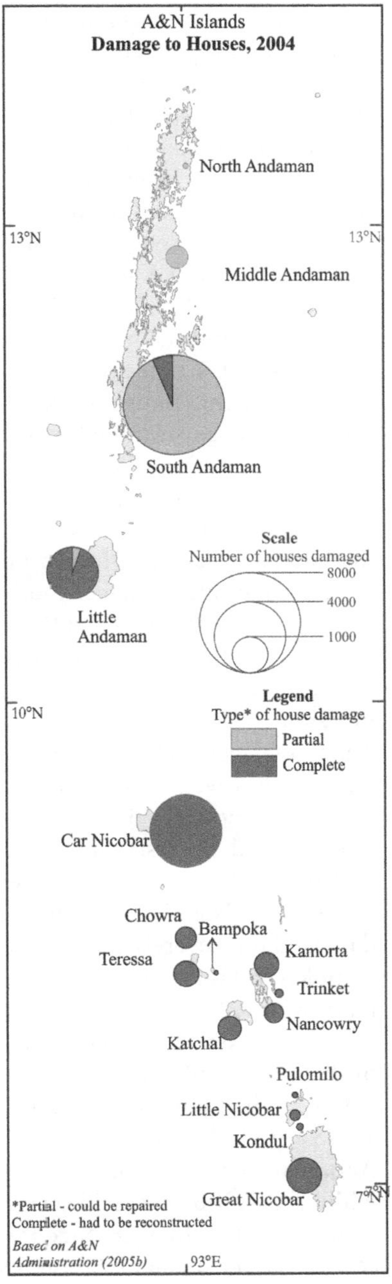

Figure 6.8 A&N Islands: damage to houses, 2004

Andaman (4 per cent) and North Andaman (0.23 per cent). No house damage was reported from any other of the remaining six Andaman Islands. Thus the damage to houses was concentrated in two islands that together accounted for 96 per cent of the total damaged houses. In contrast, the damage to houses in the Nicobar Islands was spread across all of its 12 islands. Car Nicobar accounted for the largest number, with 56 per cent of the total damaged houses, followed by Great Nicobar (13 per cent), Teressa (7 per cent) and Kamorta (6 per cent), and the islands of Katchal, Chowra, Nancowry, Little Nicobar, Trinket, Kondul, Pulomilo and Bompoka accounted for the remaining 17 per cent. Even though the number of houses damaged in the Andaman Islands was 1.4 times more than the houses damaged in the Nicobar Islands, yet this figure masks the actual impact. This is because the kind of damage sustained by the houses of Andaman Islands and the Nicobar Islands varied.

Damage to houses in the Tsunami was of two kinds: partial and complete. The kind of partial damage was further divided into two types: marginal and severe. Of the total 17,857 damaged houses, 43 per cent suffered marginal damage, another 2 per cent had severe damage, while 55 per cent of the houses suffered complete damage (Table 6.9).

Even though the Andaman Islands accounted for a higher number of houses damaged, 73 per cent of the 10,551 damaged houses sustained marginal damage, that is cracks developed in the houses. Another 4 per cent were severely damaged, as either its walls or the roof collapsed. However, these houses could be restored with some repair. It was the completely damaged houses that that could not be repaired but required reconstruction. Of the 10,551 damaged houses of the Andaman Islands, 24 per cent were completely damaged. The case of the Nicobar Islands was different. Here, there were neither marginally damaged nor severely damaged houses, as all of the 7,306 houses were completely damaged. In fact, the houses were washed away in the Tsunami waves, leaving behind no trace of their existence. One can also gauge the intensity of the damage, when the damage as a percentage of the total number of houses pre-Tsunami is considered (Figure 6.9).

Out of the total 104,238 houses in the A&N Islands in pre-Tsunami, 17,857 houses were damaged. The 10,551 damaged houses of the Andaman Islands accounted for 11 per cent of the pre-Tsunami totals while the 7,306 of the Nicobar Islands was 66 per cent of the pre-Tsunami total. In Car Nicobar and Nancowry, more than 95 per cent of the pre-Tsunami houses were damaged. In another five islands – Bompoka, Chowra, Teressa, Little Nicobar and Kondul – more than 70 per cent of the pre-Tsunami houses were damaged while in another

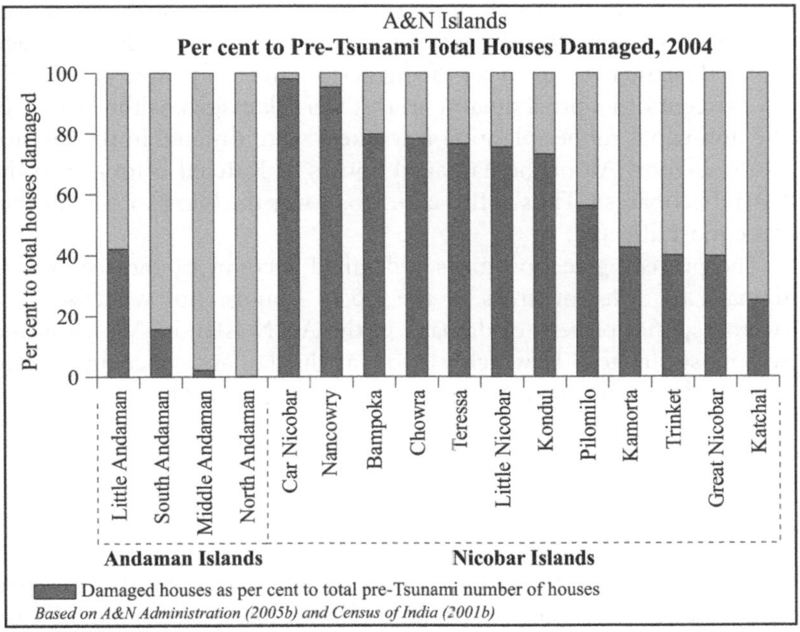

Figure 6.9 A&N Islands: per cent of pre-Tsunami total houses damaged, 2004

Table 6.9 Damage to houses, 2004

Islands	*Type of damage to houses*			*Total houses damaged*	*Per cent of total damage*
	Partial				
	Marginal[1]	*Severe*[2]	*Complete*[3]		
Andaman Islands	7,664	396	2,491	10,551	59.09
Nicobar Islands	0	0	7,306	7,306	40.91
Total A&N Islands	**7,664**	**396**	**9,797**	**17,857**	100
Per cent of total damage	42.92	2.22	54.86	100	

1 Marginal damaged houses require minor repair.

2 Severe damaged houses require major repair.

3 Complete damaged houses cannot be repaired, and require reconstruction.

Source: Based on A&N Administration (2005b)

four islands, Pulomilo, Kamorta, Trinket and Great Nicobar, the damage was more 40 per cent of the pre-Tsunami total. Even though Katchal Island accounted for the highest number of people killed, only 25 per cent of its pre-Tsunami houses were damaged as the houses of the non-tribal rubber plantation workers were situated in the interior of the islands. All of the damaged houses of Katchal belonged to the tribal Nicobarese. Thus in this case too, it was the Nicobar Islands that were most affected.

The preceding section gives a detailed account of the sector-wise damage to different areas of the A&N Islands. But what was the overall spatial pattern of damage in the A&N Islands? Which island was most affected? In which island was the damage concentrated? To answer these questions, a damage index was prepared.

Damage index

Individually the pattern of damage across physical and social infrastructure and houses clearly brings out that it was the Nicobar Islands which were most affected in terms of total damage and also as a percentage damage of the pre-Tsunami totals. However, more significant analysis is to club all these various types into a holistic picture. For this a damage index was constructed. A composite damage index was calculated to determine the overall damage in the A&N Islands.

The spatial unit of the composite damage index were the islands and not district, *tehsil* or village. This is because in the pre-Tsunami period the A&N Islands had only two districts (presently three districts); therefore, spatial variation does not emerge in district-wise analysis. The tehsils of A&N Islands are spread over two or more islands, thus, these too do not present the total picture. As for employing village as a spatial unit, in the A&N Islands the village boundaries are not clearly demarcated because the villages have encroached on the neighbouring forest. Moreover, the damage data was not available at the micro-scale of the village. Therefore individual islands were considered. The islands are different in terms of amenities, infrastructure and population composition, and together with the availability of island-wise damage detail made the islands the best spatial unit to represent the spatial pattern of damage.

The A&N Islands consist of 572 islands of which 38 were inhabited pre-Tsunami. Of these, 25 were in the Andaman Islands and 13 in the Nicobar Islands. For computing the composite damage index, only inhabited islands with a population of more than 50 people were considered (Table 6.1a). The number of such islands is 22: 10 in Andaman and 12 in Nicobar. Sixteen inhabited islands were not included. Two out of these 16 islands, namely, North Sentinel and Strait, are inhabited by the Sentinelese and Great Andamanese population, respectively. These are tribal peoples who still practice hunting and gathering and therefore the indicators selected do not apply in their case. Another two islands, Narcondam and Tillang Chang, are police outposts only, that is there are no households in these islands, only police camps are present, thus these were not included. Another six 'inhabited' islands, namely, Ross, Viper, Cinque, Aves, Curlew and Stewart, are 'accessible' during the day for tourism only and are devoid of any households. The remaining six islands, East Island, Flat Bay, Interview Island, John Lawrence, North Passage and Spike Island, have fewer than five households of the forest labourers. As the number of households is too small to compute indices, these were not considered. Moreover, the population of these 16 excluded islands is only 223 which accounts for a meagre 0.07 per cent of the total population of the A&N Islands. The 22 islands included for computing the indexes account for 99.93 per cent of the total population.

The composite damage index combined damages of various sectors to assess the extent of damage, which was not adequately represented by a single indicator (Figure 6.10). To compute the Tsunami damage index of the A&N Islands, 11 indicators were selected. These

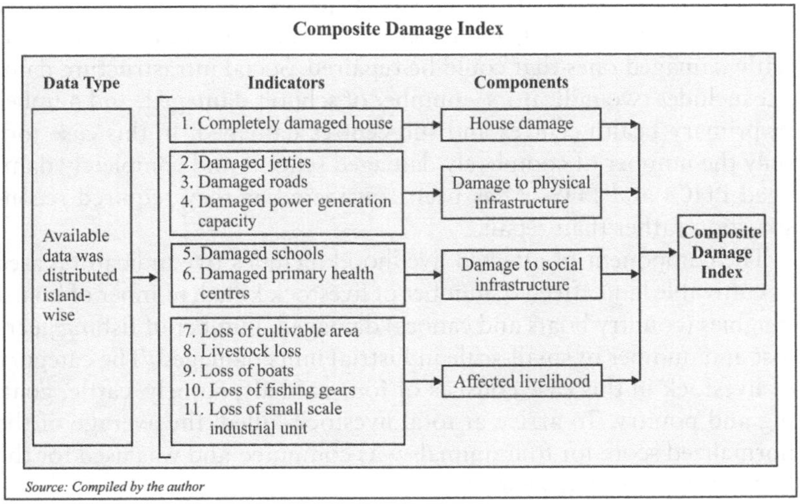

Figure 6.10 Composite damage index

indicators represent only tangible damage. The intangible impact such as sense of loss, grief and emotional turmoil of the people was not within the scope of the present research.

The 11 indicators selected are: (1) number of houses damaged, (2) number of jetties damaged, (3) road length damaged, (4) loss to power generation capacity, (5) number of schools damaged, (6) number of primary health centres and sub-centres damaged, (7) area of cultivable land affected, (8) number of livestock killed, (9) number of boats/dinghies (country boats and canoes) damaged, (10) number of fishing gears lost and (11) number of small-scale industrial units damaged. The gaps in the data were filled using suitable technique before computing the damage index. The indicators were then grouped into four components to assess which sector of damage contributed to maximum damage in an island. These are house damage, damage to physical infrastructure, damage to social infrastructure and affected livelihoods. The component of house damage has only one indicator: number of house damaged. For the purposes of damage index calculation, only the completely damaged house was considered, as this represented the actual damage. The marginally damaged or severely damaged houses could be repaired, but the completely damaged houses had to be reconstructed.

The physical infrastructure damage includes three indicators: number of wharfs/jetties damaged, road length damaged and loss to power generation capacity. For the calculation of the damage index, only the number of completely damaged wharfs/jetties was taken, as the latter had to be reconstructed, and hence its damage was more than the partially damaged ones that could be repaired. Social infrastructure damage includes two indicators – number of schools damaged, and number of primary health centres and sub-centres damaged. In this case too, only the number of completely damaged schools and completely damaged PHCs and PHSCs has been considered as they required reconstruction rather than repair.

The component of affected livelihood includes five indicators: area of cultivable land affected, number of livestock killed, number of boats/dinghies (country boats and canoes) damaged, number of fishing gears lost and number of small-scale industrial units damaged. The category of livestock in this case consists of four animals, namely, cattle, goat, pig and poultry. To arrive at total livestock killed, the average of the normalized score for four animals was computed and was used for the final composite damage index.

The data for some of the indicators was not available island-wise. For example, livestock losses were given as one unit for the islands of

Bompoka, Chowra and Teressa, also for the islands of Kamorta, Nancowry and Trinket, and also for Little Nicobar, Kondul, Pulomilo and Great Nicobar. Therefore to solve the problem, the data were distributed between the concerned islands on basis of their population distribution. The reason being that the greater size of population would require a higher livestock population, as these are domesticated and vary with the human population. Again, data for number of boats damaged and number of fishing gears lost for five islands – Trinket, Bompoka, Pulomilo, Little Nicobar and Kondul – were not given. From the available data for other islands, the number of boats damaged per 100 individuals was calculated. Correlating this with the population data of the islands, the number of boats damaged and the number of fishing gear lost was computed. The tribal Nicobarese across the various Nicobar Islands share commonalities; this gives credence to the method of completing the data set for the computation of the composite index.

For each component the indicators were normalized. Normalization refers to the division of multiple sets of data by a common variable in order to negate the indicator's effect on the data, thus allowing underlying characteristics of the data set to be compared. The 11 indicators had varied units of measurement consisting of numbers, length, area and energy units. Out of 11 indicators, eight were numbers, namely, houses damaged, jetties damaged, schools damaged, PHCs and PHSCs damaged, livestock killed, fishing boats lost, fishing gears lost and small-scale industrial units damaged. The indicator of road damaged was length (km). The indicator of cultivable land lost was area (ha), and the indicator for power capacity damaged was energy units (KW). To bring the data of the indicators on comparative scale normalization was imperative. For normalization, the formula used was

$$x_{new} = \frac{x - x_{min}}{x_{max} - x_{min}},$$

where x is given observation; x_{min} is the minimum value of the observation, x_{max} is the maximum value of the observation and x_{new} is the new normalized score. This formula scales the values of the different indicators between 0 and 1. The normalized value of different indicators was summed for the score of four components. The scores of the components were again normalized and summed up for the composite damage index. The scores of all indicators were added for the final composite score. Since each of the indicators is considered equally important, weights were not assigned to any of the indicators. The composite damage index score was arranged in ascending values

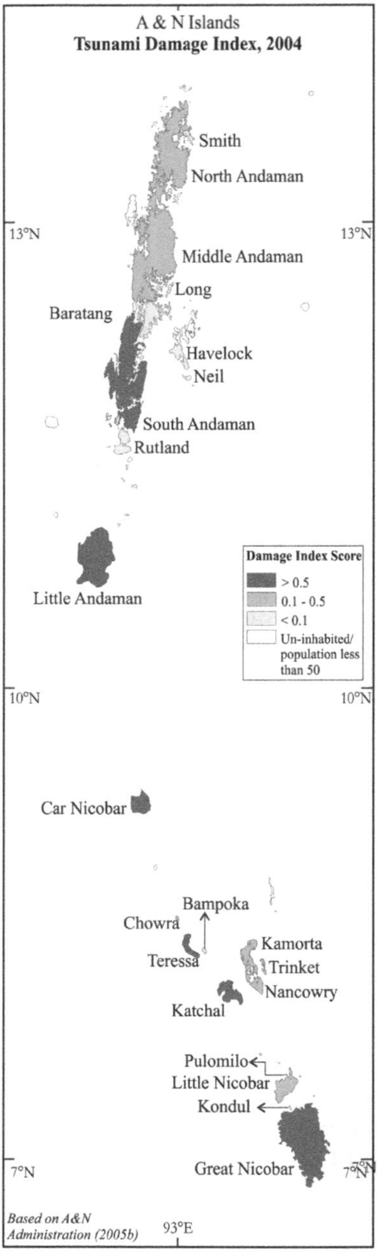

Figure 6.11 A&N Islands: tsunami damage index, 2004

and a line graph was prepared to identify the 'natural breaks'. A class interval was made using the 'natural breaks' to make three classes of high, medium and low composite damage index scores. The category of 'high' had the composite index score of > 0.5, 'medium' had the score between 0.1 and 0.5 and 'low' had the score of < 0.1. The damage index was thus calculated and mapped (Figure 6.11).

Of the 22 islands, six were in the category of high damage index score (> 0.5), another eight were in the category of medium damage index score (0.1–0.5) and the remaining eight were in the category of low damage index score (< 0.1). It is not that the islands in the category of 'low damage index score' did not have any damage, but the damage in these islands was low relative to that of the islands in other categories.

As per the composite damage index, the six islands in the category of high damage index score was Car Nicobar (3.11), followed by South Andaman (2.30), Great Nicobar (1.83), Little Andaman (1.48), Katchal (1.38) and Teressa (0.91). The eight islands in the medium damage index score (from high to low score) were Nancowry, Chowra, Kamorta, Middle Andaman, Trinket, North Andaman, Little Nicobar and Pulomilo. The remaining islands in the low damage index score category were Kondul, Bompoka,

Havelock and Neil Island. The remaining four islands had a damage index score of 0: Baratang, Long, Rutland and Smith.

From the map it is evident that the islands towards the south have more damage than islands in the north. Out of the 10 Andaman Islands, only two, South Andaman and the Little Andaman, were in the high damage index score category; another two were in the medium damage index score category; and the remaining six were in the low damage index score category. In fact, the four islands with damage index score 0 belong to the Andaman Islands.

In contrast, out of the 12 Nicobar Islands, four islands were in the high damage index score category, six in the medium category and two in the low category. Thus from the composite damage index it can be concluded that the damage was concentrated in the Nicobar Islands. Even though the composite damage index gives the overview of the damage in the A&N Islands in the Tsunami, it too has its flaw. Most of the islands with a low damage index score in the Nicobar Islands did not have infrastructure that could be damaged. Thus islands that had maximum infrastructure emerged as the islands in which the damage was concentrated. Again, the damage index does not provide the consequences of the damage. A case may be taken of jetty damage between South Andaman Island and Car Nicobar Island. The number of completely damaged jetties in the former was four, while for the latter it was two. Therefore in terms of total damage, it was the former which had a higher score than the latter. But the case is, in the pre-Tsunami period South Andaman had 19 jetties of which only four were completely damaged, thus 15 jetties could be made operational, while out of the three pre-Tsunami jetties of Car Nicobar, two were completely damaged. Thus the impact of the jetty damage was more severe for Car Nicobar than South Andaman even though the latter had more number of completely damaged jetties. This was the case with all the indicators related to infrastructure. Thus, the damages in the A&N Islands clearly reflect a disaster divide between the Andaman Islands and the Nicobar Islands. It was the vulnerability of the Nicobar Islands that led to the disaster when the Tsunami hit its shore while the Andaman Islands remained comparatively less impacted. The next chapter explores the vulnerability of the A&N Islands.

Appendix

Table 6A.1 Islands included for calculation of damage and vulnerability index

District	*Island*	*Population*	*Area (km²)*
Andaman	Baratang Island	6,062	297.8
Andaman	Havelock Island	5,354	113.9
Andaman	Little Andaman Island	17,528	731.6
Andaman	Long Island	2,199	17.9
Andaman	Middle Andaman Island	54,385	1,535.5
Andaman	Neil Island	2,868	18.9
Andaman	North Andaman Island	42,163	1,376
Andaman	Rutland Island	688	137.2
Andaman	Smith Island	676	24.7
Andaman	South Andaman Island	181,949	1,347.97
Nicobar	Bompoka Island	55	13.46
Nicobar	Car Nicobar	20,292	126.9
Nicobar	Chowra Island	1,385	8.2
Nicobar	Great Nicobar	7,566	1,045.1
Nicobar	Kamorta	3,412	188.2
Nicobar	Katchal	5,312	174.4
Nicobar	Kondul	150	4.6
Nicobar	Little Nicobar Island	353	159.1
Nicobar	Nancowry	927	66.9
Nicobar	Pulomilo	145	1.3
Nicobar	Teressa	2,026	101.4
Nicobar	Trinket	432	36.3
	Total included	355,927	7,527.33
	Per cent of total population	**99.94**	**95.9**

District	*Island*	*Population*	*Area (km²)*
Islands not included for calculation			
Andaman	Aves Island	2	0.2
Andaman	Cinque Island	6	9.53
Andaman	Curlew Island	2	0.03
Andaman	East Island	17	6.1
Andaman	Flat Bay Island	10	0.14
Andaman	Interview Island	16	133.4
Andaman	John Lawrence Island	25	41.98
Andaman	Narcondam Island	17	6.8
Andaman	North Passage Island	11	22
Andaman	North Sentinel Island	39	59.67
Andaman	Ross Island	0	0.65
Andaman	Spike Island	19	11.7
Andaman	Stewart Island	2	6
Andaman	Strait Island	42	6
Andaman	Viper Island	4	0.5
Nicobar	Tillang Chong Island	13	16.83
	Total	225	321.53
	Per cent of total population	**0.06**	**4.1**

Source: Census of India (2001a)

7 Vulnerability and the Tsunami

It is often stated that proximity of the A&N Islands to the epicentre of the Tsunami generated from the earthquake of 26 December 2004 was responsible for the damage and number of people killed. The epicentre was at 3°19′12″ north latitude and 95°51 east longitude, 300 km south-south-east of Great Nicobar Island. Given the reasoning, Great Nicobar Island should have recorded maximum killed and highest intensity of damage. To test this relationship, that proximity to epicentre causes more damage and killed, the statistical test called the Pearson's Product Moment Correlation (PPMC) coefficient (developed by Karl Pearson in the 1880s) was calculated. PPMC was chosen because it measures the degree of linear correlation (dependence) between two independent variables. It is denoted by the Greek letter 'ρ' (rho) or by the Roman letter 'r'. It is the covariance of the two variables divided by the product of their standard deviations; that is,

$$r = \frac{\sum(x - \bar{x})(y - \bar{y})}{\sqrt{\sum(x - \bar{x})^2 \sum(y - \bar{y})^2}}$$

where
x – variable 1
y – variable 2
$\bar{x}$ – mean of variable 1
$\bar{y}$ – mean of variable 2.

The value of rho ranges between +1 and –1 inclusive, where 1 is total positive correlation, 0 is no correlation and –1 is total negative correlation. Through this test the relation between the distance from the epicentre and damage and the killed could be established. Therefore, this statistical test was computed between the distances of the islands from the epicentre with two variables, namely, composite damage index score and the number of people killed in each of the island.

The PPMC coefficient was –0.15 between the distance of the island from the epicentre and the composite damage index, while it was –0.34 between the distance of the island from the epicentre and the number of people killed. Therefore, both the values indicate that there is small to medium negative correlation between the variables. A negative correlation indicates an inverse relationship, that is if one variable increases the other decreases. Thus, as the distance from the epicentre increases, the composite damage index score and the number of persons killed decrease. This relationship was comparatively stronger in case of the latter. However, it is a very weak correlation, especially in case of the composite damage index score. Thus, one can conclude that there is no strong role of proximity to the epicentre in determining either the damage or number of people killed.

Therefore, what was the reason for the concentration of numbers killed and the damages if not the distance from the epicentre? It is in this context that vulnerability was drawn into the frame of analysis. Vulnerability refers to pre-disaster conditions which make a population 'prone' to damage, destruction and death. It is not inherent, but is rendered over time through socio-economic structures and government policies which fail to remove inadequacies of society.

Vulnerability is a multidimensional concept and thus has been defined in various ways by different scholars. There is no consensus on a universal definition of vulnerability. The dictionary defines vulnerability as being 'capable of or susceptible to being wounded or hurt, as by a weapon'. Taking from the dictionary definition, Gabor and Griffith (1980) defined vulnerability 'as a threat to which a community is exposed'. In a similar vein, Cutter (1996) defined vulnerability 'as the likelihood that an individual or group will be exposed to and adversely affected by a hazard. It is the interaction of the hazards of the place (risk and mitigation) with the social profile of the communities'. Blaikie et al. (1994) too in their 'pressure and release' model postulated that physical hazards represent one aspect of force, while human factors of social and economic differentiation is another force. Together they create vulnerability like in a nutcracker, and when these two operate simultaneously they culminate in a disaster. But, Timmerman (1981) defined vulnerability as the degree to which a system reacts adversely to the occurrence of a hazardous event. The degree and quality of the adverse reaction are conditioned by the resilience of the system (a measure of its capacity to absorb and recover from the event). Granger (1997) suggested that vulnerability is in a form of a continuum from total resilience at one end and total susceptibility at the other. It implies that there can be a ranking of people in terms of vulnerability. The population that is most resilient is also the least vulnerable. Thus, the definition of vulnerability ranges from a focus on physical exposure, measures of socio-economic status, access to resources, sociological investigations of the differential ability of groups to resist harm and to recover afterwards and also to discussions of how the 'hazard of place' is linked to social profiles (United Nations Environmental Programme 2003).

Disasters are created by vulnerability. Kapur (2010) defines vulnerability as the state of being vulnerable, wherein 'vulnerable' means to wound. The Oxford Dictionary defines 'vulnerable' as 'exposed to the possibility of being attacked or harmed, either physically or emotionally'. Vulnerability is not inherent, but is rendered over time through socio-economic structures and government policies which reinforce and fail to remove existing inequalities in the society. The concept of vulnerability in this research draws from the definition given by Quarantelli (2005), where he defines vulnerability as 'weaknesses in the social structures or social systems' that are the cause of a disaster. In reality, disasters are the manifestations of these latent societal vulnerabilities. Accordingly, Quarantelli emphasizes neither an event nor a physical place or time as relevant to disasters. Instead, the entire conception is social: vulnerability is socially constructed by relationships in the social system (Rodríguez et al. 2007). Associated with this is the concept of rendered vulnerability. Kapur (2010) distinguishes between 'vulnerability simpliciter' and 'rendered vulnerability'. The former 'is like mortality, an inescapable fact, entailed by one being born as a human being'. On the other hand, 'rendered vulnerability' implies to be killed rather than to die. The social and economic marginalization and collective apathy of the society renders a particular group vulnerable. It is this vulnerability that manifests as disasters.

Though difficult to define, vulnerability is not that difficult to identify. Kapur (2010) has identified four features of vulnerability. First, vulnerability is an in-between condition: 'this "in-between condition" is akin to a rope walker who is neither secure nor has he yet fallen . . . it is a precarious state of affair' (Kapur 2010). Second, vulnerability is a warning as it points out what 'may happen'. This feature lends it the virtue of it being predictive. Therefore,

> given the present vulnerability one can forecast the outcome with a reasonable degree of accuracy. Since the state of collapse is 'yet' to occur, vulnerability carries the inherent score and hope to work out plans and policies for a better future.

A third feature of vulnerability is that it has elasticity of use as 'there can be vulnerable individuals, population, institutions and societies and places'. Because of this flexibility, vulnerability spreads across various disciplines and thus has applicability in varied streams of both physical and social sciences. Finally, the fourth feature of vulnerability

> is a state that stands not in isolation but in relation to some other component or variable which could be a substance, power, agent or process. It could be natural, manmade or both. A society could be vulnerable to globalization, climate change, food insecurity, poverty, crime, disease or for that matter 'natural forces'. Whatever be the 'other', the latter is critical, for without it vulnerability has neither meaning nor existence.
>
> (Kapur 2010)

Its facets differ according to the 'other'. Therefore vulnerability to health would need to be measured against indicators of age, food intake, nutrition, work engagements, among others. Vulnerability to war would need the length of the international borders, strength of the armed forces, efficiency of the armaments, to name a few. It is in relation to 'the other' that the vulnerability is measured.

Measuring vulnerability

To measure the vulnerability of the A&N Islands to disasters, a composite vulnerability index was computed. For computing this, the same 22 islands, for which the composite damage index was calculated, were taken and the same methodology was used. Vulnerability is an elusive and multidimensional concept that incorporates a range of issues. To grasp and measure the vulnerability of the A&N Islands to the Tsunami, there is a need to identify set of indicators that are holistic, critical and quantifiable. The choice of indicators was guided first by the aspect of vulnerability it represented, and second by the availability of data. After weighing and considering many indicators such as island-wise total income and expenditure, contribution of the island in the gross state domestic product (GSDP) of the A&N Islands, import and export of goods in each of the islands, employment in the government sector, frequency of travel to mainland, poverty, social inequality, well-being and quality of life, among others.

One key indicator, for which data were available, was access to amenities because lack of basic amenities creates vulnerability. Infrastructure providing the amenity is 'the basic physical and organizational structures needed for the operation of a society or enterprise, or the services and facilities necessary for an economy to function' (Goel 2002). The UN/ISDR (2009) termed these amenities as 'critical facilities' and defined them as

> the primary physical structures, technical facilities and systems which are socially, economically or operationally essential to the functioning of a society or community, both in routine circumstances and in the extreme circumstances of an emergency. [. . .] Critical facilities are elements of the infrastructure that support essential services in a society. They include such things as transport systems, air and sea ports, electricity, water and communications systems, hospitals and health clinics, and centres for fire, police and public administration services.
>
> (UN/ISDR 2009)

Poor civic infrastructure is an indicator of vulnerability. Functional and quality infrastructure reduces vulnerability first by providing safety in times of disaster and second, by facilitating the relief, rescue, reconstruction, rehabilitation and recovery after a disaster. The condition of infrastructure also reflects the economic development of the area. There is a negative correlation of quality of infrastructure with poverty, that is the areas with low quality infrastructure tend to have high poverty and vice versa. The village directory of the Census of India (2001b) is a source that lists the presence of the amenity such as schools, medical facilities, drinking water, post and telegraph, and electrified villages, among others. But it does not provide the quality of the services these infrastructures provide. For example, it documents the presence of the medical centre, but not of the doctors, or availability of medicines, or kind of cases treated in these. The presence or absence of primary school is listed but not the availability of teachers, or details on the functioning of these schools. Similarly, if a village is 'electrified' (see Chapter 6 for definition of electrified villages) is given, but the hours for which electricity is not provided is not available. As the quality of the service provided is more significant than the mere presence of the amenity, these indicators were not used for the vulnerability analysis.

After carefully weighing finally five key indicators were selected. These are percentage to total population of (1) scheduled tribes (STs); (2) illiterate population; (3) percentage of marginal workers to total workers; and islands without (4) cargo handling jetties and (5) metalled roads. The first three indicators were sourced from Census of India (2001a) and the remaining two from the *Island-wise Statistical*

Handbook 2006 (Directorate of Economics and Statistics 2006). The Census of India 2001 was the last census before the Tsunami in December 2004 and hence was the only authentic source of population data of the pre-Tsunami period. The data for total illiterate population were not directly available. Hence, the indicator given in the census had to be 'inverted'. The 'Primary Census Abstract' of the Census of India (2001a) lists the number of literate people. The number of illiterate population was calculated by subtracting total literate population from the total population above the age of seven years. The *Island-wise Statistical Handbook 2006* lists the availability of different kinds of jetties and the length of metalled roads for all the inhabited islands of the A&N Islands. Accordingly the indicators were tabulated.

It was seen that these five indicators encompassed two broad facets of vulnerability. Therefore these indicators were grouped into components representing a dimension of the vulnerability: (1) socio-economic vulnerability and (2) lack of transport facilities (Figure 7.1).

Since the indicators had varied units, to make them comparable for calculating the composite vulnerability index, the indicators were normalized, using the following formula:

$$x_{new} = \frac{x - x_{min}}{x_{max} - x_{min}},$$

where x is given observation; x_{min} is the minimum value of the observation and x_{max} is the maximum value of the observation and x_{new} is the new normalized score.

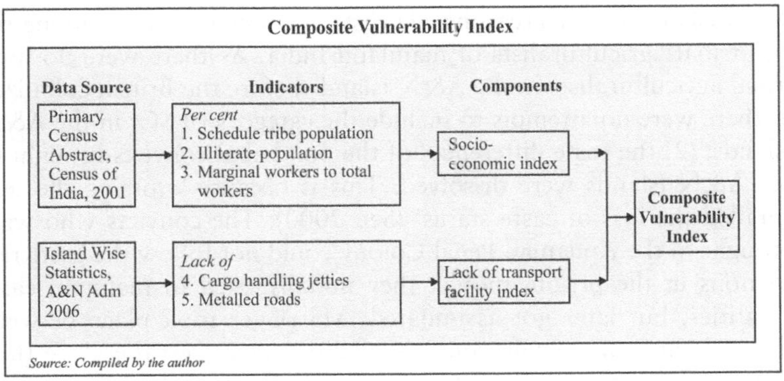

Figure 7.1 Composite vulnerability index

This formula scales the values of the different indicators between 0 and 1. The normalized value of different indicators was summed for the score of three components. The scores of the components were again normalized and summed up for the composite vulnerability index. The final scores were grouped into the category of high, moderate and low, and were mapped.

Socio-economic vulnerability

Low socio-economic status can create vulnerability to disasters. People belonging to a low socio-economic group have no social security and are most often poor with no employment and therefore are engaged in the informal sector. They are impoverished, neglected, oppressed, exploited and live mostly in squalor and are under huge debts. They have a hand-to-mouth existence and thus have low resilience for any emergency situation and therefore have very high vulnerability. Low socio-economic status itself carries the characteristics which disallow the people under this category to combat disasters. The component of socio-economic vulnerability includes three indicators, namely, percentage schedule tribe population, percentage of illiterate population and percentage of marginal workers to total workers.

Scheduled tribe population: The scheduled castes (SC) and schedule tribe (ST) population by and large are socially, economically and educationally backward and hence are 'scheduled' as per the Constitution of India (Kapur 2010). They were scheduled by the colonial government to provide them with benefits and reservations. As per the 2001 Census of India, the ST population in the A&N Islands accounted for 8 per cent of the total population while there was no SC population. The Census of India does not record the status of SC in the A&N Islands for two reasons: (1) caste is a social division among the non-tribal agriculturalists of mainland India. As there were no non-tribal agriculturalists in the A&N Islands before the British colonized it, there were no attempts to include the category of SCs in the A&N Islands; (2) the caste differences of the non-tribal convicts brought to the A&N Islands were dissolved. This is because 'crossing the seas entailed the loss of caste status' (Sen 2000). The convicts who were brought to the Andaman Penal Colony could not follow the caste restrictions in the prison, though they initially tried to maintain caste identities, but later got assimilated. Marriages took place between the male and the female convicts of different castes; therefore their progeny did not adhere to the customs of any caste and religion strictly (Dass 1937). Inter-caste and inter-religion marriages were common among the convicts of the Andaman Penal Colony (Dhingra 2005). After the resettlement scheme, most of the Bengali settlers belonged to the lower caste. But as these people were not indigenous to the A&N Islands, their castes were never recorded under the scheduled caste category. However in 2011, the pre-1942 settlers and the post-1942 Bengali resettlers were included in the Central OBC list (Roy 2012). Therefore, in the Census of India, the SC population is returned as '0' and the disadvantaged group is the ST population of the A&N Islands.

Prior to British colonization, the entire A&N Islands were inhabited by the tribal population. But in 2001 the tribal population accounted for only 8 per cent of the total population (Census of India 2001a). There are six scheduled tribes notified from the A&N Islands, four from the Andaman Islands and two from the Nicobar Islands. The tribes of the Andaman Islands are hunter-gatherers and include the tribes of Great Andamanese, Jarawas, Sentinelese and Onges. The two tribes of the Nicobar Islands are the Shompens and the Nicobarese. The former are hunter-gatherers while the latter are settled in well-defined villages and pursue subsistence copra production, fishing and hunting. The tribal Nicobarese population accounts for 90 per cent of the total tribal population of the A&N Islands.

The tribal population of the A&N Islands are disadvantaged when placed vis-à-vis the non-tribal population. The hunter-gatherers which include the Great Andamanese, Jarawas, Sentinelese, Onges and Shompens require a large area of forest for survival as it is the source of food and all other necessities. With the influx of non-tribal population, the forest area available for the tribes has sharply declined, thereby putting their survival at stake. The tribal Nicobarese, though, are not hunter-gatherers, yet their subsistence living makes them vulnerable. They are unable to compete with the 'developed' non-tribal population who exploit them. The percentage of STs is 0.9 per cent in the Andaman Islands and 63 per cent in the Nicobar Islands. Out of the 10 Andaman Islands, seven islands have insignificant tribal population. The tribal population is limited to Little Andaman (7 per cent), South Andaman (0.7 per cent) and Middle Andaman (0.4 per cent). The tribal population of the A&N Islands is concentrated in the Nicobar Islands. Eight out of the 12 Nicobar Islands have more than 90 per cent tribal population. Car Nicobar has 78 per cent, Kamorta and Katchal have 50 per cent and Great Nicobar has 12 per cent of tribal Nicobarese population. Not only do the Nicobar Islands have high tribal population, but the illiteracy too is high.

Illiterate population: Illiterate population is the second indicator of the component low socio-economic status. The Census of India defines literacy as the ability to read and write any language (Census of India 2001c). Taking from this definition, an illiterate can be defined as someone who does not possess the ability to read and write with comprehension in any language. The ill effects of illiteracy are numerous. First, the inability to read and write limits the ability to obtain and understand essential information. Illiteracy affects the access to knowledge as well as skills and thereby limits employment opportunities. They end up having low and menial jobs with very low income. They are exploited by their employers. Illiteracy is also the cause and the effect of poverty, malnutrition, health problems, child labour, migration and lack of access to continuous teaching and learning environments. These inequalities influence the social vulnerability of individuals (Martínez and Fernández 2010). With only 'labour' to sell, the illiterate population do not have any savings to fall back on in times of disaster. They constitute the poor population and precariously live on the marginal areas, and thus are the first ones to be impacted. In a disaster, not only are the illiterates are worst affected, but their recovery is also delayed, as neither are they aware of nor do they have access to the schemes and programmes post-disaster. They are mostly cheated by the middlemen and sometimes the government officers too. In the post-disaster period the situation is worse as they fail to receive compensation and benefits provided under various government schemes. It is very difficult for an illiterate to pursue the documentation which is required in claiming and receiving the compensation. It deters them in seeking the same. Since illiteracy is also representative of poverty and marginalization, it satisfies its inclusion as an indicator of vulnerability to disasters.

The percentage of illiteracy is low for the A&N Islands as compared to the national average. In 2001, the percentage of illiteracy for India was 35 per cent, while for the A&N Islands it was 19 per cent. However, the entire islands do not have the same illiteracy. The illiteracy is comparatively low for the Andaman Islands than the Nicobar Islands. It is 32 per cent for the former while the latter has 42 per cent illiterate population. Among the Andaman Islands, Rutland has the highest illiteracy with 54 per cent and the lowest is South Andaman with 28 per cent. But out of the Nicobar Islands, Little Nicobar has 92 per cent illiterate population, followed by Chowra and Bompoka with 75 per cent and 72 per cent, respectively. The lowest is in Great Nicobar where 37 per cent of the population is illiterate, followed closely by Car Nicobar with 39 per cent of illiterate population. Illiteracy leads to employment in the informal sector. The high illiteracy of the Nicobar Islands is also reflected in the high percentage of marginal workers.

Marginal workers population: The percentage of marginal workers to total workers is the third indicator of the component of the low socio-economic status. The Census of India defines a marginal worker as 'those who worked any time at all in the year preceding the enumeration but did not work for a major part of the year, i.e. those who worked for less than 183 days (or six months)' (Census of India 2001d). The marginal workers have one of the minimum wages among all the sectors of the economy. Most of the marginal workers are illiterate and therefore are unable to have a secure source of income. They have hand-to-mouth existence and thus they possess no savings or assets. Therefore, in a disaster like the Tsunami, on one side they tend to lose the little possession they have and it also limits their chances to work and earn their daily wages. Not only do they suffer in a disaster in terms of asset loss, but their livelihoods are also destroyed. In the A&N Islands, of the total working population, 17 per cent are marginal workers, but the percentage varies between the Andaman Islands and the Nicobar Islands. Marginal workers as a percentage of total workers are 13 per cent in the Andaman Islands and 34 per cent in the Nicobar Islands. The tribal Nicobarese livelihoods are diverse as they pursue copra production, fishing, plantation of fruit plants and areca nuts, while some are engaged in government jobs. Most of these activities are subsistence in nature. The tribal Nicobarese largely depend on copra production for earnings, therefore only copra production is considered as 'economic activity' which is produced as and when required. As the households are not engaged daily in the process of production, the percentage of marginal workers among the tribal Nicobarese is high.

The socio-economic vulnerability index was calculated from the three indicators, scheduled tribe population, illiterate population and percentage of marginal workers to total workers (Figure 7.2). The socio-economic vulnerability score ranged from 0.18 to 2.98. Natural breaks in the scores were used to make three classes: high, medium and low. Out of 22 islands, 11 islands are in the low score, nine in the medium score and two in the high score categories. Out of 10 Andaman Islands, Smith Island with a score of 1.27 is in the medium socio-economic vulnerability category while the remaining nine islands are in the low score category. The lowest score, 0.18, is for South Andaman Island. None of the Andaman Islands is in the high socio-economic vulnerability score category.

In contrast, out of 12 Nicobar Islands, two are in the low, eight in the medium and two in the high socio-economic vulnerability score categories. The two islands with the highest score are Little Nicobar

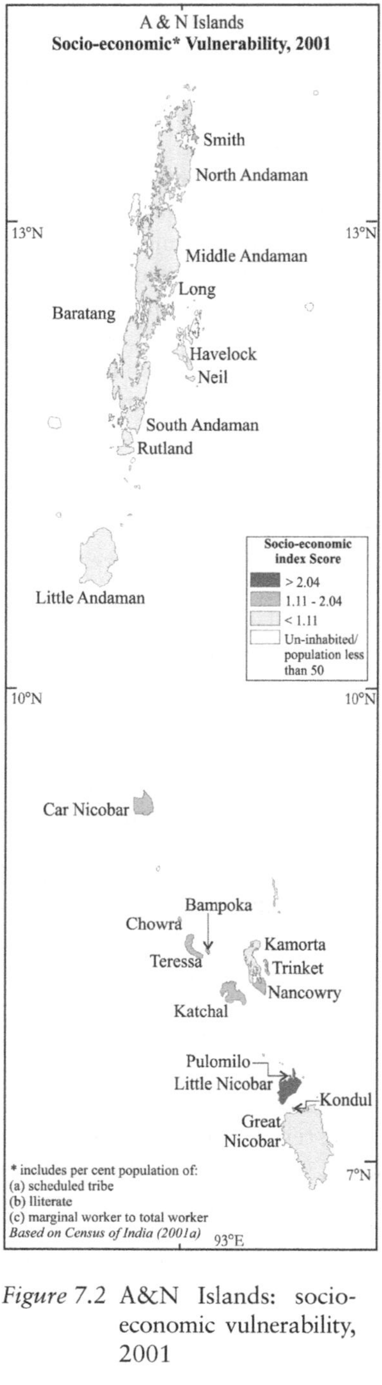

Figure 7.2 A&N Islands: socio-economic vulnerability, 2001

and Pulomilo with scores of 2.98 and 2.30, respectively. The Nicobar Islands in the medium socio-economic vulnerability are, in order: Teressa, Car Nicobar, Chowra, Bompoka, Nancowry, Trinket, Kondul and Katchal, while Kamorta and Great Nicobar Islands are in the low socio-economic vulnerability category. The scores indicate that Nicobar Islands have comparatively low socio-economic status than the Andaman Islands. It can be concluded that islands with dominant tribal population scored high on the vulnerability component of low socio-economic status and are more vulnerable to disasters.

Lack of transport facilities

The third component of vulnerability pertains to the non-availability of transport systems in the A&N Islands. During and in the aftermath of a disaster, transport facilities are of critical importance for the movement of people, services, goods and relief supplies for reconstruction and for rehabilitation. The lack of a good transport network leads to vulnerability. Therefore, for the third component – lack of transport facility – two indicators were selected: first, lack of cargo handling jetties and second, lack of metalled roads.

Jetties are of various kinds. Not all jetties can handle cargo. In the absence of cargo handling jetties, the cargo is unloaded at another island that has the facility. The goods are then transferred to smaller boats that can berth at the smaller jetties. A second option is that ships are anchored in mid-sea and goods are unloaded onto a pontoon (floating dock) which is towed to the concerned island. Therefore, the absence of jetties that can handle cargo reflects the vulnerability of the island as it limits the transportation of goods and services.

The second indicator was lack of metalled road. A metalled road connects villages and settlements on the same island and is required for transporting goods and services. It is also essential for movement of vehicles. In the absence of metalled road, no vehicles can operate. Metalled road also facilitates the movement of medical services and transportation of the sick to the health centres. As seen in the previous section, most of the amenities, especially the primary health centre and sub-centres, are not present in all villages. Therefore in case of sickness and injury, a metalled road is required for efficient movement of the ambulance. In the absence of metalled roads, the sick and the injured are carried on foot in locally made stretchers to the health centre, thereby delaying the response. It also hampers the evacuation and relief work, as the rescuers cannot reach a site promptly in the absence of metalled road.

The number of jetties and the length of metalled road could not be taken as indicators because the islands are of varied sizes. Large islands have more than one jetty and the length of metalled road is more when compared to smaller islands. Hence, a score was assigned to the islands on the basis of the presence of the transport facility. As the presence of cargo handling jetties improves connectivity and lowers vulnerability, islands with a cargo handling jetty were assigned a score of '0' and the other islands without it were assigned '1', as the latter are more vulnerable than the former. Out of 22 A&N Islands, 11 islands did not have cargo handling jetties. Out of 10 Andaman Islands, only two islands, Rutland and Smith, did not have cargo handling jetties while the other eight islands had one. In the case of the 12 Nicobar Islands, the cargo handling jetty was present only in three islands, Car Nicobar, Kamorta and Great Nicobar. The remaining nine islands did not have cargo handling jetties. These islands are Bompoka, Chowra, Katchal, Kondul, Little Nicobar, Nancowry, Pulomilo, Teressa and Trinket.

Similarly, a score of '0' was assigned to islands that had metalled roads and a score of '1' was assigned to islands without metalled roads. Out of 22 A&N Islands, 10 islands did not have metalled roads. In the Andaman Islands, three islands – Long, Rutland and Smith – did not have metalled roads, while the remaining seven islands had

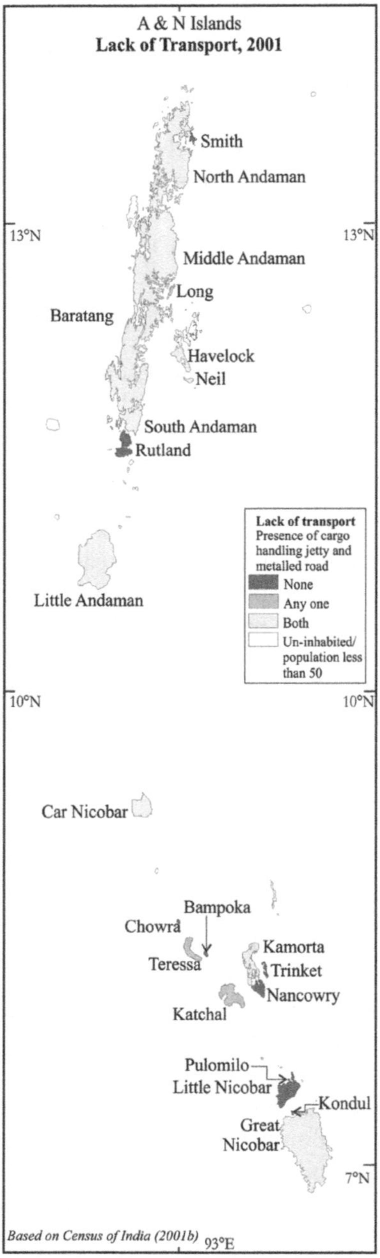

Figure 7.3 A&N Islands: lack of transport, 2001

metalled roads. Seven of the Nicobar Islands did not have metalled roads and roads were limited to five islands.

For the component lack of transport facilities, the index score was calculated by summing up the assigned scores, and three categories were identified (Figure 7.3). The three categories were first, islands with both cargo handling jetty and metalled road; second, islands with either cargo handling jetty or a metalled road; and third, islands with neither cargo handling jetty nor a metalled road. The first category had the least vulnerability, the second had moderate vulnerability while the third was the most vulnerable in the component of lack of transport facility.

Out of 22 A&N Islands, 10 islands have both transport facilities, three of the islands have only one or the other form of transport facility, while nine islands do not have any of the transport facilities. In the case of the Andaman Islands, out of the 10, seven had both cargo handling jetties and metalled roads. Long Island had only one of the transport facility, and two islands, Rutland and Smith, had none. Out of the 12 Nicobar Islands, three islands – Car Nicobar, Great Nicobar and Kamorta – had both the transport facilities. Another two islands, Katchal and Teressa, had metalled road only. The remaining seven islands did not have either a cargo handling jetty or metalled roads. Therefore in this component too, the Nicobar Islands had a higher vulnerability score than the Andaman Islands.

To assess the composite vulnerability score of the A&N Islands, the scores of the two components were normalized and added. Class interval was calculated by dividing the difference between the highest and the lowest scores by three, as 22 scores can be optimally represented through three classes. Thus the category of low, medium and high vulnerability was made and mapped. The scores are relative. It does not indicate that islands with low vulnerability score are not vulnerable. It indicates that islands with low vulnerability score are comparatively less vulnerable than the ones in the category of medium score or high score. From the composite vulnerability scores, the following facts distinctly emerge.

First, of the 22 A&N Islands, 11 are in low, three are in medium and eight are in high score category (Figure 7.4). The least vulnerable island was South Andaman with score '0' while Little Nicobar emerged as most vulnerable with score '2'. The scores are distributed unevenly across the A&N Islands. The islands in the low score categories are concentrated in the Andaman Islands while the islands with moderate and high score dominate the Nicobar Islands. Out of the 10 Andaman Islands, eight are in low and one

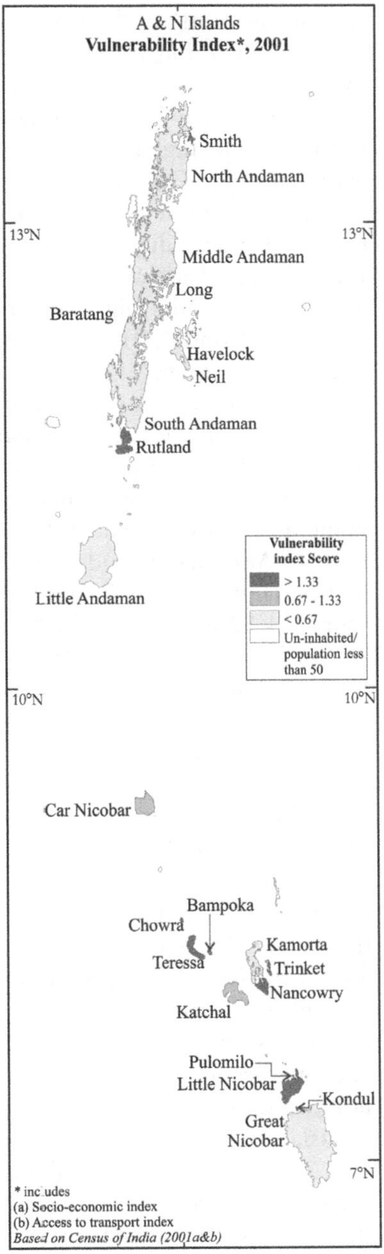

Figure 7.4 A&N Islands: vulnerability index, 2001

each in medium and high vulnerability score category. The islands in the low vulnerability score category in ascending order are South Andaman, Little Andaman, Neil, Middle Andaman, North Andaman, Havelock, Baratang and Long. Rutland Island has medium score while Smith Island is in the high score category. In the case of the Nicobar Islands it is the opposite. Out of the 12 Nicobar Islands, three are in low, two in medium and seven in high vulnerability score categories. The islands in the low vulnerability score category are Great Nicobar, Kamorta and Car Nicobar. Katchal and Teressa are in the medium score category. The islands in the high vulnerability score categories, in increasing order, are Kondul, Trinket, Nancowry, Bompoka, Chowra, Pulomilo and Little Nicobar.

Second, the contrast between vulnerabilities between the islands is very high. Among the Andaman Islands, on the one hand there are islands with low vulnerabilities, South Andaman, Little Andaman and Neil; on the other hand there are Smith and Rutland with very high vulnerability scores. The difference between the highest and the lowest score of Andaman Islands is 1.39. Similarly, the difference between the highest (Little Nicobar) and the lowest (Great Nicobar) vulnerability score is 1.82 for the Nicobar Islands.

Third, the difference between the highest vulnerability score, i.e. 2 for Little Nicobar and the lowest score, i.e. 0 for South Andaman, is 200 times. This shows that there is a stark contrast between vulnerabilities of the islands. Little Nicobar Island has an area of 159 km^2 and had a population of 353 while South Andaman Island has an area of 1,348 km^2 and had a population of 181,949 (Census of India 2001a). The city of Port Blair is situated in South Andaman which accounts for nearly one-third the population of the total A&N Islands. It has also the highest number of tourist population, as all tourists who visit the A&N Islands first have to land in Port Blair before proceeding to any of the islands. It is the most developed part of the A&N Islands. In contrast, Little Nicobar lies in neglect. It had 98 per cent of tribal Nicobarese population, one health sub-centre, two primary schools and one branch post office. There was no metalled road and it was not connected with boat service either. It's a similar pattern among the islands, where the revenue-generating islands, such as Havelock and Neil, have low vulnerabilities in contrast to the one from which the economic returns are low. The former is dominated by the non-tribal population while the latter has a dominant tribal population. The tribal dominated islands cannot be 'developed' for economic purposes as they are protected under the Andaman and Nicobar Islands (Protection of Aboriginal Tribe) Regulation, 1956. Under this garb of 'protection', not even basic services have been provided for the far-flung Nicobar Islands.

It emerges that the islands with high tribal population lacked the transport network. Both the components are positively related and dictated the vulnerability scores. A well-developed transport network aids in the spread of economic development and assimilation of the population. In times of globalization, without a transport network the areas remain aloof from development and thus have high vulnerability. The critical dominant factor which gives high vulnerability scores to so many of the Nicobar Islands is its tribal population and lack of a well-developed transport network.

The spatial picture that emerges from the composite vulnerability score based on pre-Tsunami secondary data confirms that the Nicobar Islands were more vulnerable than the Andaman Islands. But how many people inhabited these vulnerable islands is the next question. Therefore, the population of the islands was grouped according to their composite vulnerability scores to assess the quantum of the vulnerable population.

The total population of the A&N Islands as per the Census of India 2001 was 356,152, of which the Andaman Islands account for 88 per cent and the Nicobar Islands account for the remaining 12 per cent. Of this total population, 90 per cent is in the category of low vulnerability, 8 per cent in the medium and 2 per cent in the high vulnerability category. But, when the Andaman Islands and Nicobar Islands are viewed separately, the picture changes. Of the total population of the Andaman Islands, 98.9 per cent are in the category of low vulnerability, 0.7 per cent population is in the medium and 0.4 per cent in the high vulnerability category. In contrast, in the Nicobar Islands, 26 per cent are in the low, 61 per cent in medium and 13 per cent in high vulnerability categories. Thus, about three-fourths of the population of the Nicobar Islands were vulnerable before the Tsunami in 2004.

Does spatial pattern of vulnerability correspond to spatial patterns of the killed and the damage? To assess the correlation of deaths and damages ascribed to Tsunami with the vulnerability PPMC coefficient was computed. The PPMC coefficient between the vulnerability index and the number of persons killed was –0.03, which indicates insignificant negative correlation. This is despite the fact that the Nicobar Islands emerged more vulnerable in the vulnerability analysis based on secondary data, and when juxtaposed with the data of the killed it confirmed that higher the vulnerability, the higher the killed. The coefficient between the vulnerability and the damage index was –0.45. This indicates a medium negative correlation, i.e. islands with higher vulnerability had low damage and vice versa. The macro level assessment of the vulnerability index does not correspond with the numbers killed or with the damage index. Again, the spatial distribution of vulnerability in the Nicobar Islands also does not correspond with the pattern of the killed in the Nicobar Islands. Car Nicobar Island has a low vulnerability score, but it was second in the number of people killed. Again, Katchal Island had a medium vulnerability score, yet it ranked '1' in terms of number of people killed. Similarly, Nancowry Island is in the category of high vulnerability, but the number of people killed was low as compared to the other islands. What was the reason for these discrepancies?

There are two prominent reasons for these discrepancies. First, the islands with the transport infrastructure had low vulnerability index score and vice versa. In the Tsunami the infrastructure was damaged and thus these islands with low vulnerability score ranked high in damage index scores. In contrast, islands with high vulnerability score had a low damage index score because these islands did not have any infrastructure that could be damaged. A second reason for this contrast is that the vulnerability analysis was based on secondary data which did not take into consideration the quality of the transport infrastructure present in the islands. The islands with the infrastructure providing the amenities had a low vulnerability index score. But the pre-Tsunami condition of these structures was so poor that it collapsed, and hence accounted for a high damage index score. The vulnerability index based on secondary data sources has its drawbacks. However, for an objective analysis it was the only available data set. Therefore, an attempt was made to assess vulnerability based on fieldwork carried out in 2010 and qualitative secondary sources.

Assessing vulnerability

Fieldwork was undertaken in 2010 to assess vulnerability substantiate the indicators selected for measuring vulnerability. In the A&N Islands, after the Tsunami households were resettled in the post-Tsunami houses that were constructed under the Tsunami Rehabilitation Programme. These post-Tsunami houses were constructed across 10 islands, of which 100 households were selected from two islands, 65 from South Andaman and 35 from Car Nicobar (rationale of the selection of these households is discussed in Chapter 9). Pre-Tsunami characteristics of these 100 households were documented as all of these households were directly affected in the Tsunami and were residing in the post-Tsunami houses during the fieldwork.

The three indicators of measuring socio-economic vulnerability were percentage schedule tribe population, percentage of illiterate population and percentage of marginal workers to total workers. The composition of these 100 affected households based on these indicators was determined. Among the 100 households, 65 households from South Andaman were non-tribal while the 35 households from Car Nicobar Island belonged to the tribal Nicobarese and thus were scheduled tribes.

On the second indicator, illiterate population, the formal education level of the head of the households was documented. Out of the 100 head of the households, 26 were illiterate; 16 had dropped out before completing primary school; 16 completed primary school; 21 dropped before secondary school; nine completed secondary school; two completed higher secondary; and only 10 were graduates. There was a difference in the formal education level among the heads of the households of non-tribal households of South Andaman and tribal Nicobarese from Car Nicobar. Among the 65 heads of the households of South Andaman, 20 per cent were illiterate; 17 per cent dropped out before completing primary school; 15 per cent completed primary school; 17 per cent dropped before secondary school; 12 per cent completed secondary school; 5 per cent completed higher secondary; and 14 per cent completed their undergraduate degree. In Car Nicobar Island, 37 per cent of the head of the households were illiterate, 14 per cent dropped out before completing primary school; 17 per cent completed primary school; 29 per cent dropped before secondary school; 3 per cent completed secondary school, which was the highest level of formal education among the 35 heads of household. Therefore, out of the 100 households 79 per cent did not complete their secondary education. These 79 per cent of the households are literate for namesake, and are as good as an illiterate. It is the remaining 21 per cent that can actually reap the benefits of literacy. Thus, it is concluded that the level of education was low for the affected heads of household.

The third indicator was percentage of marginal workers to total workers. Out of the 65 non-tribal households, five (8 per cent) are non-workers and are dependent on family members, while four (6 per cent) are sustained by regular pension, 18 (28 per cent) are in the government sector, 12 (18 per cent) have their own business which operates for the entire year and the remaining 26 (40 per cent) head of the households are marginal workers. The remuneration of 26 marginal workers per head ranged between Rs. 150 and Rs. 200 a day. In India the poverty line is Rs. 816 per capita per month in rural areas and Rs. 1,000 per capita per month in urban areas (Planning Commission 2008). Therefore, to sustain a five-member family (consumption expenditure – Rs. 4,080 per month in rural areas and Rs. 5,000 per month in urban areas), these households barely meet the minimum requirement. These heads of household are daily wage labourers whose earnings are dependent on the availability of work. Securing a job daily is every day's struggle. The case of the tribal Nicobarese is different. The entire tribal Nicobarese household, in their respective capacity, is engaged in the subsistence economy of copra (dried coconut) production for entire year. But activity is not carried out every day. Copra is produced as and when required. Thus, it is difficult to categorize the tribal Nicobarese as main or marginal worker. A small proportion, nearly 5 per cent, of the tribal Nicobarese population is engaged in the government sector.

Thus, the vulnerability characteristics of the affected households that emerged from the fieldwork were that the households belonged to the low socio-economic backgrounds, had very little formal education and most of the households did not have a secure source of earnings. These affected households were also questioned about the quality of infrastructure and services pre-Tsunami in their localities. Based on their perception, the state of physical infrastructure pre-Tsunami could be drawn.

State of physical infrastructure pre-Tsunami

A clear distinction existed between the state of infrastructure in South Andaman and Car Nicobar. As the affected households from South Andaman Island were living in Port Blair, they had access to good infrastructure, which was not the case with the households from Car Nicobar Island. The households informed that even though piped water was there, the supply of water was erratic and therefore the households depended on the wells rather than on the piped water. The access to medical infrastructure in the Nicobar Islands was equally dismal. The households from South Andaman had access to the G. B. Pant hospital in Port Blair, which is the only multi-speciality of the A&N Islands. It was the households from the Nicobar Islands that lacked access to medical facilities. In every village of the Nicobar Islands there was either a primary health sub-centre (PHSC) or a dispensary. The households recalled that these PHSC and the dispensaries were operational only for few hours during the day and were not equipped to handle serious cases. In Car Nicobar Island, for emergencies, the households rushed to the Bishop John Richardson Hospital. This hospital came into existence in 1946 under the Japanese. Post-Independence, a larger hospital was constructed and inaugurated in 1960 and was known as the civil hospital. In 2003, it was named as Bishop John Richardson (BJR) Hospital in honour of the influential Nicobarese leader Bishop John Richardson who worked for the development of the tribal Nicobarese. He was also the one to introduce the tribal Nicobarese to Christianity and was also the first Member of Parliament from the A&N Islands. At the time of the Tsunami, it was a 112-bed hospital that was equipped to deal with only basic services such as obstetrics, fractures, animal bites and viral fevers, to name a few. This hospital was not equipped to handle serious cases, for which patients were and still are referred to the G. B. Pant Hospital at Port Blair or further to mainland India.

As a greater number and variety of schools were available in South Andaman Island, the number of children going to school was higher for them. In contrast, the school infrastructure and the quality of education were appalling in the Nicobar Islands. The per capita public expenditure on education in the A&N Islands is the highest in the country and it claimed eighth rank in literacy among 35 states and union territories in 2001. Yet, the majority of the heads of household of the Nicobar Islands stated that 'there are no teachers in the school, why should we send children there. They waste time at school'. For the tribal Nicobarese households, education is important yet the heads of household are not much in favour of higher education because they feel the younger generation does not want to come back to the islands after completing their education and are distanced from the community. The younger members do not want to go to school since for them study is 'boring'. Thus, very few of the Nicobarese children, who are interested in studies, complete higher secondary education. There is no college in the Nicobar Islands, therefore the ones who want to pursue studies further have to leave their islands and relocate to Port Blair or to mainland India. The initiative to begin a college in the Nicobar Islands was a failure. The present-day Mahatma Gandhi Government College (presently known as Mayabunder College), situated at Mayabunder in Middle Andaman Island, was initially started at Car Nicobar Island in 1992. But because of the paucity of students as well as teachers, it was relocated to Mayabunder in 1994 (fieldwork, 2010). This incidence shows that even though the Nicobar Islands had a number of schools, the quality of education is poor and unrelated to the tribal way of life.

The Post and Telegraph (P&T) facility pre-Tsunami was also rudimentary in the Nicobar Islands. The households recalled that phone services were weather and electricity dependent. In the A&N Islands, even though 67 per cent households are 'electrified', for most of the day there is no electricity. The power generation is dependent on diesel, and delays in supply of diesel leads to power cuts for several hours in a day. In far-flung islands, when diesel is in short supply, the bus service also gets affected. It also affects phone service as the cellular phone towers require electricity to function. During fieldwork in Car Nicobar in 2010, there was power failure in the villages for 12–15 hours in a day. The fieldwork was conducted during the monsoon. Diesel was in limited supply as it could not be transported due to the rough sea. This resulted in power failures.

The transport network was in poor condition in the pre-Tsunami period. From secondary sources based on qualitative surveys, it was evident that in pre-Tsunami period waterways, airways and roadways infrastructure of the A&N Islands were in dismal condition:

> travel on the islands and inter-island transit involves relatively long waiting time due to infrequent and irregular availability of modes of transport and lack of synchronisation between the alternative and (often) complementary modes. Moreover, the speed of travel between the islands averages between 12 km and 18 km per hour. Not only are the vessels old and slow, but also they cannot be utilised to their reported capacity due to frequent break down and repair and maintenance work.
>
> (Planning Commission 2008)

The ships are the lifeline of the A&N Islands, yet out-of-date ships are used for transport of goods and passengers. As per the Planning Commission (2008) reports, among the three important vessels that the A&N Islands possessed, two vessels (MV *Sentinel* and MV *Yerewa*) were supposed to be decommissioned in 2001 and MV *Chowra* in 2004. However, all three ships are still in use after the Annual Passenger Service (APS) in 2011.

The jetties too were in poor condition and hence collapsed in the Tsunami. A reconnaissance survey to assess the damage in infrastructure to the A&N Islands concluded that the damage to jetties was caused because of 'inadequate shear design of piles, improper detailing (mainly inadequate lapping of longitudinal bars) and inadequate anchorage length. Prior deterioration due to age, poor construction quality or prior faults was also responsible for the damage' (Tang et al. 2006). The structures of the jetties are in constant contact with the sea water. This results in weathering both mechanical (force of the waves and the daily tides) and chemical (work of seawater on the concrete piles), leading to deterioration of the structure. The wharfs and jetties require regular maintenance as they are situated in a high-seismic zone. It was the case of lack of maintenance that led to their collapse.

The road conditions were equally appalling: 'the east-west stretch of the Great Nicobar Trunk Road was almost non-existent in many parts because of the frequent heavy landslides' (American Society of Civil Engineers 2007). Again, 50 per cent of the villages in the Nicobar Islands did not have metalled roads; another 20 per cent had either metalled road or mud-road: 'condition of many existing roads is poor having an adverse impact on the speed of road traffic and life of vehicles . . . the condition of the existing roads is less than satisfactory. All weather connectivity for many villages is limited' (Planning Commission 2008). Since 70 per cent of the villages in the Nicobar Islands had 'temporary *kutcha*' roads, these were washed out.

After the 2004 earthquake, the middle section of Chengappa Bridge connecting Middle Andaman Island and North Andaman Island fell from its bearing, making the bridge unusable. The construction design of the bridge was not suited for an area lying in seismic zone V as reported by the engineers from IIT Kanpur in 2002. They carried an evaluation study of the Chengappa Bridge after the earthquake of 14 September 2002. The report concluded that

> inadequate seating of bridge deck over piers and abutments is a serious concern for its safety during a stronger earthquake in future. The bearings are simple neoprene pads, which are far from satisfactory for a bridge located in seismic zone V. Bridge deck restrainers are the minimum that need to be provided to ensure that the spans are not dislodged from the piers in future earthquakes. As the impact of the earthquake of 26 December 2004, the spans dislodged from its piers. It remained out of operation for more than six months.
>
> (Rai et al. 2006)

In contrast to the state of roads and jetties, the airports were well maintained as they were under the Indian Navy. These sustained only minor damage and were operational immediately after the Tsunami. The A&N Islands have four airports/airstrips. The Port Blair and Diglipur airstrips are made of flexible pavement and hence suffered minor damages at the time of Tsunami. The Campbell Bay airstrip did not suffer any major damage even though it was closest to the epicentre, because the airstrip was at a height of 5 m above sea level (Rai et al. 2006).

After measuring vulnerability based on the secondary data, the functionality of these indicators was looked at during the fieldwork. What emerged was that the affected households indeed were vulnerable. A part of the affected households belonged to the scheduled tribe who were largely illiterate or did not complete secondary education, while most of the remaining non-tribal households were marginal workers. It was also concluded from the fieldwork that the amenities provided in the pre-Tsunami period in the Nicobar Islands were rudimentary and were non-functional for most of the year. Similarly, in the case of transport network, it was poor construction design and lack of maintenance that made these structures fragile and vulnerable to the Tsunami. It was the vulnerability of the island that led to the disaster in the Nicobar Islands. Another aspect of vulnerability that was assessed in the fieldwork was the condition of houses of the affected households in the pre-Tsunami period.

State of houses pre-Tsunami

The Census of India (2001e) classifies houses into permanent, semi-permanent and temporary. The permanent houses have walls and roofs made of permanent materials like galvanized iron sheets, asbestos sheets, burnt bricks, stone or concrete. In semi-permanent houses, either wall or roof is made of permanent material and the other made of temporary material like grass, thatch, bamboo, plastic, polythene, mud, unburnt brick or wood. A temporary house has walls and roofs made of temporary materials. Permanent houses, if constructed with proper guidelines and with regular maintenance, are the most durable kind of houses. Only 28 per cent of the houses in the pre-Tsunami A&N Islands were of the permanent type, while the remaining 72 per cent (about two-thirds) were semi-permanent and temporary. The pattern was similar for both the Andaman Islands as well as the Nicobar Islands (Table 7.1).

The vulnerability analysis based on secondary data sources brought out the fact that the Nicobar Islands were more vulnerable in terms of socio-economic conditions, had no access to amenities as well as lacked good transport network, and hence emerged as the most vulnerable islands. The fieldwork conducted among the Tsunami-affected households confirmed that they indeed belonged to poor socio-economic conditions, had no access to the basic amenities and they also suffered because of the lack of transport network. In the fieldwork, it also emerged that these affected households lived in frail houses on the coasts. These pre-Tsunami conditions made the people vulnerable to the Tsunami.

Thus, in the vulnerability analysis it emerged that the Nicobar Islands lacked infrastructure, the little infrastructure that existed were ill-maintained and most of the households lived in frail and fragile houses. However, the above analysis did not look into the nuances

Table 7.1 Vulnerable pre-Tsunami house type in A&N Islands

Islands	*Per cent of total pre-Tsunami residential houses*				
		Non-permanent			
	Permanent	*Semi-permanent*	*Temporary*	*Unclassified*	*Total*
Andaman	28.48	39.12	32.16	0.24	71.52
Nicobar	27.66	33.45	38.84	0.06	72.34
Total	28.40	38.55	32.83	0.22	71.60

Source: Based on Census of India (2001a)

of the vulnerability that emerged from the number of people killed as they were disproportionately distributed. Why was it that Nicobar Islands accounted for the maximum number of killed? Who were the killed? What is the reason for the uneven distribution of the killed even in the Nicobar Islands?

Identifying the vulnerable population

The people killed in the Tsunami in the A&N Islands were disproportionately distributed (Table 7.2). The Andaman Islands accounted for 2 per cent while the Nicobar Islands accounted for 98 per cent of the total 3,513 persons killed. There also was a difference in the age of the persons killed, as one-third of the total killed were children. There was slight gender disparity. Women and men accounted for 51 per cent and 49 per cent of the total killed, respectively. The disproportionate distribution was reiterated in fieldwork. In Car Nicobar Islands, a total of 732 persons were killed. Of this total, 401 persons killed were from villages of Arong and Kimius, which accounted for 55 per cent of the total killed. A profile of the killed revealed that the number of children and young adults was high among the killed (Table 7.3).

Of the total killed, 23 per cent were of age less than 10 and another 30 per cent were between 11 and 30 years. Thus more young adults were killed. The children were vulnerable because they could neither run nor climb trees efficiently. In several cases they were carried away from their parents and were drowned in the Tsunami waves. Children were also most affected by the trauma. The children suffered greater adjustment disorder, depression, panic disorder and post-traumatic stress disorder (Math et al. 2008).

More women were killed as compared to the men. In the overall numbers killed, the number of killed women was slightly higher than the

Table 7.2 Vulnerability differential of the killed

	Numbers killed	*Per cent of total*
Age		
Adult (> 18 years)	2,336	66.50
Children (< 18 years)	1,177	33.50
Total number killed	3,513	100
Gender		
Male	1,729	49.22
Female	1,784	50.78
Total number killed	3,513	100
Tribal status		
Tribal	2,955	84.12
Non-tribal	558	15.88
Total number killed	**3,513**	**100**

Source: Based on Digal (2006)

Table 7.3 Vulnerable killed in Car Nicobar Island*

Age group	*Total*	*Per cent of total killed*	*Numbers killed*		*Per cent of total killed*	
			Male	*Female*	*Male*	*Female*
0–5	47	11.72	16	31	34.04	65.96
6–10	47	11.72	19	28	40.43	59.57
11–20	64	15.96	17	47	26.56	73.44
21–30	58	14.46	12	46	20.69	79.31
31–40	52	12.97	15	37	28.85	71.15
41–50	52	12.97	19	33	36.54	63.46
51–60	38	9.48	15	23	39.47	60.53
61–70	28	6.98	13	15	46.43	53.57
71–80	6	1.50	3	3	50.00	50.00
More than 80	9	2.24	5	4	55.56	44.44
Total	**401**	**100**	**134**	**267**	**33.42**	**66.58**

* In the villages Arong and Kimius.

Source: Based on Fieldwork (2010)

men. But at the village level, the difference between the deaths of women and men was larger. Of the 401 persons killed, 33 per cent were men while 67 per cent killed were women. Of the total 267 women killed, 83 per cent were less than 50 years. The 'fertile' age group of 11–40 years accounted for 49 per cent of the women killed. During the fieldwork, a young unmarried Nicobarese man remarked: 'there is scarcity of girls in the island [referring to Car Nicobar Island] to get married'. An explanation put forth by the Nicobarese themselves as to why more women were killed is that women put the needs of family, especially children, before their own. When the Tsunami waves approached, the women attempted to save children and other family members and in the process were killed. As children and women accounted for the most killed, they were the most vulnerable group. However, the starkest difference was between the deaths of the non-tribal and tribal Nicobarese.

The most vulnerable: tribal Nicobarese

To seek answers for the concentration of those killed in the Nicobar Islands, the characteristics of the 14 islands were analyzed where deaths were registered. These islands were grouped into three categories on the basis of their population composition – (1) the predominantly non-tribal, (2) the predominantly tribal and (3) the mixed non-tribal and tribal. In the predominantly non-tribal islands, the non-tribal population was more

Table 7.4 Killed according to non-tribal and tribal population

Islands	*Per cent of total non-tribal population*	*Per cent of total tribal population*	*Per cent of total killed*
1. Predominantly non-tribal islands > 90 per cent of the total population is non-tribal			
South Andaman	99.27	0.73	0.31
Little Andaman	93.02	6.98	1.77
2. Predominantly tribal islands > 90 per cent of the total population is tribal			
Nancowry	4.96	95.04	0.08
Bompoka	0	100.00	0.39
Kondul	6	94.00	1.04
Teressa	9.87	90.13	1.50
Pulomilo	4.83	95.17	2.47
Chowra	0.65	99.35	2.58
Trinket	0.69	99.31	3.31
Little Nicobar	2.27	97.73	3.39
3. Mixed non-tribal and tribal islands 10 < non-tribal population < 90 per cent of the total population			
Katchal	49.89	50.11	37.29
Car Nicobar	21.65	78.35	28.20
Kamorta	45.66	54.34	9.75
Great Nicobar	87.38	12.62	7.94

Source: Based on Census of India (2001a) and A&N Administration (2009)

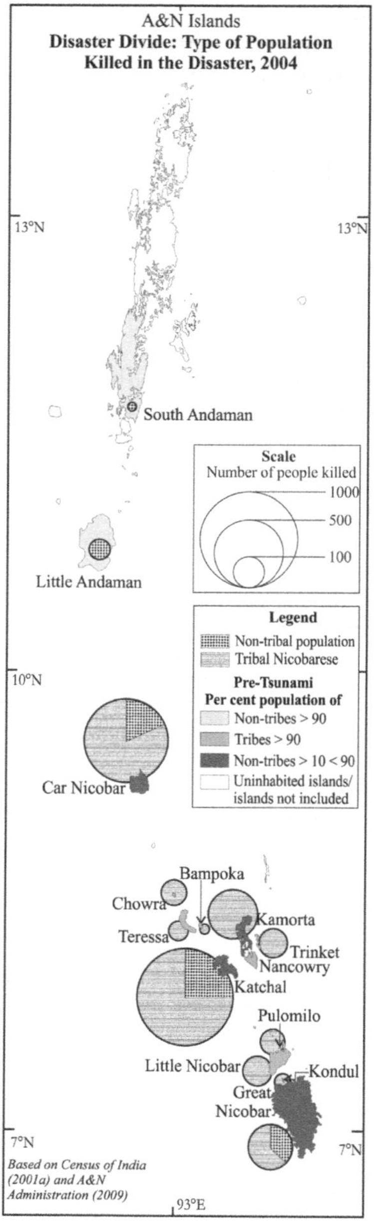

Figure 7.5 A&N Islands: disaster divide: type of population killed in the disaster, 2004

than 90 per cent of the total population of the island. Out of the 14 islands that registered the killed, two fell in this category, namely, South Andaman Island and Little Andaman Island. Another eight of the islands belonged to the category of predominantly tribal islands where the population of tribes account for more than 90 per cent of the total. These islands are Bompoka, Chowra, Little Nicobar, Trinket, Pulomilo, Nancowry, Kondul and Teressa. In the third category – mixed non-tribal and tribal islands – the percentage of the non-tribal population was more than 10 per cent. A population of more than 10 per cent is considered a sizeable population that can make a difference to the economy and the ecology of a place. Therefore, the cutoff of 10 per cent of the non-tribal population was taken for the third category. There are four islands in this category, namely, Katchal, Kamorta, Car Nicobar and Great Nicobar. When the percentage distribution of people killed in the Tsunami was pitched alongside these categories, it became evident that the islands with mixed non-tribal and tribal population account for 83 per cent of the total population killed (Table 7.4).

It was not the islands with dominant tribal or dominant non-tribal population that recorded the maximum number of people killed; rather it was the islands that had dominant tribal population along with sizeable non-tribal population where the deaths were concentrated (Figure 7.5). Therefore, it can be concluded that the islands with mixed non-tribal and tribal population were vulnerable and hence reported the deaths. Nevertheless, the non-tribal of Nicobar too were vulnerable, however their vulnerability differed from that of the tribal Nicobarese.

Vulnerability of the non-tribal population from Nicobar Islands

In the pre-Tsunami period, although the Nicobar Islands were protected under the Protection of Aboriginal Tribes Regulation (ANPATR), 1956, they had a sizeable non-tribal population which accounted for 37 per cent of the total population of the Nicobar Islands. Under the ANPATR parts of the A&N Islands dominated by aboriginal tribal population were declared as protected areas, in which the entry of the non-tribal population is prohibited (barring few exceptions). Nevertheless the non-tribal population increased in the Nicobar Islands as the ANPATR was amended from time to time to accommodate the needs of the non-tribal population.

As per the initial Regulation of 1956, the protected areas of Nicobar Islands included 'the entire area comprised in, and enclosed within the coast line of each of the following islands: (1) Car Nicobar; (2) Batti Malv; (3) Chowra; (4) Tilangchong; (5) Teressa; (6) Bambooka; (7) Camorta; (8) Trinkat; (9) Nancowry; (10) Katchall; (11) Moreo; (12) Trak; (13) Treis; (14) Menchal; (15) Little Nicobar; (16) Pulo Milo; (17) Great Nicobar; (18) Kondul and (19) Kabra, but excluding the area comprising air-field in Car Nicobar and ports of Camorta and Car Nicobar' were declared reserved (Andaman and Nicobar Gazette 1956). (The spellings of the islands' names are given here as per the ANPATR, which is different from the spellings as used by the *Islandwise Statistical Handbook* 2006 published by the Directorate of Economics and Statistics, Andaman and Nicobar Administration.)

However, the ANPATR of 1956 has been amended 12 times from 1956 till before the Tsunami, and areas of the Nicobar Islands were 'de-regularized' for resettlement purposes. Of these 12 amendments, three pertain directly to the Nicobar Islands, in which the areas of the Nicobar Islands were de-regularized and the rules for obtaining tribal passes were modified. Under Section 7 of the Regulation, in some occasions, such as for research or for some government-identified projects for the welfare of the tribal population, a non-tribal person can be permitted to enter the tribal areas for which a tribal pass is required and is issued by the deputy commissioner's office.

One of the amendments was of January 1969, under which the employees of the Indian Rubber Board that were on duty to reserved areas were exempted from taking a tribal pass. This amendment affected the Katchal Island. Rubber plantations were set up in Katchal Island and 47 families of the Sri Lankan Tamil repatriates were 'employed' in this plantation. This served two purposes: first, resettlement of the Sri Lankan repatriates; second, initiating rubber plantation in the Nicobar Islands. Another amendment came in April 1972, in which the area located between longitude 93°50′ to 93°57′ east and latitude 6°53′ to 7°2′ north, in the eastern coast of Great Nicobar Island, was de-regularized and 300 families of ex-servicemen were resettled from 1969 to 1980. The last amendment before the Tsunami came in July 2004, under which the employees of public sector banks and members of their family were exempted from taking a tribal pass to enter the Nicobar District during the duty period.

Thus, the four islands with sizeable non-tribal population in Nicobar Islands were Car Nicobar, Kamorta, Katchal and Great Nicobar. The first two islands had the presence of a non-tribal population prior to the implementation of the ANPATR while the latter two were added in 1969 and 1972, respectively. Again, the initial Regulation permitted the armed forces, forest officers, civil police and emergency service officers and their families to enter the 'protected' areas on duty. The later amendments of 1969 and 2004 granted permissions to more government employees to enter and reside in the Nicobar Islands. Though the permission was granted for the duty period only, some of the families settled in the Nicobar Islands after the retirement of the employees. The deregulation and amendments encouraged the growth of population of non-tribal which increased 176 times between the decade of 1961 and 1971 (Table 7.5).

Table 7.5 Population growth in Nicobar Islands (1961–2001)

Years	*Total population*		*Per cent growth of population*	
	Non-tribal	*Tribal Nicobarese*	*Non-tribal*	*Tribal Nicobarese*
1961	21	14,461	–	–
1971	3,696	17,969	17,500.00	24.26
1981	9,244	21,183	150.11	17.89
1991	14,355	24,403	55.29	15.20
2001	15,503	26,565	8.00	8.86

Source: Based on Census of India (1961–2001)

The population growth rate has remained high for the non-tribal population as compared to the tribal Nicobarese population. Between 1961 and 1971, the percentage growth of population of the non-tribal population was as high was 17,500. It was only in the decade of 1991–2001 that the growth rate for both non-tribal and the tribal Nicobarese became near equal.

The non-tribal population of the Nicobar Islands has grown on five accounts. First, the government resettled non-tribal families in the Nicobar Islands from the late 1960s to mid-1980s. The Sri Lankan Tamil repatriated families were resettled in Katchal Island and the ex-servicemen families from Punjab were resettled in Great Nicobar Island. The second factor was the natural population growth of the resettled population. Third, the government servants that were posted in the islands brought their families along with them, which too contributed to the natural increase in population. Fourth, these families settled permanently in the Nicobar Islands and did not return back. Lastly, as the population of the non-tribal grew in the Nicobar Islands, it attracted small traders and labourers, who came as opportunists and settled in the Nicobar Islands and also brought in their families. It is the last two groups of the non-tribal population of the Nicobar Islands that were vulnerable.

Fieldwork confirmed that the non-tribal households from the Nicobar Islands are not a homogenous group. They differ in many ways. First, the state of origin of the 38 non-tribal households from Nicobar Islands varied: Tamil Nadu – 16; Andhra Pradesh – 12; Ranchiwalas (Chotanagpur Plateau area, present-day Jharkhand) – 5; West Bengal – 3; and Uttar Pradesh – 2. A second difference was in the duration of stay in the Nicobar Islands. The average residence period of the 38 non-tribal households was 27 years, but varied from less than two years to nearly 50 years. Most of the early migrants, the ones who went to the Nicobar Islands in the early 1950s, were small traders who set up shops for the tribal Nicobarese. They carried on with their businesses even after the implementation of ANPATR in 1956. The third difference was in terms of economic status. The heads of household with government services were economically better than ones with petty businesses and the daily wage labourers, as the former had a secure source of income.

The legal status was the fourth and the most important difference noticed. All non-tribal population who were not officially 'resettled' or were sent on duty by the government, but continued to live after retirement, were classified as 'illegal residents'. They formed the bulk of the non-tribal households in the Nicobar Islands and were vulnerable because they had no rights and privileges. Under the ANPATR of 1956 it was forbidden for them to reside in the Nicobar Islands. They were without financial security in the first place and hence had come to the Nicobar Islands in search of employment opportunities. In the Nicobar Islands the tribal Nicobarese considered the non-tribal population as a 'necessary evil' since they were largely traders and provided the Nicobarese with goods and services, but also cheated most of the time. The tribal Nicobarese rented their houses to the non-tribal households. An oral rent agreement existed between the non-tribal population and tribal Nicobarese. Thus the non-tribal households lived on Nicobarese lands and houses which were close to the coast. They lived in relative poverty and had no investments. Again, they were 'illegally' residing in the Nicobar Islands and owned neither land nor houses, thus they were not 'eligible' for compensation in case of a disaster. Their situation was so grave that after the Tsunami, they were resettled in South Andaman Island and not at their pre-Tsunami island. Though the 'illegal' non-tribal population was vulnerable, they lived there out of their own choice.

A high number of killing was reported from islands in which there was a sizeable non-tribal population. But in these islands it was the tribal population that was killed in large numbers. Of the total killed in these four islands, 79 per cent were tribal population. In each of these islands the share of tribal population among the killed was high. In Kamorta Island, all killed were tribal Nicobarese. In Car Nicobar Island, Katchal Island and Great Nicobar Island, of the total killed tribal population accounted for 82 per cent, 75 per cent and 62 per cent, respectively (Table 7.6).

To understand why the tribal population was killed in such large numbers, it would be worth investigating the conditions which have rendered them vulnerable.

Table 7.6 Disaster divide: more tribal population killed in islands with mixed non-tribal and tribal population

Islands	*Per cent population killed*	
	Non-tribal	*Tribal*
Kamorta	0.00	100.00
Car Nicobar	18.03	81.97
Katchal	24.90	75.10
Great Nicobar	38.35	61.65

Source: Based on Census of India (2001a) and A&N Administration (2009)

Vulnerability of the tribal Nicobarese

The tribal Nicobarese have been rendered vulnerable primarily because of two major reasons: (1) due to the amendments to ANPATR of 1956 that led to the juxtaposition of non-tribal and tribal Nicobarese population and (2) because of the exclusion of the tribal Nicobarese from the development process.

To regulate the movement of the non-tribal population in tribal-dominated islands, the ANPATR was implemented in 1956. But because of the deregulation of areas and amendments to the ANPATR, the non-tribal population increased. At the time of Independence, the Nicobar Islands had a meagre non-tribal population, that too in Car Nicobar Island and Kamorta Island only. But after Independence the non-tribal population infiltrated the Nicobar Islands, so much so that in 2001 in Great Nicobar Island, the non-tribal population accounted for 87 per cent of the total population, while the tribal Nicobarese were reduced to 13 per cent.

The continuous influx of the non-tribal population had an adverse impact on the tribal population. The traders of the Nicobar Islands consisted of the non-tribal population. The tribal Nicobarese narrated incidents where they were cheated by the non-tribal population, as they are always charged more for the product. Another adverse impact of the juxtaposition of the non-tribal population along with the tribal Nicobarese was that the latter began imitating the former.

One prominent example was the changes in the traditional Nicobarese houses. The building materials and the design of traditional Nicobarese houses were modified to imitate the non-tribal houses, which made it susceptible to the earthquake and the Tsunami. In this process of assimilation, the tribal Nicobarese became 'advanced' as compared to the other five aboriginal tribes of the A&N Islands. However, in the process of development, the tribal Nicobarese became detached from their old customs and traditions that had evolved over centuries and were adapted to the environment. Neither did they remain completely 'tribal' as the Jarawas, Onges, Sentinelese and Shompens, nor were they 'advanced' as the non-tribal population. This in-between situation created the vulnerability. The tribal Nicobarese were left out while the Nicobar Islands were being developed.

As pointed out earlier, infrastructure growth is an indicator of development. Both the physical and the social infrastructure were concentrated in islands dominated by non-tribal population, while the islands with dominant tribal Nicobarese population were left out (Table 7.7).

Table 7.7 Concentration of infrastructure in the Nicobar Islands with sizeable non-tribal population, 2003

Islands	*Per cent non-tribal population*	*Physical infrastructure*			*Social infrastructure*	
		Road length (km)	*No. of cargo handling jetties*	*Power generation capacity (KW)*	*Total schools*	*Total medical facilities*
Great Nicobar	87.38	102.4	1	2,706	12	9
Katchal	49.89	26.66	0	579	11	5
Kamorta	45.66	10	1	858	6	5
Car Nicobar	21.65	72.8	1	3,400	17	7
Teressa	9.87	4.5	0	428	9	5
Kondul	6.00	0	0	0	1	0
Nancowry	4.96	0	0	295	7	3
Pulomilo	4.83	0	0	24	1	1
Trinket	2.7	0	0	0.194	2	1
Little Nicobar	2.27	0	0	50	2	1
Chowra	0.65	0	0	165	1	1
Bompoka	0.00	0	0	0	1	0
Total	**35.72**	**216.36**	**6**	**8,505.194**	**70**	**38**

Source: Based on Directorate of Economics and Statistics (2006)

Take the case of roads. Of the total metalled roads in the pre-Tsunami Nicobar Islands, 98 per cent were concentrated in islands with a dominant non-tribal population. The pre-Tsunami Nicobar Islands had three cargo handling jetties which were confined to three islands, namely, Car Nicobar, Kamorta and Great Nicobar; all three are islands with a dominant non-tribal population. Similarly, 89 per cent of the power-generating capacity was installed within these four islands only. Out of the total schools and total medical facilities, 66 per cent and 68 per cent, respectively, are confined to these four islands. Again, in the Nicobar Islands, the piped water supply was concentrated in non-tribal pockets of Campbell Bay, Govind Nagar, Shastri Nagar and Vijay Nagar settlements of Great Nicobar Island, Kamorta headquarters in Kamorta Island, West Bay and Kapanga in Katchal Island and Safed Balu in Teressa Island.

Not only was the infrastructure concentrated, all infrastructure constructions were initiated after the resettlement of the non-tribal population began in the Nicobar Islands. The east-west and the north-south roads in Great Nicobar Island were constructed in 1971, after the ex-servicemen were resettled in 1969. So was the case in Katchal Island and Kamorta Island as well. Even the jetties of Nicobar Islands were constructed in the 1960s after the non-tribal population was resettled, and regular shipping services were established only after 1965, i.e. after the resettlement of non-tribal rubber plantation workers in Katchal Island.

Therefore, in the total scheme of development the tribal population was bypassed and marginalized while non-tribal population pulled the amenities, services and facilities towards themselves. On the one hand, schemes were introduced to develop the Nicobar Islands so that the non-tribal population could be resettled; on the other hand, the tribal population continued to have a subsistence economy. More than 75 per cent of the tribal population remained engaged in subsistence copra production and fishing even in 2004, 57 years after Independence (Planning Commission 2008). No initiatives were taken in order to harness the resources of the tribal Nicobarese that includes coconut, forest, livestock rearing and the sea. Instead of harnessing and developing the local resources, new practices were introduced such as rubber plantation which was completely alien for the tribal population, nor were they employed in it.

Coconut is the mainstay of Nicobarese livelihood. The Nicobarese sell copra to obtain money with which they fulfil the cash requirements. During the fieldwork it was apparent that the method of copra production was subsistent. Ripe nuts are collected, de-husked, split into two and dried in bamboo shack kilns. After they are sufficiently heated, copra is scooped and is sun-dried. In 2010 copra was sold at a rate of Rs. 36.60 per kg to Ellon Hinengo Limited (EHL), a co-operative society of the tribal Nicobarese. In a week, a tribal Nicobarese household, known as a *tuhet*, produced approximately 20 kg copra and generated a weekly income of about Rs. 800. The figures vary with the size of tuhets. Besides copra, Nicobarese depend on forest to obtain fruits of keveri trees (pandanus), yams and tubers, wood and leaves for house construction, weaving baskets and mats. Forests are also used to rear pigs. The latter are first domesticated and then left in the forest and finally hunted and consumed. An annual pig festival known as 'Ossuary Feast' is celebrated in each village. In Car Nicobar Island there are 15 tribal villages. Therefore every month there is a pig festival in any of the villages, and at times there are months with more than one pig festival. On average in the pre-Tsunami period, a tuhet owned approximately 150 pigs.

Besides the coconut, forest and pig, the important source of livelihood for the tribal Nicobarese is the sea. Seafood such as fish, lobsters, dugongs and turtles add protein to the diet. The sea was also used for maintaining sanitation of the village, as waste generated of biodegradable products were thrown in the sea. The sea cools the islands and provides a moderating milieu within an equatorial location. Tribal Nicobarese thus have a livelihood and sustenance closely bound with coconut, forest and the sea. As the tribal population's living depended on copra, pig and fishing, they were left with little choice except to build settlements on/close to the coast. They continued to live on the coasts until the Tsunami (Figure 7.6).

When a location of settlements was mapped for the representative Nicobar Islands, Car Nicobar Island and the Katchal Island, it became

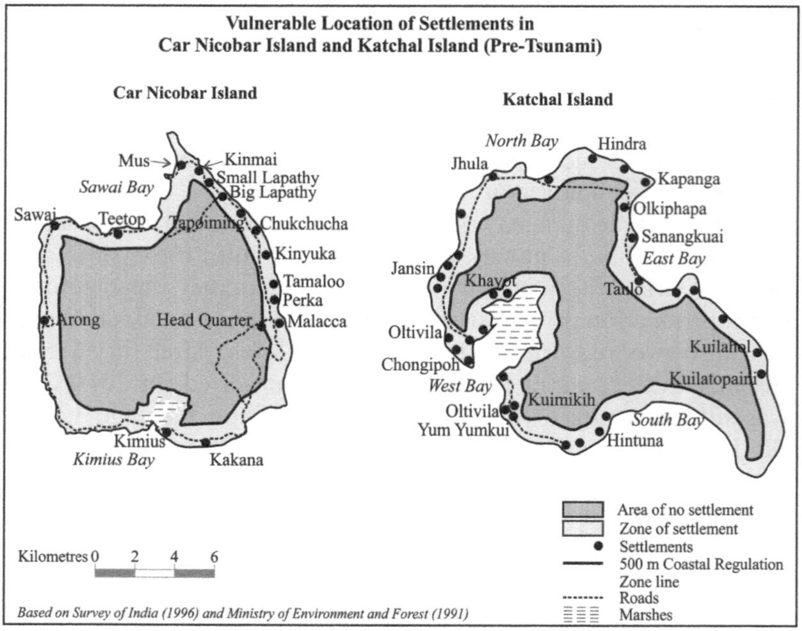

Figure 7.6 Vulnerable location of settlements in Car Nicobar Island and Katchal Island (pre-Tsunami)

evident that all villages were located on the coast. In fact, settlements clearly dotted and hugged the coastline even though it came under the Coastal Regulation Zone. Under the Environment Protection Act, 1986 of India, notification was issued in February 1991 for regulation of activities in the coastal area by the Ministry of Environment and Forests.

The coastal zones demarcated for various activities came to be known as the Coastal Regulation Zone (CRZ). The regulation states that

> the Central Government hereby declares the coastal stretches of seas, bays, estuaries, creeks, rivers and backwaters which are influenced by tidal action (in the landward side) upto 500 m from the High Tide Line (HTL) and the land between the Low Tide Line (LTL) and the HTL as CRZ; and imposes with effect from the date of this Notification, the following restrictions on the setting up and expansion of industries, operations or processes, etc. in the said CRZ.

For the purposes of the CRZ Notification, the High Tide Line means the line on the land up to which the highest water line reaches during the spring tide. The High Tide Line was demarcated uniformly in all parts of the country by the demarcating authority or authorities so authorized by the central government, in accordance with the general guidelines issued in this regard. The coastal stretches were classified in four zones (Table 7.8).

Almost all of the A&N Islands were placed in CRZ-IV. Category IV includes the 'coastal stretches in the Andaman & Nicobar, Lakshadweep and small islands, except those designated as CRZ-I, CRZ-II or CRZ-III'. Port Blair came under CRZ II, inhabited island of the Andaman Islands, and the Nicobar Islands came under CRZ-III and uninhabited islands came in CRZ-IV. The tribal Nicobarese villages continued to reside on the coasts even though their village location was declared vulnerable. The tribal Nicobarese villages came under CRZ-III. As per the CRZ-III, 'construction/reconstruction of dwelling units between 200 and 500 metres of the High Tide Line permitted so long it is within the ambit of traditional rights and customary uses such as existing fishing villages and gaothans' (in Maharashtra, gaothan is portion of the land of the village which is ordinarily used for settlement; it is the land included within the site of a village, town or city).

Table 7.8 Coastal regulation zones in the A&N Islands, 1991

Coastal Regulation Zones 1991	*Definition*	*Areas of A&N Islands*
Zone I	Areas that are ecologically sensitive and important, such as national parks/marine parks, sanctuaries, reserve forests, wildlife habitats, mangroves, corals/coral reefs, areas close to breeding and spawning grounds of fish and other marine life, areas of outstanding natural beauty/historically/heritage areas, areas rich in genetic diversity, areas likely to be inundated due to rise in sea level consequent upon global warming and such other areas as may be declared by the central government or the concerned authorities at the state/union territory level from time to time; and area between the low tide line and the high tide line.	NA
Zone II	The areas that have already been developed up to or close to the shoreline. For this purpose, 'developed area' is referred to as that area within the municipal limits or in other legally designated urban areas which is already substantially built up and which has been provided with drainage and approach roads and other infrastructural facilities, such as water supply and sewerage mains.	Port Blair Bambooflat
Zone III	Areas that are relatively undisturbed and those which do not belong to either Zone I or II. These will include coastal zone in the rural areas (developed and undeveloped) and also areas within municipal limits or in other legally designated urban areas which are not substantially built up.	Some villages of the Andaman Islands and all villages of the Nicobar Islands
Zone IV	Coastal stretches in the Andaman and Nicobar, Lakshadweep and small islands, except those designated as CRZ-I, CRZ-II or CRZ-III.	Uninhabited islands

Source: Based on Ministry of Environment and Forest (1991)

Therefore no action was taken to relocate the tribal Nicobarese villages even though the two Disaster Management Plans of 1996 as well as 2004 noted:

> flat islands like Car Nicobar, Chowra, Teressa, Katchal, Nancowry, Pilomillow (spelling as per the Disaster Management Plans), Little

> Andaman, Neil, Havelock, Strait, Long Island, etc. are vulnerable to tsunamis in Andaman and Nicobar Islands. The Nicobarese tribal are particularly vulnerable to tsunamis since their villages are found on the coastal belt only.
>
> (A&N Administration 1996, 2004)

Since the tribal Nicobarese were within 'the ambit of traditional rights and customary uses', no initiative or action was taken to move the population to safe ground even though it was vulnerable, and so on 26 December 2004 when the Tsunami hit the Nicobar Islands, the villages were completely demolished (Figure 7.7).

The height of the Tsunami was 5 m above the traditional tribal Nicobarese huts and the surrounding coconut groves. The Tsunami waves inundated more than a kilometre in Car Nicobar Island. As the tribal Nicobarese over time had lost the traditional wisdom of interpreting the movement of the sea, they were caught unaware, while the remaining five aboriginal tribes survived as they used their traditional knowledge, and moved inland when the water of the sea receded. The infrastructure of the tribal villages of the Nicobarese was also not well developed, thus they could not escape either and were caught in the wave.

The vulnerability of the tribal Nicobarese is a case of rendered vulnerability. The short-sighted development agenda for the Nicobar Islands, the amendments to the ANPATR of 1956, the neglect of the warnings of the successive Disaster Management Plans and the marginalization of the tribal Nicobarese rendered the Nicobar Islands and

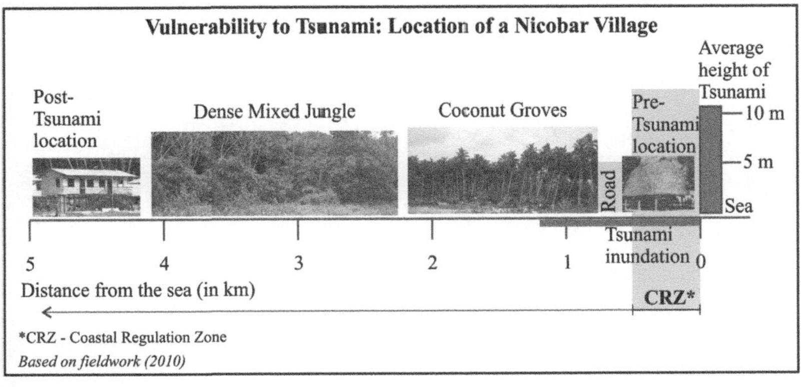

Figure 7.7 Vulnerability to Tsunami: location of a Nicobar village

especially the tribal Nicobarese vulnerable. The Tsunami was followed by an equally massive response that spread over seven years. What has been the response to the Tsunami in the A&N Islands? Did the response lead to recovery? Was the response effective in reducing the vulnerability of the Nicobar Islands? The next chapter investigates the immense response to the Tsunami in the A&N Islands.

8 Response and the Tsunami

Response to disaster refers to ways, modes, measures and actions taken after a disaster to restore the pre-disaster conditions. Response has many phases which can be clubbed into three stages – immediate relief, reconstruction, and rehabilitation and recovery. Immediate relief provides assistance to the affected people immediately after the event and includes emergency aid and search and rescue operations, setting up of relief camps and temporary shelter. Reconstruction includes rebuilding of infrastructure, both physical and social, as well as post-disaster houses in many cases. The most difficult and long-lasting aspect of the response is rehabilitation. The trauma of the disaster and the pain of the killing of loved ones can never be recovered. However, to rehabilitate the affected, livelihoods need to be restored. This involves various support systems, aids and incentives such as providing free ration until households could earn a living, providing opportunities for employment and even psychological counselling to rebuild lives and livelihoods.

The A&N Islands are a union territory. Therefore the responsibility to manage disasters in the A&N Islands rests with the central government. The main lineaments of disaster management in the A&N Islands are similar if not identical to the progress of mainland India. As India was a signatory to the United Nations 'International Decade for Natural Disaster Reduction' (IDNDR) from 1990–99, the Government of India entrusted the state governments and the administration of the union territories to prepare an integrated disaster management plan. In the light of this, a disaster management plan for the A&N Islands was prepared in 1996 by the then deputy commissioner, S. A. Awaradi of the Nicobar District.

Pre-Tsunami disaster management system

Even though India was a signatory to the IDNDR in 1990, action on disaster management took place only after the super cyclone hit Orissa in 1999, and the earthquake occurred in Bhuj in 2001, in which

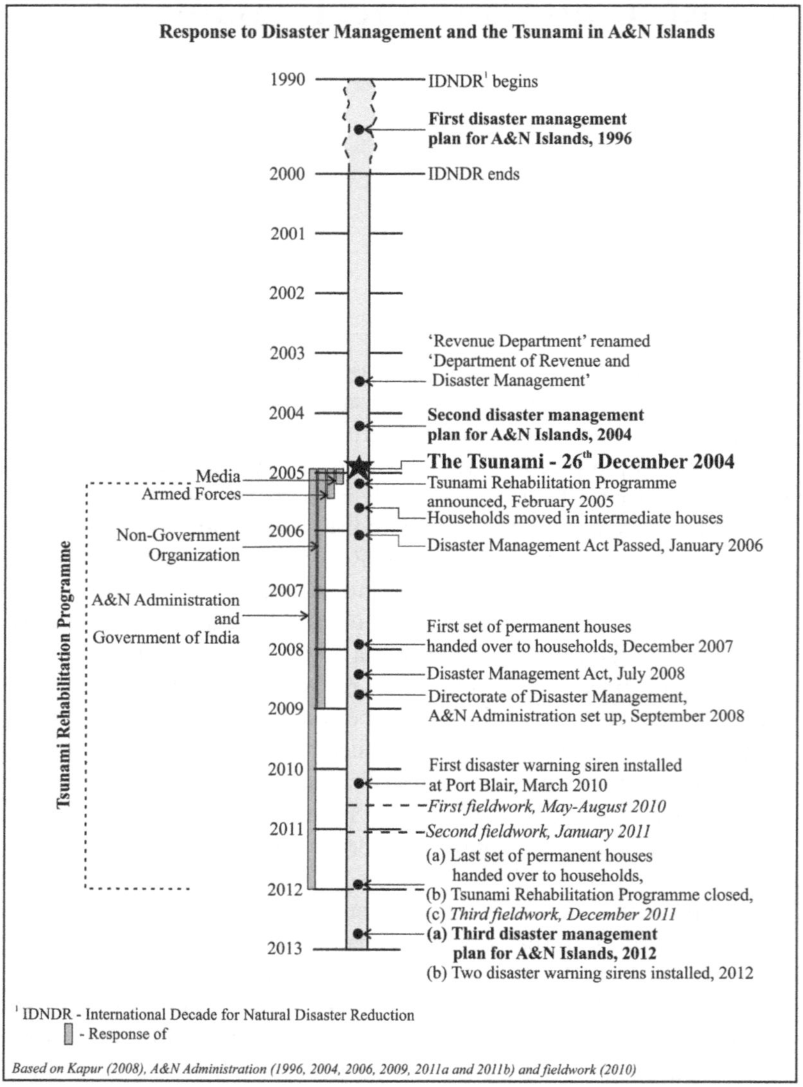

Figure 8.1 Response to disaster management and the Tsunami in A&N Islands

9,000 and 20,000 people were killed, respectively (Figure 8.1). On 20 August 1999, a High Powered Committee (HPC) under the chairmanship of Shri J. C. Pant was constituted to review the then existing mechanisms for preparedness and mitigation of natural disasters, recommend measures for strengthening organizational structures, to formulate a comprehensive modal plan for Natural Disaster Management at the National, State and District levels and to draw a framework for disaster management in India (Ministry of Agriculture 2001). As per the recommendations of the HPC post-2001, some changes were ushered in the field of disaster management across India including the A&N Islands.

The first major change was renaming of the department to include the term 'disaster management'. In the A&N Islands under the Order No. 2116 of the Andaman and Nicobar Administration, dated 21 May 2003, the 'Revenue Department' was renamed as the 'Department of Revenue and Disaster Management' and 'all aspects of disaster management, namely, prevention, mitigation, preparedness, response, relief and rehabilitation' came under its purview. The whole idea of the name change was to coordinate the procedure of disaster management under various departments of the administration.

In 2003, key departments such as Police, Health Services, Andaman Public Works Division (APWD) and Agriculture constituted various teams which were given responsibility to co-ordinate with each other for efficient management of relief works at the time of the disaster.

A Disaster Management Plan was prepared in 1996 by the then deputy commissioner of the Nicobar Islands. However, a need was felt to reformulate the Plan in view of the recommendations of the High Powered Committee on Disaster Management. Therefore, in 2004, a special committee under the chairmanship of the deputy commissioner prepared a second disaster management plan for the A&N Islands. Accordingly, the administration had to identify potential disaster-risk zones and lay out the mechanism of operation in case of a disaster. Five potential threats were identified which could lead to disasters in A&N Islands. These were, first, cyclone and high velocity winds; second, floods; third, volcanoes; fourth, earthquakes; and fifth, tsunami and maritime disasters that included shipwrecks on the high seas, collision of ships with submerged rocks, ships in distress due to mechanical/navigational failures and cyclones and oil slicks from oil tankers leading to environmental hazards (Andaman and Nicobar Administration 2004).

A paragraph is devoted to each, in which the last significant occurrences and potential areas under threat were outlined. The plan also stated that the nodal agency for disaster management is the Department of Revenue and Disaster Management and is responsible for 'all aspects of disaster management viz. prevention, mitigation, preparedness, response, relief and rehabilitation'. It is also responsible for 'coordination on steps taken by different departments of the Administration

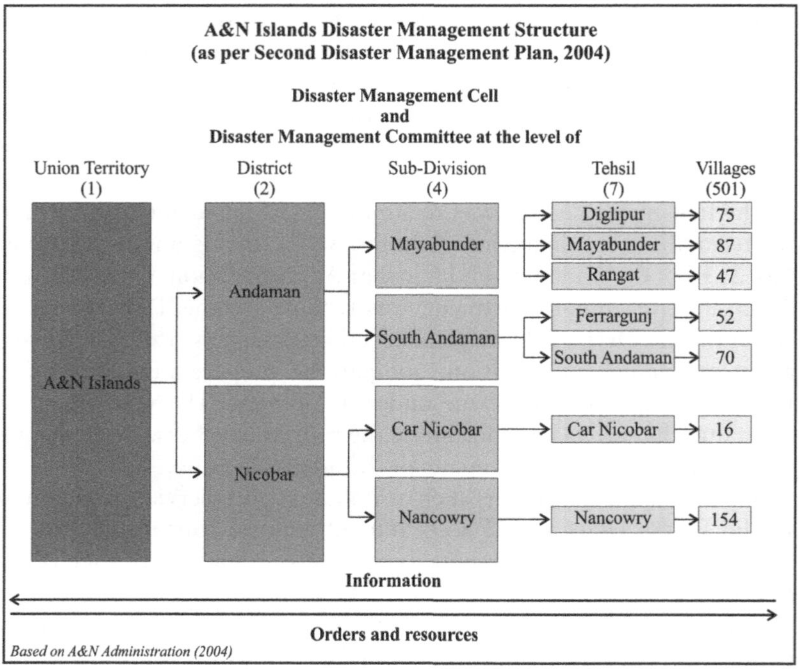

Figure 8.2 A&N Islands disaster management structure (as per second disaster management plan, 2004)

on the above (disaster) subject'. The disaster management plan proposed for creation of disaster management machinery at five levels – union territory (UT) level, district level, sub-division level, tehsil/island level and village level (Figure 8.2). The levels were to be interconnected whereby information would flow from lower to the highest level and each lower level would report to the level immediately above it. At each level a disaster management committee and a disaster management cell had to be made. The former would make plans and review the preparedness in case of disasters while the latter would have a role in times of crisis. The disaster management cell of each level would be responsible for disseminating warning, preparing people by informing them about 'dos and don'ts', assessing the scale of disaster after its occurrence and monitoring rescue, relief and rehabilitation operations. Provision for island-level committee was not there. Even though the islands constitute a geographical unit, for the administration the mainland model is followed.

The personnel for the disaster management committee and cell were to be drawn from the existing administrative setup; the lieutenant governor was to be the chairman of the union territory disaster management committee; and the chief secretary was to be the chairman of the union territory disaster management cell. The plan proposed the deputy commissioner of the district as the chairman of the district disaster management committee and also the district disaster management cell. The assistant commissioner of the district was to be given the responsibility to chair the sub-division disaster management and disaster management cell. The tehsil level disaster management committee and cell were chaired by the corresponding officers posted in the tehsil. The village level committee was to be headed by the patwari in case of revenue villages and the first captains in case of tribal Nicobarese villages. In urban areas, the councillor was to be the chairman of the municipal area disaster management committee and cell.

The plan also proposed for mitigation of disaster through awareness generation, capacity-building, preparedness, response and rehabilitation. The preparedness plan provided the procedure to be followed before the disasters, and included maintaining an updated list of the functioning body, construction of a control room within the premises of the deputy commissioner's office, identifying hazard-prone areas and devising a co-ordination method of the machinery. Procedures to be followed at the time of disaster were also provided. This included activation of the control room, evacuation of the affected area, releasing a press note, providing relief and shelter to the affected people and monitoring the situation. In case of a disaster, the areas to be evacuated and the duration of evacuation would be defined by the district disaster management cell. The disaster management cell was made the nodal authority to co-ordinate and implement the rehabilitation plan. For the affected people, the plan laid out that aids like rehabilitation schemes, ex-gratia, loans and subsidies would be given. For the psychosocial rehabilitation, the tehsil disaster management cell would arrange for doctors to treat post-traumatic stress disorder.

In the rehabilitation phase, the plan places responsibility on the disaster management cells to appraise, document and report the rehabilitation process. However, no details are given regarding the procedure of how the appraisal is to be done. Only a format of reporting is suggested which would include history (details regarding how the disaster started, the trauma and the suffering of the people, and the remedial action taken in the district) as well as success stories, like successful interventions by the district administration, public and private persons, philanthropists, social welfare organizations and religious institutions. The effort of the people to save others' life and property may be recognized so that the people shall be publicly honoured.

The management of high disaster-prone setups, such as the Indian Oil Corporations compound and the wharfs and jetties with cargo, do not come under the purview of the union territory disaster management cell. A separate plan is provided for each of these high disaster-prone setups. The A&N Command consisting of personnel from the army, air force, navy and the coast guard would be the operating agencies.

In addition to the disaster management plan, the A&N Administration in 2004 has constituted a mobile search and rescue team to deal with the various types of disasters in the A&N Islands. This team consists of the staff of police, fire services, medical department and the APWD.

Apart from this, the Ministry of Home Affairs had provided very high frequency (VHF) sets for search and rescue to the A&N Administration. VHF is the International Telecommunication Union (ITU) designated range of radio frequency and electromagnetic waves from 30 MHz to 300 MHz. Common uses for VHF are FM radio broadcasting, television broadcasting, land mobile stations (emergency, business, private use and military), long-range data communication up to several tens of kilometres with radio modems, amateur radio, and marine communications. Hence the VHF sets are very useful for search and rescue in a disaster. The A&N Administration had also procured three satellite phones for search and rescue and imparted training to select officers for its use. The administration notified all departments to orient the plan schemes to include disaster mitigation and preparedness.

Thus, even with two disaster management plans in existence with detailed procedure of preparedness and operation in a disaster, the Tsunami was followed by a trail of damage and destruction. The areas 'prone to' tsunami were identified by both the disaster management plans of 1996 and 2004, yet no measure was taken to ensure the safety of the islanders. Both the disaster management plans state that

> flat islands like Car Nicobar, Chowra, Teressa, Katchal, Nancowry, Pillowmillow, Little Andaman, Neil, Havelock, Strait Island, Long Island, etc., are vulnerable to tsunamis in Andaman and Nicobar Islands. The Nicobarese tribals are particularly vulnerable to Tsunamis since their villages are found on the coastal belt only.
>
> (A&N Administration 1996)

A similar sentiment and fear is repeated in the 2004 disaster management plan,

> these [tsunami] waves may reach enormous size and have been known to travel across entire oceans. Flat islands like Car Nicobar, Chowra, Teresa, Katchal, Nancowry, Pilowmillow, Little Andaman, Neil, Havelock, Strait, Long Island, etc., are vulnerable to tsunamis. The Nicobarese tribals are particularly vulnerable since their villages are mostly on the coastal belts.
>
> (A&N Administration 2004)

The 1996 plan proposed,

> it is necessary to locate high grounds in each village and island where the people can move to escape from the effects of tsunamis . . . a thick belt of plantation on the coastal regions of the islands, particularly the flat ones may help reduce the sweeping effects of the tsunamis. Suitable species of the plants including mangrove could be planted as a tidal wave breaker.
>
> (A&N Administration 1996)

Even after eight years of the plan, no steps were taken to protect the coastal villages of the Nicobar Islands from the Tsunami. The 2004 plan highlighted the same situation and suggested similar measures: 'absence of coastal plantation/mangroves and non-availability of high altitude grounds/buildings can enhance the vulnerability of the people to tsunamis' (A&N Administration 2004).

Neither warning of the disaster management plan of 1996 nor 2004 was followed. Even though a disaster management cell and a control room existed in Port Blair, integration with remote areas was not established in 2004. The administration was unprepared to respond effectively to a disaster of the magnitude of the Tsunami. The phase immediately after the Tsunami was marked with unpreparedness and chaos. It was barely a few months before the Tsunami that the disaster management plan of 2004 came into existence, and it remained a document on paper. It was yet to gear up its machinery, however, the disaster management plan of 1996 was in existence for nearly eight years. The plans failed to address the disaster in the A&N Islands because both plans were mechanical plans. The provisions enlisted focused on technological development and the bureaucratic setups but failed to address the deeper issues of vulnerability. The plans identified the islands 'prone to' natural occurrences such as cyclones, earthquakes and tsunamis, but it did not identify the vulnerable population who lacked choices and were forced to inhabit the 'vulnerable islands'. The plans were there in principle but not in practice. Had the plan been implemented and put into effect, the number of killed in the A&N Islands would surely have been far less. But this wasn't done. When the Tsunami came in 2004, destruction and damage was rampant. It was a global Tsunami and foreigners were killed. The response was equally massive. In the A&N Islands alone, the post-Tsunami response lasted for seven years and was declared closed in December 2011. It changed the islands completely.

The response to the Tsunami in the A&N Islands can be divided into two segments: (1) immediate phase and (2) long-term phase. In the immediate phase, the focus was on saving lives and evacuating the stranded islanders, providing for their basic necessities such as food, clothing and shelter and restoring critical transport and communications infrastructure. Therefore, this phase can also be called the relief phase. In the long-term response phase, the focus was on reconstruction of the damaged infrastructure and rehabilitating the affected households by providing them with compensation for the killed, and the losses, new houses and assistance in restoring livelihoods; hence this phase can be called the rehabilitation phase. The phases are not exclusive, as there were overlaps. For example, reconstruction of transport infrastructure began immediately after the Tsunami to restore the communication with the Nicobar Islands. Again, compensation for the household items was distributed immediately after the Tsunami whereas the other compensation was released over the seven-year period. Again, relief camps were provided in the immediate phase while different kinds of shelters were provided later. Hence, the six-month period that the households spend in the relief camps has been taken as the immediate phase, while the remaining six and a half years consists of the long-term response phase.

Immediate phase of response

The first action in the immediate phase of the Tsunami response in the A&N Islands was the meeting of the Union Territory Disaster Management Committee under the chairmanship of the lieutenant governor at Port Blair on 26 December 2004 itself. Other members included the head of A&N Command, the chief secretary, the inspector general of police, the member of parliament, secretaries of various departments and other officers. The crisis management team was called for assistance from the armed forces, as all connections with mainland India were severed. The news about the disaster in the Nicobar Islands reached Port Blair and the lieutenant governor, the member of parliament and the commander in chief of the A&N Command flew to Car Nicobar Island to assess the damage. They reached Car Nicobar Island by 4 p.m. Contact with the mainland could be re-established only by 5 p.m. and help was sought from the Ministry of Home Affairs. Satellite phones were flown in from mainland India and sent to the affected Nicobar Islands to aid in evacuation and rescue.

An apex coordination cell with members from all government departments and A&N Command was formed on 27 December 2004. A control room was set up at Port Blair in the office of the deputy commissioner to co-ordinate search, evacuation and relief-related activities. All secretaries to the administration and senior administrative officers were deputed to affected islands for the co-ordination of relief and rescue operations. Deputy resident commissioners of Kolkata and Chennai were designated as special relief officers for the purpose of co-ordinating procurement and dispatch of relief materials through ships, airlines and air force planes from Kolkata and Chennai, respectively. Simultaneously, the crisis management group headed by the cabinet secretary met in Delhi immediately after the Tsunami hit Tamil Nadu by 9 a.m. and keeping in view of the magnitude and scale of the killed and damage, the armed forces were mobilized.

Response of armed forces to Tsunami in A&N Islands

As the A&N Islands have a 1,962 km coastline which accounts for 30 per cent of India's Exclusive Economic Zone (EEZ), the armed forces were already stationed for its protection. The defence of the A&N Islands is the responsibility of the armed forces, i.e. army, air force, navy and the coast guard. The Indian Coast Guard was established on 18 August 1978 as an armed force for the protection of India's maritime interests and enforcement of maritime law with jurisdiction over both territorial (including contiguous zone and exclusive economic zone) and international waters. It operates under the Department of Defence of the Union Ministry of Defence. The presence of the defence forces and the Coast Guards at the A&N Islands during the time of the Tsunami helped rescue operations. They came into action immediately.

At the time of Tsunami, there was an air force base in Car Nicobar Island, while Port Blair, Campbell Bay (Great Nicobar Island) and Diglipur (North Andaman Island) had naval bases. Apart from them, the army had its garrison in Port Blair, as did the Coast Guard. Although the Coast Guard is stationed at Port Blair, it patrols the entire coastline of the A&N Islands. To co-ordinate the efforts of the four forces, the A&N Command was set up in 2001. The Command was created to safeguard India's strategic interests in Southeast Asia and the Strait of Malacca and to counter any threat from China, which was reported to have set up surveillance posts in Myanmar's Coco Islands, 40 km off the northern tip of the Andaman Islands (www.ids.nic.in). The A&N Command helps in preventing smuggling, piracy, drug and gun trafficking, poaching and illegal immigration in the region and especially in the Strait of Malacca. It is the only unified command in the country. Immediately after the Tsunami, due to the advantage of the unified structure of the A&N Command, relief operations could be carried out systematically with co-ordination.

The Tsunami hit Car Nicobar Island at around 7:00 a.m. The first information about the Tsunami was relayed by Air Force Base in Car Nicobar Island at 7:30 a.m. through a mayday message. Mayday is an emergency procedure word used internationally as a distress signal in voice procedure radio communications. It derives from the French *venez m'aider*, meaning 'come help me'. It is used to signal a life-threatening emergency primarily by mariners and aviators, but in some countries' local organizations it is used by the police forces, firefighters, and transportation organizations. The call is always given three times in a row ('Mayday Mayday Mayday') to prevent mistaking it for some similar-sounding phrase under noisy conditions and to distinguish an actual Mayday call from a message about a Mayday call. The Mayday message to the Oceanic Control Centre, Chennai read: 'Mayday, Mayday, Mayday, Chennai, Chennai, Chennai. This is Victor Juliet Golf on ground at Carnicobar. Carnicobar hit by a severe earthquake and flooding of the island by tidal waves. Island is sinking, request immediate rescue and relief' (Mittal 2009). Immediately after the message, the Air Force Headquarters at New Delhi put two AN-32 aircraft on alert at the Southern Air Command at Thiruvananthapuram, Kerala. Other commands were also instructed to keep aircraft ready for relief and rescue. On 26 December 2004 at 10:30 a.m. and 10:36 a.m., AN-32 aircraft were launched for Car Nicobar which reached at 4:46 p.m., with food and relief items.

The Ministry of Home Affairs mobilized the armed forces and the Operation Sea Wave was launched for relief and rescue of the stranded islanders of the A&N Islands. The Operation Sea Wave, with nearly 20,000 personnel involved in rescue and relief, became the biggest peace time relief operation. The last war before the Tsunami was the Kargil War of May–July 1999, in which 30,000 soldiers and airmen were deployed (Mittal 2009). Approximately 18,500 personnel were sent to the A&N Islands for the 'emergency' period. It included personnel from the armed forces, coast guards, central reserve police

Table 8.1 Number of armed forces deployed in A&N Islands

Forces	*Number of*			
	Personnel	*Ships*	*Aircraft*	*Helicopters*
Army	8,000	–	–	–
Navy	5,500	21 + 34 (LCU*)	4	12
Air Force	2,000	–	34	42
Coast Guard	2,000	9	–	–
CRPF[1]	1,000	–	–	–
Total	**18,500**	**64**	**38 (+34 civil)**	**54**

* LCU is a type of boat used by amphibious forces for transportation.
1 Central Reserve Police Force.

Source: Based on Planning Commission (2007b)

force, A&N Police Force, A&N Police Radio and A&N Fire Services. Apart from personnel, the armed forces also provided ships, aircraft and helicopters for the relief work in the A&N Islands (Table 8.1).

The armed forces remained engaged with the Tsunami response in the A&N Islands for six months and pulled out in June 2005.

The first task of Operation Sea Wave was to rescue stranded islanders and evacuate small islands that were completely devastated. The coast guard and the navy helicopters searched for survivors on land while ships looked for them in the sea. Nine navy ships, five coast guard ships and two merchant vessels were involved in search and rescue. Six islands – Chowra, Bompoka, Trinket, Kondul, Pulomilo and Little Nicobar – were evacuated.

Within 10 days of the Tsunami, more than 50 per cent of the Nicobar Islands had been evacuated. Islanders from Chowra and Bompoka were taken to Teressa, from Trinket they were taken to Kamorta while islanders from Kondul, Pulomilo and Little Nicobar were evacuated to Great Nicobar Island. From Car Nicobar a total of 20,000 persons were evacuated. Of this, nearly 6,000 people were evacuated to mainland India to Chennai and Kolkata. The latter consisted of Tsunami-affected family of the air force officers posted in Car Nicobar Island. The remaining 14,000 (tribal Nicobarese and the non-tribal) were evacuated to Port Blair. The armed forces also cleared the debris and cremated the dead bodies.

To accommodate evacuated people, relief camps were set up in eight of the A&N Islands: two in the Andaman Islands and six in the Nicobar Islands. In the Andaman Islands, relief camps were located in two islands, namely, South Andaman Island and Little Andaman Island, while in the Nicobar Islands relief camps were set up in six islands – Car Nicobar, Teressa, Katchal, Kamorta, Nancowry and Great Nicobar. The relief camps of the Andaman Islands were maintained by the Port Blair Municipal Council and the Gram Panchayats, but the relief camps of the Nicobar Islands were set up and maintained by the armed forces. The Nicobar Islands are protected under the Andaman and Nicobar Aboriginal Tribes Protection Regulation of (ANATPR) 1956: only the armed forces, paramilitary forces, police and administration officers were permitted to go to Nicobar Islands. Thus in the Nicobar Islands the armed forces took up the charge of setting up and maintaining the 118 relief camps using 3,000 tents.

The medical teams of armed forces were the first to provide medical help to the injured islanders. The doctors' team from mainland India arrived after a week of the Tsunami in the Nicobar Islands. During this period the medical team of the armed forces along with the resident doctors of the Primary Health Centres treated the injured people in the relief camps.

The army engineers restored the essential infrastructure of A&N Islands. They restored water supply in Port Blair, repaired the Suryachakra Power Plant of Bambooflat in South Andaman Island, and the power plant at the Car Nicobar Air Force Base. The repair of Car Nicobar Air Force Base runway and the Great Nicobar airstrip was also undertaken by the army engineers during the night, as during the day there was heavy air traffic bringing in relief supplies. These two airstrips became the base for the relief operations carried out in the Nicobar Islands. While the Campbell Bay airstrip served Great Nicobar Island that had rescued population from the islands of Kondul, Pulomilo and Little Nicobar, Car Nicobar Island became the base for the relief supplied to islands of Teressa, Katchal, Kamorta and Nancowry. New roads were also constructed by the army engineers in Car Nicobar Island.

After evacuation and setting up of relief camps, the next important task of the Armed Forces was to transport relief items and materials to six Nicobar Islands housing the relief camps. Ships were used to send relief items from the mainland. The ships came in from Kolkata, Chennai, Vishakhapatnam and Thiruvananthapuram. Food and drinking water cans, relief materials collected from various parts of the country and other items such as blankets, generators, pumping sets and tents were brought to the ports via rail and road. The ships first came at Port Blair and after breaking bulk moved to the Nicobar Islands in smaller vessels. As the jetties of Nicobar Islands were damaged, assault vessels of the armed forces and barges were used to transfer material to the islands that had the relief camps. Almost 15,000 tonnes (1 tonne = 1,000 kg) of relief material was transported from the mainland of which 2,800 tonnes was ration; 8,000 litres of aviation fuel was also transported to Car Nicobar Island for the helicopters engaged in rescue and transporting relief material by the armed forces.

In the Nicobar Islands the armed forces were also involved in the construction of intermediate houses. The construction material was transported from mainland to Port Blair, from where it was taken to the Nicobar Islands. The armed forces were involved in relief work till June 2005. The immediate response phase ended in June 2005 after the relief camps were closed and people moved to the intermediate houses. The armed forces played a crucial role in the period immediately after the Tsunami. The discipline of the armed forces helped the 'emergency' period. The six months intensive work by the armed forces after the Tsunami was the biggest peace time operations. The relief camps set up across the A&N Islands became the focus of all activities during the immediate response phase. A hallmark of response in the immediate phase was the establishment of the relief camps.

The post-Tsunami relief camps

People became homeless after the Tsunami and therefore had to be evacuated. Post-evacuation the households were accommodated in relief camps that were set up across the A&N Islands. The number of relief camps changed as and when required. The maximum number of camps was registered on 25 January 2005, when there were 207 camps with more than 46,000 people in eight islands (Figure 8.3). Car Nicobar had the maximum number of relief camps followed by South Andaman Island. The Andaman Islands accounted for 43 per cent of the total of 207 relief camps and the remaining 57 per cent were in the Nicobar Islands. The relief camps of the Andaman Islands accommodated 39 per cent of the 46,000 people while the latter housed 61 per cent (Table 8.2).

The 84 relief camps of Car Nicobar Island accommodated 35 per cent, while the 80 relief camps of South Andaman Island accommodated 25 per cent of the total population living in the relief camps. This clearly indicates that the relief camps at the Nicobar Islands were under more stress. Coupled with this is the fact that accessing the camps of the Nicobar Islands became even more difficult because of the breakdown of jetties. On the other hand, accessing the relief camps of the Andaman Islands was comparatively easier as these were located in the administrative headquarters of the A&N Islands, i.e. around Port Blair.

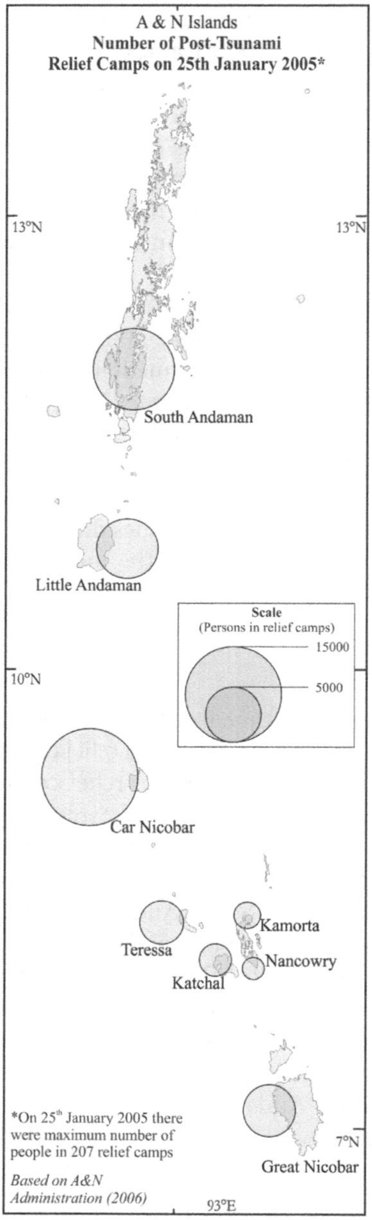

Figure 8.3 A&N Islands: number of post-Tsunami relief camps on 25 January 2005

The relief camps looked different across the A&N Islands as these were constructed with different materials in different sizes. Based on the fieldwork interviews it was known that in South Andaman Island, there were three kinds of relief camps. First were camps set up in situ, that is, at the site, where houses were damaged. This was the case of Junglighat – Dairy Farm, Bambooflat jetty, Panihati, Wandoor, Ograbraj, Teylorabad, Sippighat, Biblitan, Mithakhari, Burman Nallah and Chidiya Tapu. These camps were made of plastic sheets and bamboo poles which had been distributed by the panchayat leaders. These first type of relief camps housed people from South Andaman from the respective locality. The second type of relief camps were set up in the schools, where a classroom was shared by two to three families. In the playground of schools, tents and community kitchens for meals were made. The school turned into relief camps housed the non-tribal households evacuated from the Nicobar Islands. The third type of relief camp was large tents set up in the Island Tourism Festival Ground (approx. 30,000 m^2) at Port Blair, and these housed the tribal Nicobarese from the various Nicobar Islands. The segregation was done so that the families/communities could be together.

In Little Andaman Island and the Nicobar Islands, relief camps were made of tents. One tent was

Table 8.2 Number of relief camps with affected persons

Island	*Relief camps*		*Persons in relief camps**	
	Number	*Per cent of total*	*Number*	*Per cent of total*
Car Nicobar	84	40.58	16,194	35.19
South Andaman	80	38.65	11,395	24.76
Little Andaman	9	4.35	6,569	14.27
Great Nicobar	14	6.76	4,690	10.19
Teressa	9	4.35	3,296	7.16
Katchal	4	1.93	1,818	3.95
Kamorta	4	1.93	1,227	2.67
Nancowry	3	1.45	834	1.81
Total	**207**	**100**	**46,023**	**100**

*As on 25 January 2005; on this day the camps had maximum number of people.

Source: Based on A&N Administration (2006)

allotted for one family of four to five members. In six out of 12 of the Nicobar Islands, 118 relief camps with 3,000 tents were set up (Mittal 2009). The total time taken to set up relief camps varied across islands. In South Andaman, within six hours of the Tsunami relief camps were set up, whereas in Car Nicobar and Great Nicobar, it took three days to set up the relief camps and more time was taken to set up relief camps at Katchal and Kamorta. Based on the fieldwork interview with households that lived in these relief camps, it was inferred that though these relief camps varied in appearance and were distributed across the islands, they shared one common characteristic. All the relief camps were crowded and lacked even the basic facilities such as toilets and bathrooms. These were constructed almost a week after the Tsunami.

Relief camps were an important shelter for the affected people for the six months following the Tsunami. While in relief camps, the immediate compensation of Rs. 2,000 was distributed to each household to purchase items of basic necessities. About 300 doctors were brought in from mainland India to treat the affected people. Three types of medical teams were deployed in the affected islands. First were general physicians deployed by the Ministry of Home Affairs and the doctors from the armed forces. The focus of this team was to provide immediate medical service to the injured and affected. The second were medical teams from the National Institute of Communicable Diseases, New Delhi, which specifically looked after the prevention of the outbreak of communicable diseases. The third team consisted of psychologists and psychiatrists from National Institute of Mental Health and Neuro Sciences (NIMHANS), Bangalore. Their focus was to treat post-traumatic stress disorder (PTDS).

The National Institute of Communicable Diseases, Delhi established a network for epidemiological surveillance. The central unit was established at Port Blair and six other peripheral units were established in six of the Nicobar Islands where relief camps were set up. Essential medicines, life-saving drugs, reagents, consumables including anti-malarial and anti-diarrheal drugs and instruments were shipped or airdropped to all affected islands. To control the outbreak of measles, children between the ages of six months and five years were immunized and vitamin A supplements were also administered to them. To prevent malaria and cholera outbreak, mosquito and fly control measures were adopted. These included application of larvicide, indoor residual spray and thermal fogging of insecticides. Personal protection measures were instituted; adequate numbers of insecticide-treated bed nets were distributed. Action was taken to minimize fly breeding. Water purification measures were taken and provision of safe drinking water was ensured. Halogen tablets were distributed at every relief camp and village. Oral rehydration solution (ORS) depots were established at relief camps and ORS packets were distributed to all diarrhoea cases for preventing dehydration.

The medical team from NIMHANS, Bangalore was in the A&N Islands till February 2005. It set up 74 counselling camps, 12 in the relief camps of Port Blair and 62 in the relief camps of Car Nicobar Island. As the counselling was not compulsory for inmates of the relief camps, only 475 persons (out of 46,000 in relief camps) sought psychiatric help. Medicines were administered in serious trauma cases while others were encouraged to speak out about the Tsunami. Group discussions were held to ease the grief of the affected people. This was also the time when various non-governmental organizations (NGOs) from the island itself and from different parts of the country and the world came to the A&N Islands to provide relief.

Response of the NGOs to the Tsunami in A&N Islands

A total of 72 NGOs participated in the post-Tsunami relief and rehabilitation in the A&N Islands (Table 8.1a). The NGOs that came can be classified on the basis of their area of operations into international

NGOs, national NGOs and local NGOs. The international NGOs have centres across the world, whereas the national NGOs work primarily within India and the local NGOs are from the A&N Islands itself. Out of these 72 NGOs, 10 were international, 45 were national and 17 were local. Out of the 10 international NGOs, eight had centres located in mainland India, while three – Ramakrishna Mission, Chinmaya Mission and Nehru Yuva Kendra – had a pre-Tsunami presence in the A&N Islands. Out of the 44 national NGOs, one, i.e. Rajasthani Yuva Manch had a pre-Tsunami presence in the A&N Islands. The bulk of NGOs, i.e. 51 (seven international NGOs and 44 national NGOs) came to the A&N Islands for the first time after the Tsunami. Their representatives from regional centres located at Delhi, Chennai, Kolkata, Mumbai and other cities of mainland India came to the A&N Islands. The seven international NGOs that came to the A&N Islands were Action Aid, Adventist Development Relief Agency (ADRA), Friends Society in Social Service, Oxfam, Red Cross, Save the Children and the United Nations International Children's Emergency Fund (UNICEF). Therefore NGOs in large number descended on Port Blair after the Tsunami with the intention of providing relief to the affected households. These reached the A&N Islands within two days of the Tsunami.

The NGOs that sprung in action first were the NGOs that were present in Port Blair at the time of the Tsunami. Immediately after the Tsunami, the NGOs such as Sangathan Manch, Ramakrishna Mission, Chinmaya Mission and Vanvasi Kalyan Ashram distributed food in a common kitchen at Dairy Farm and Junglighat region of Port Blair. By day two, people evacuated from Nicobar Islands arrived in Port Blair and to accommodate them, schools, community halls and open grounds were converted into relief camps. The NGOs from mainland India joined in to 'manage' the relief camps by distribution of relief items and maintaining the community kitchens at the camps.

As per the ANATPR 1956, NGOs were not permitted to go to the 'worst affected' Nicobar Islands. Dr Stuart Zimble of the Médecins Sans Frontièrs (Doctors without Borders) said 'we have the expertise to help people but are handicapped by lack of access' (Gonsalves 2005). To resolve this, the 'Tsunami NGO Co-ordination Cell for Resettlement and Rehabilitation of Survivors' was set up by the Planning Commission consisting of representative members of 25 NGOs and officials of the Planning Commission (Khosa 2005). Following this, 38 NGOs went to Hut Bay in Little Andaman Island and 15 NGOs were permitted to work in Campbell Bay (in Great Nicobar Island). Both these islands are dominated by the non-tribal population. Only the NGOs of Christian missionaries were permitted to carry relief work in

Table 8.3 Type of assistance by the NGOs

Type of assistance provided by NGOs	*Number of NGOs*
Only relief	46
Only reconstruction	3
Only rehabilitation	4
Relief + reconstruction	5
Relief + rehabilitation	4
Reconstruction + rehabilitation	1
Relief + reconstruction + rehabilitation	5
Other (environment and animal protection)	2
Not available	2
Total	72

Relief: Providing food, clothing and shelter in the immediate phase of response.
Reconstruction: Construction of damaged buildings, infrastructure and houses.
Rehabilitation: Livelihood and skill training programme.

Source: Based on Digal (2006)

Car Nicobar and Nancowry Island as 98 per cent of tribal Nicobarese are Christians. The NGOs assisted the Administration in three ways – in relief efforts, reconstruction and rehabilitation (Table 8.3).

The latter two are part of a long-term response while relief efforts were carried out in the immediate phase of response. Fifty-eight NGOs were engaged in providing relief items, 12 NGOs joined in reconstruction and 14 in assisting the people in rehabilitation. Of the total 72 NGOs, 50 left either during or at the end of the relief phase. A total of 22 NGOs were engaged in reconstruction and rehabilitation. The duration of stay of these NGOs ranged from six months to five years.

Work done by the NGOs

Based on the fieldwork enquiry, it was concluded that the works of NGOs in the A&N Islands were of three kinds – (1) distribution of relief items, (2) trauma counselling and initiating training programmes for livelihood restoration and (3) reconstruction of houses, toilets, health centres and schools.

Many relief items were provided by the NGOs for the immediate relief of the affected households. The most significant was the Oxfam package which contained items of daily needs (Table 8.2a). The list of items can be grouped into eight categories, namely, items for personal hygiene and sanitation, clothing, bedding, kitchen items, first-aid kits, school kits, hardware and other items. These packages were given to the households while they were in the relief camps. After six months of stay at the relief camps, the households moved to intermediate houses. This house consisted of only one room made of corrugated galvanized iron (CGI) sheets fastened together to either wooden or iron poles. In the intermediate houses, the households were given cots, cupboards, sewing machines, tables and chairs by the NGOs. In the Andaman Islands, each intermediate shelter site was 'adopted' by an NGO, while the intermediate houses of the Nicobar Islands were overseen by the A&N Administration. In the Andaman Islands each of the NGOs distributed different items. Therefore, the households who lived in Chouldari and Bambooflat intermediate shelter got cots, cupboards and sewing machines, while the households from Brichganj intermediate shelter did not get these items. Instead, they got tables, chairs, bicycles and school books. There was discontent among the households as some households felt that they missed out on 'items' distributed at the other site. For the households of the Andaman Islands, the NGOs represented organizations that were in the shelter to distribute 'items' free of cost. It is evident that NGOs focused more on delivering a count of the goods, but did not focus either on the quality or utility. In the Nicobar Islands, the 'items', though donated by the NGOs, were delivered by the armed forces through the captains of the tribal villages. Hence, in the Nicobar Islands there was equality in distribution of relief items.

The second kind of work of NGOs among the affected households was trauma counselling and imparting training for livelihood. In the Andaman Islands, among the non-tribal households, trauma counselling was organized for the affected people by three NGOs – Voluntary Health Association of India, West Bengal Voluntary Health Association and Swami Vivekananda Yoga Anusandhan Sansthan. The NGO Butterflies helped in setting up of the *anganwadi* centres in the affected villages. Seven NGOs imparted vocational training. Care India initiated classes for sewing lessons. The Adventist Development Relief Agency (ADRA) began an eco-club, where participants were taught kitchen gardening. When the project was completed, certificates were issued to the participants. Repair workshops were also conducted to train islanders to repair cycles, mobiles and laptops. NGOs such as World Vision (India), PRAYAS, CARITAS – CRS and Hindustani Covenant Church gave new boats to the fishermen. Cash for work was initiated by the World Vision (India). A microcredit programme was also initiated by the Sustainable Environment and Ecological Development Society.

In the Nicobar Islands, the NGOs of Christian missionaries, such as Church of North India and Church Auxiliary for Social Action (CASA) provided trauma counselling in the villages of the Nicobar Islands. Oxfam and UNICEF engaged the tribal Nicobarese in the construction of water harvesting structures. In Car Nicobar Island, Action-Aid helped in village-by-village documentation of the resources available with the villagers. These reports were prepared by the villagers and is available with the village captains.

The third important contribution of the NGOs was their support in the reconstruction process. Seventeen NGOs participated in the reconstruction process. Twelve participated in reconstruction of intermediate houses, and four participated in construction of permanent shelter, namely, Hindustan Convent Church (HCC), CARE India, Mata Amritanandamayi Math (MATA) and ACANI. The NGOs constructed houses in the Andaman Islands only. Another NGO, the Bharatiya Jain Sangathan, constructed 20 schools and 30 primary health sub-centres across the A&N Islands. Oxfam, Church of North India and CASA constructed schools in Car Nicobar Island. UNICEF and Oxfam were engaged in construction of toilets for the intermediate shelters.

Thus it is evident that the NGOs were engaged in various relief and rehabilitation activities. However, during the fieldwork in 2010, no substantial impact of NGO response was visible apart from the logos of the NGOs on the donated items, on the permanent houses constructed by each of these and the certificates issued after the completion of sponsored courses.

Critique of NGO response: a fieldwork-based analysis

At the time of the fieldwork for this study, during May–August 2010, there were no NGOs found in the permanent shelters in A&N Islands as their projects of tsunami rehabilitation had finished. So the respondents were asked to recollect and assess the level of satisfaction of NGOs' response. As for receiving the relief items, all households acknowledged that they did receive the 'items'. However, none of the households was satisfied with the quality of the items provided. They informed that the clothes tore after two to three washes and utensils began to leak within a few months of use. The cots and cupboards started to rust. As for the training programme conducted by the NGOs such as sewing lessons, eco-clubs, cycle and mobile phone repairing, kitchen gardening, to name a few, an analysis of the feedback from 65 non-tribal households showed that men and women from

52 households had taken the 'training'. However, only one woman made use of the training to earn a living by making household snacks. The remaining participants went back to their pre-Tsunami occupations. This is for three reasons.

First, the islanders did not get a feeling of 'connection' with the NGOs. As the NGO workers visited the A&N Islands for the first time, after the Tsunami, they had no prior briefing or knowledge about the islands. They were unaware about the local people and their needs. Relief items distributed by them reflect the unawareness of the NGOs about the A&N Islands and their lack of effort to enquire and learn about it. They provided unsuitable items. One such example was the saris given to the Nicobarese women. The Nicobarese women mostly wear long skirt with blouse and not saris. The distributed saris were thus used as mosquito nets and curtains. Another useless item for the households was the blankets. The A&N Islands lie in the tropical region, closer to the equator, and thus experience no winter and blankets are not at all needed, yet they received blankets from NGOs. Similarly, motor boats were distributed to tribal Nicobarese, but diesel for running these was not provided. Therefore, people made their traditional canoes known as hoodies instead and new boats lay idle and wasted along the coast. Radios and ceiling fans were distributed, but there was no electricity. Bicycles lost importance when the roads were damaged after the Tsunami. By the time of completion of the reconstruction of the roads, bicycles had rusted and became useless. The households stated that items received in the name of relief were poor in quality. Clothes tore after two washes and utensils rusted and became unusable. The people simply 'collected' what was offered free of cost and found a use for each of the items. For example, saris became curtains and mosquito nets, and the cans of the water filters and other leaking utensils were used as flower pots. The NGOs focused more on delivering goods; they failed to provide items that the people really needed.

Second, NGOs were not permitted to visit the Nicobar Islands with the dominant tribal Nicobarese population. The ANATPR (1956) prohibits entry of non-tribal people into the tribal reserve area. So NGOs' response towards the tribal Nicobarese was limited. The relief items supplied by the NGOs were distributed by armed forces in the Nicobar Islands. The tribal Nicobarese came in contact with the Christian missionaries NGO which worked in close association with the village church. Here it needs to be brought forth that fieldwork confirmed that the strong social bonds of the tribal Nicobarese gave them resilience. Thus, tribal Nicobarese were indifferent to the NGO presence, and hence, the impact of the NGOs on tribal people was limited.

The third reason for the limited NGO impact was the scarcity of dedicated field workers (Rawal et al. 2006). NGOs changed their workers frequently over the five years during the reconstruction and rehabilitation period, and thus failed to create trust with the people. This hampered training programmes like sewing, snack-making, kitchen gardening, laptop and mobile phone repairing and other such programmes which were initiated by the NGOs. Thus NGOs' work in post-Tsunami failed to make a holistic impact on the lives of the tsunami-affected people of the A&N Islands.

By the end of the immediate phase, the armed forces had moved out and so did many of the NGOs. Only 22 NGOs remained and assisted administration in reconstruction and rehabilitation. The long-term response phase was dominated with large-scale reconstruction and efforts towards rehabilitation of the Tsunami-affected people.

Long-term phase of response

The long-term response phase began when the immediate phase of response waned. It aimed at restoring the A&N Islands to its pre-Tsunami condition, albeit reducing the vulnerability. Therefore, it entailed reconstructions of basic infrastructure, damaged houses and rehabilitation of the affected people. For all these works funds were provided by the Government of India and partly from the NGOs.

Allocation of funds for Tsunami relief, reconstruction and rehabilitation

The Tsunami of 2004 was a global Tsunami and hence it also generated global response. The Tsunami generated unprecedented and unanticipated aid efforts. The Asian regional governments and communities and the international community raised a sum total of USD 17 billion that was utilized for relief, rehabilitation and reconstruction projects following the Tsunami (Jayasuriya and McCawley 2010).

The damage due to the Tsunami in the A&N Islands was estimated at Rs. 38.27 billion (Ministry of Home Affairs 2005b). The methodology of these estimates could not be ascertained through any of the reports of either the Government of India or the A&N Administration. The financial damage of the A&N Islands was 33 per cent of the estimated total damage (Rs. 115.45 billion) across India (especially Tamil Nadu). The A&N Islands received funds for the relief and rehabilitation from three main sources – the Prime Minister Relief fund, the Lieutenant Governor (of A&N Islands) Relief Fund and the Government of India.

The then Prime Minister Shri Manmohan Singh sanctioned Rs. 581.8 million for the Tsunami relief from the Prime Minister Relief Fund in the A&N Islands. Out of this fund an ex-gratia payment to the next of the kin at rate of Rs. 100,000 and a Tsunami Rehabilitation Relief of Rs. 298.5 million was sanctioned and disbursed. Besides this from the Prime Minister's Fund, three more schemes (Tsunami Child Assistance, Tsunami Scholarship Scheme and Social Infrastructure Scheme) totalling Rs. 283.3 million were sanctioned. Along with this, the lieutenant governor of the A&N Islands announced the gratuitous relief that was to be paid out of the Lieutenant Governor's Relief Fund. A financial relief amount of Rs. 36.1 million was distributed to 18,073 affected households at the rate of Rs. 2,000 per household (A&N Administration 2009).

The central government fund came in two packages – (1) Rajiv Gandhi Rehabilitation Package (RGRP) and (2) Tsunami Rehabilitation Package (TRP). Within a month after the Tsunami, the RGRP was announced for the A&N Islands. It had two components – first relief and second rehabilitation. The relief package was for Rs. 2.06 billion and the remaining Rs. 4.54 billion was for rehabilitation. The latter was added to the TRP which was announced later. Therefore the A&N Islands received Rs. 6.60 billion under the RGRP. The Rs. 2.06 billion was spent as payment of ex-gratia under different heads, immediate relief and reconstruction of intermediate shelter. Out of the total ex-gratia, 54 per cent was paid to the next of kin of the deceased at rate of Rs. 100,000 per person killed. Ex-gratia was also paid for the injured, household articles, temporary relief, and assistance to orphans, widows, unmarried girls and small businessmen.

The TRP was announced by the Empowered Group of Ministers after two months of the Tsunami. It was based on the damage report prepared by the expert committee that visited the A&N Islands after the Tsunami to assess the damage. For reconstruction and rehabilitation, Rs. 26.55 billion under the TRP was sanctioned for the A&N Islands. This included funds for rehabilitation under the RGRP amounting to Rs. 4.54 billion.

The NGOs too participated in the relief, reconstruction and rehabilitation. The share of the NGOs is approximately Rs. 2 billion. This is an approximate figure based on information available in the government documents, internet and the fieldwork in 2010. NGOs participated in construction of the 3,850 intermediate shelters, 706 permanent shelters, 20 schools, 30 PHSCs and the four PHCs. Apart from this, they also contributed in providing ex-gratia relief to 10,000 affected households and conducting livelihood restoration programmes. The approximate cost of construction of one unit was immediate shelter – Rs. 120,000; permanent shelter – Rs. 1,400,000; school – Rs. 1,200,000; PHSC – Rs. 1,798,000; PHC – Rs. 5,674,968; Rs. 10,000 ex-gratia for each household; along with this approximately Rs. 350 million were spent on the rehabilitation programmes. Therefore the total expenditure of the NGOs is the sum of each of these sectors which amounts to about Rs. 2 billion.

The total amount received by the A&N Islands from all the three sources is estimated at Rs. 29.225 billion (Table 8.4). Therefore, when the funds available from government sources (Rs. 29.22 billion) and NGOs (Rs. 2 billion) were put together, they amounted to Rs. 31.22 billion.

The funds were used for four main purposes – funding for immediate relief, ex-gratia compensation, reconstruction of damaged infrastructure and houses and for restoring livelihoods of the affected population. The Rs. 26.55 billion for reconstruction was thus allocated:

Table 8.4 Monetary response to the Tsunami

Source of fund	*Amount sanctioned (in Rs. billion)*	*Activities funded*
Lieutenant Governor Relief Fund	0.036	Immediate relief
Prime Minister Relief Fund	0.582	(i) Ex-gratia for the killed (ii) Tsunami Child Assistance Scheme (iii) Tsunami Scholarship Scheme (10 years) (iv) Support for Social Infrastructure Scheme
Government of India		
Rajiv Gandhi Rehabilitation Package	2.06	(i) Immediate relief (ii) Ex-gratia under different heads (iii) Reconstruction of intermediate shelter
Tsunami Rehabilitation Package*	26.55	(i) Reconstruction of houses (ii) Reconstruction of infrastructure (iii) Restoration of livelihood (iv) Protection of environment
Total	**29.225**	

*The package was revised in February 2010 to Rs. 27.93 billion, an addition of Rs. 1.383 billion.

Source: Based on A&N Administration (2006)

reconstruction of houses (46 per cent); reconstruction of wharfs/jetties (11 per cent); restoration of agriculture (8 per cent); reconstruction of social infrastructure such as schools and medical facilities (8 per cent); reconstruction of power supply infrastructure (8 per cent); and reconstruction of roads/bridges (8 per cent). The remaining Rs. 3.01 billion or 11 per cent of the funds was allocated for revival of fisheries, restoration of tourism, reconstruction of water and sewerage lines, for coastal protection and other miscellaneous purposes. The allocated fund was released annually. The first instalment was released in the financial year 2005–06 and was carried out through 2010–11. The TRP was declared closed in December 2011, seven years after the Tsunami.

The question which gets raised is, where and how was the total of more than Rs. 30 billion (from Government of India and NGOs) that came into the A&N Islands as a response to the Tsunami spent? To gather this data, Right to Information (RTI) was requested from 10 departments involved in the Tsunami reconstruction for island-wise allocation and expenditure on Tsunami reconstruction for financial years 2005–12. RTI had to be pipelined because the expenditure data, sector-wise and island-wise, was unavailable among the existing sources such as the government reports and the internet. Though the department-wise annual expenditure under the TRP was available in the annual plans of the A&N Islands, it did not provide the detail of island-wise work done. Thus, RTI seemed the most authentic source for the expenditure data as the departments maintain a record of all activities carried out in different islands. Different departments supplied information in different formats some furnished it island-wise while some provided it circle-wise. As the records provided by different departments were detailed, from these records data was compiled accordingly and was used for the analysis.

Reconstruction and rehabilitation

The post-Tsunami period in the A&N Islands was characterized by intense reconstruction activity for which a total Rs. 21.75 billion out of total TRP of Rs. 26.55 billion was allocated (Table 8.5). The reconstruction work that began immediately after the Tsunami focused on restoring the basic communication network in the affected islands.

These were not planned expenditures but were carried out in the emergency situation. Therefore these activities were not reconstruction per se but were 'restoration'. For example, the communication link was restored between the A&N Islands and mainland India.

Table 8.5 Sector-wise allocation and expenditure of funds under Tsunami Rehabilitation Package (2005–11)

Sectors	*Amount (Rs. in billion)*	
	Allocated[1]	*Expenditure*[2]
House reconstruction		
Intermediate houses	NA	1.15
Permanent houses	12.21	16.73
House reconstruction	*12.21*	*17.88*
Infrastructure reconstruction		
Ports and jetties	3.00	2.90
Power and ICT	2.08	2.67
Roads and bridges	2.00	1.83
Water and sewerage	0.38	0.46
Social infrastructure and welfare	2.08	1.68*
Infrastructure reconstruction	*9.54*	*9.54*
Livelihood restoration		
Agriculture + livestock	2.17	NA
Fisheries	1.48	0.38
Tourism	0.67	0.14
Environmental and coastal protection	0.28	0.02
Livelihood restoration	*4.60*	
Others		
Miscellaneous	0.10	NA
Technical assistance	0.10	NA
Others	*0.20*	
Total	**26.55****	**> 27.96**

*Does not include the expenditure incurred by the NGOs which constructed 21 schools and 31 health service infrastructures.

**The package was revised in February 2010 to Rs. 27.93 billion.

NA: Not available.

Source: Based on the following:

1 A&N Administration (2006)

2 RTI from the departments – Andaman Lakshwadeep Harbour Works, Electricity Department, Andaman Public Works Department, Department of Social Welfare, Department of Fisheries, Department of Tourism, Department of Forest and Environmental Protection

Again, the Ministry of Home Affairs sent 715 diesel sets from the mainland India to restore power supply in the affected islands. Similarly, telephone exchanges were restored so was the water supply in Port Blair city. Repairable jetties were made functional such as the Mus jetty of Car Nicobar and the jetty of Kamorta, even though both were damaged. Temporary helicopter landing facilities were created at Teressa, Katchal, Nancowry and Great Nicobar as the relief camps were located in these islands. Therefore in the immediate response period the critical infrastructures were made serviceable, not reconstructed. The reconstruction began after the TRP was announced in February 2005. The post-Tsunami reconstruction can be grouped into two categories: reconstruction of infrastructure and reconstruction of houses. The former encompasses all reconstruction of physical and social infrastructure, while the latter includes the reconstruction of the intermediate and the permanent houses.

Reconstruction of infrastructure

Reconstruction of infrastructure accounted for 36 per cent of the total funds allocated under the TRP and included the reconstruction of social infrastructure and the physical infrastructure. The expenditure on social infrastructure included reconstruction of schools, primary health centres, the primary health sub-centres and the state transport infrastructure as it included purchase of buses for various islands. The latter included the reconstruction of water supply, sewerage lines, power supply infrastructure, communication networks, roads/bridges and wharfs/jetties.

Expenditure on infrastructure reconstruction

The total expenditure on infrastructure reconstruction was Rs. 9.54 billion (as per the replies of the RTIs). Of this, Rs. 1.68 billion was for reconstruction of social infrastructure and Rs. 7.87 billion was for physical infrastructure. At the onset it is seen that the expenditure for reconstruction of physical infrastructure was 4.7 times the amount of social infrastructure. Within the amount sanctioned for physical infrastructure, Rs. 2.90 billion was spent for reconstruction of wharfs/jetties, which accounts for 30 per cent of the total expenditure on infrastructure reconstruction. It was followed by power supply, roads and bridges and water supply and sewerage with 28 per cent, 19 per cent and 5 per cent of the total allocation, respectively. The remaining 18 per cent was expended on social infrastructure. The TRP aimed at reconstruction of the damaged infrastructure, and therefore with the amount allocated only the damaged infrastructure was reconstructed; new additions could not and were not made.

Expenditure on infrastructure reconstruction spreads over the entire seven-year period of the TRP (Figure 8.4). From 2005–06 to 2007–08 it ranged around Rs. 1.5 billion per year, exceeding Rs. 2.2 billion in year 2009–10 when many ongoing reconstruction projects were completed. The following section charts the reconstruction of social infrastructure followed by the reconstruction of physical infrastructure.

Reconstruction of schools: Before the Tsunami there were 369 schools across the A&N Islands; 34 were partially and 33 were completely damaged. A total of 57 schools were reconstructed, of which 16 were in the Andaman Islands and 41 were in the Nicobar Islands (Table 8.6).

In the Andaman Islands the number of schools reconstructed was less than the number of schools damaged as most of the schools were partially damaged and were repaired. On the other hand, schools of the Nicobar Islands had to be reconstructed at the new settlement location,

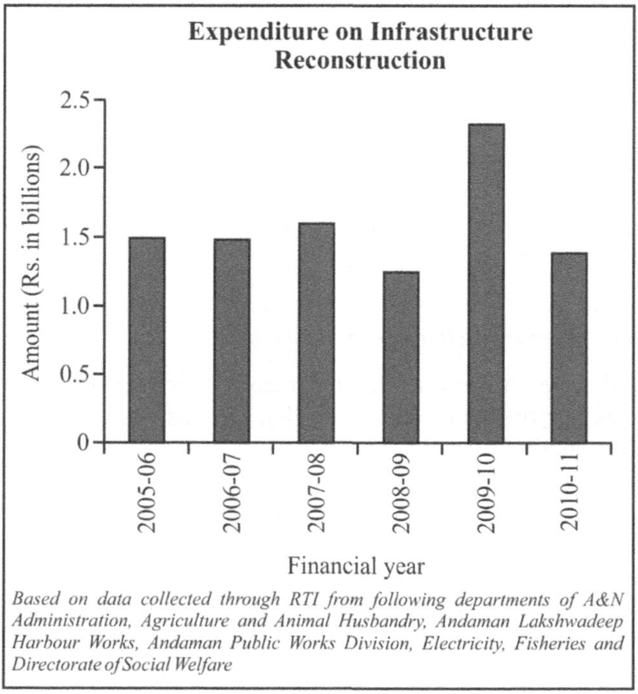

Figure 8.4 Expenditure on infrastructure reconstruction

Table 8.6 Reconstruction of schools (2005–11)

Name of the Island	*Number of schools*	
	*Damaged**	*Reconstructed*
Little Andaman Island	6	6
Middle Andaman Island	17	6
North Andaman Island	1	3
South Andaman Island	8	1
Sub-total Andaman Islands	*32*	*16*
Car Nicobar Island	10	14
Kamorta Island	2	10
Katchal Island	6	6
Teressa Island	7	5
Nancowry Island	2	3
Chowra Island	1	1
Little Nicobar Island	2	1
Great Nicobar Island	0	1
Bompoka, Kondul, Pulomilo and Trinket	5	Islands declared unfit for habitation
Sub-total Nicobar Islands	*35*	*41*
Total	**67**	**57**

* Includes both partially and completely damaged schools.

The schools of A&N Islands were reconstructed by four NGOs, namely Bharatiya Jain Sangathan (BJS), Oxfam, Church of North India (CNI) and Church Auxiliary for Social Action (CASA) and the Andaman Public Works Division (APWD).

Source: Based on A&N Administration (2005b) and fieldwork (2010)

even though some schools were partially damaged and undamaged. Hence, the number of schools reconstructed in Nicobar Islands was more than the number of schools damaged.

Reconstruction of PHCs and PHSCs: A total of four PHCs and 30 PHSCs were reconstructed, which was near equal to the damage (Table 8.7). The four reconstructed PHCs were located in the islands Little Andaman, Great Nicobar, Katchal and Teressa. Out of 30 PHSCs reconstructed, 29 were in the Nicobar Islands. Car Nicobar accounted for the 10 out of these 29. It was followed by the islands of Kamorta and Great Nicobar in which seven and four PHSCs were reconstructed, respectively. The islands of Katchal, Nancowry, Chowra and Little Nicobar accounted for the remaining eight PHSCs.

Fieldwork in 2012 in Car Nicobar Island confirmed that though schools and primary health centres/sub-centres have been reconstructed, there is a paucity of 'skilled personnel'. Adding to the constant transfers of teachers, another problem faced by the schools of the Nicobar Islands is the non-availability of adequate teachers.

The teachers of the schools are posted for a tenure of two years. Lack of dedicated teachers and constant rotation of teachers leads to an inefficient teaching system. By the time the teachers and the students get accustomed to each other, the teachers are transferred. Sometime the new teacher is not compassionate, or may pick on some students (Chattopadhyay and Bhattacharya 2007). Similarly, the PHSCs are managed by compounders. The doctor visits it every day at a particular hour only. Thus, function of PHSC has been restricted to provide first aid only. In times of emergency, the patient needs to be carried to BJR Hospital in Car Nicobar or to the PHCs in case of other Nicobar

Table 8.7 Reconstruction of medical facilities (2005–11)

Name of the island	*Number of PHC*		*Number of PHSC*	
	Damaged	*Reconstructed*	*Damaged*	*Reconstructed*
Little Andaman	1	1	1	1
Sub-total Andaman Islands	*1*	*1*	*1*	*1*
Car Nicobar	NA	0	5	10
Kamorta	0	0	0	7
Great Nicobar	1	1	6	4
Katchal	1	1	4	3
Nancowry	NA	0	3	3
Teressa	1	1	4	1
Chowra	NA	0	1	1
Little Nicobar	0	0	1	0
Bompoka, Kondul, Pulomilo and Trinket	NA	Islands declared unfit for habitation	2	Islands declared unfit for habitation
Sub-total Nicobar Islands	*3*	*3*	*26*	*29*
Total	4	4	27	30

NA: Not available before the Tsunami.

The primary health centres (PHC) and the primary health sub-centres (PHSC) were reconstructed by the NGO Bharatiya Jain Sangathan (BJS).

Source: Based on A&N Administration (2005b) and Bharatiya Jain Sangathan (2005)

Islands. The PHCs too are not equipped to handle serious illness. The case is then referred to the G. B. Pant Hospital in Port Blair.

The spatial pattern for the reconstruction of social infrastructure followed the damage pattern. The damage to social infrastructure was concentrated in Nicobar Islands, and so was the reconstruction. After the Tsunami, the location of the settlements was changed, hence infrastructure had to be constructed at the new settlement sites. Therefore, the number of social infrastructure reconstructed was more than the numbers damaged, because at the new site, each of the social infrastructures had to be reconstructed, even though it was partially damaged or was not at all damaged. Schools and health centres are the dominant social infrastructures that were reconstructed across A&N Islands. The TRP focused on reconstruction of social infrastructure but it did not facilitate the quality of the services. The success of social infrastructure depends on the manpower handling it. In the post-Tsunami period, the damaged social infrastructure was reconstructed but the lacuna of trained individual still remains. Another important sector of the TRP was the reconstruction of physical infrastructure.

In the Tsunami when the critical infrastructure such as power supply, roads and wharfs/jetties were damaged, it hampered not only relief and rescue but also reconstruction of houses and rehabilitation. Therefore before any construction could take place on the affected islands, it was essential to reconstruct the basic infrastructure. Adding to the difficulty was the fact that in the Nicobar Islands the reconstruction material had to be 'imported', thereby increasing the delay and the cost of reconstruction.

In A&N Islands, the total expenditure on the post-Tsunami physical infrastructure reconstruction was Rs. 7.87 billion. It had four components – water and sewerage, roads and bridges, power supply and wharfs/jetties, accounting for 5 per cent, 23 per cent, 34 per cent and 38 per cent of the total, respectively. Reconstruction of water pipelines and sewerage included works such as laying down of pipelines for drinking water supply and for drainage of waste water, and amounted to Rs. 0.46 billion, which was least among the four components. Reconstruction of the remaining three components accounted for the lion's share of the total expenditure on physical infrastructure reconstruction.

Reconstruction of Roads and Bridges: Roads and bridges were reconstructed by the Andaman Public Works Department (APWD), except in Great Nicobar Island where reconstruction of roads and bridges is being carried out by the Border Road Organization. Total expenditure on reconstruction of roads and bridges was Rs. 1.87 billion. The pattern of expenditure on reconstruction of roads and bridges has been fluctuating and peaked to Rs. 0.4 billion in 2007–08 and 2009–10. In the intervening years it was less than Rs. 0.3 billion (Figure 8.5).

The Andaman Islands accounted for 17 per cent and the Nicobar Islands accounted for 83 per cent. The Great Nicobar Island alone accounts for 36 per cent of the total expenditure. Here 30 km of 50 km of trunk road (known as north-south road), which connected Campbell Bay to Indira Point (the southernmost tip of India), was washed away and had to be reconstructed around 5–6 km inward from the coast. This involved clearing of the equatorial rainforests in the hilly terrain, which has added to the cost of reconstruction.

Reconstruction of power supply and communication network: In the post-Tsunami period the Electricity Department of the A&N Administration carried out reconstruction of power supply and the other communication network. It expended Rs. 2.67 billion for the reconstruction work. The Andaman Islands accounted for 72 per cent while the Nicobar Islands accounted for 28 per cent of this total (Figure 8.6).

The three islands, namely, South Andaman, Little Andaman and Car Nicobar accounted for 76 per cent of the total expenditure. The year 2009–10 accounted for 38 per cent of the total while for other years it did not exceed 20 per cent. The dominant share of the expenditure

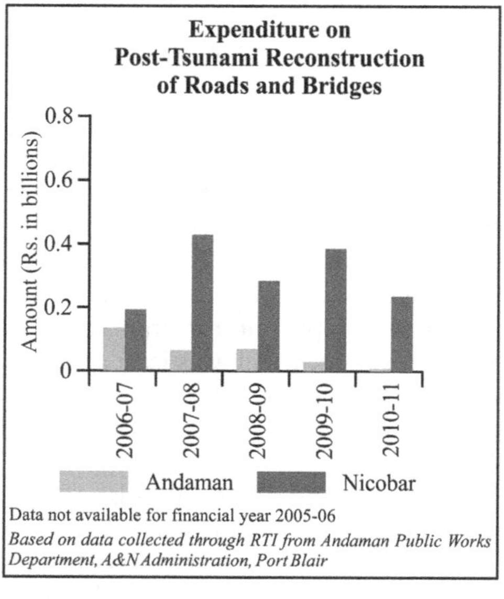

Figure 8.5 Expenditure on post-Tsunami reconstruction of roads and bridges

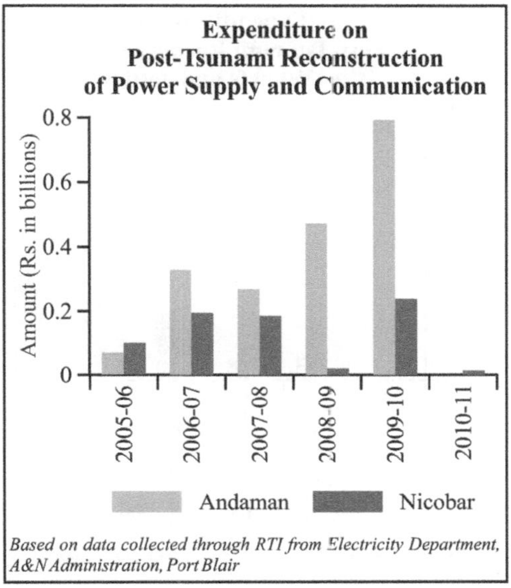

Figure 8.6 Expenditure on post-Tsunami reconstruction of power supply and communication

on physical infrastructures was on the fourth component, the wharfs/jetties which are the lifeline of the A&N Islands and were severely damaged in the Tsunami.

Reconstruction of wharfs/jetties: Reconstruction of wharfs/jetties amounted to Rs. 2.87 billion which is 38 per cent of the total expenditure on physical infrastructure reconstruction. The reconstruction of wharfs/jetties of A&N Islands was carried out by the Andaman Lakshwadeep Harbour Works (ALHW). The expenditure on the reconstruction of wharfs/jetties increased over the financial years (Figure 8.7).

It was highest in 2010–11 at Rs. 0.96 billion. The two financial years, 2009–10 and 2010–11 accounted for 64 per cent of the total expenditure on reconstruction of wharfs/jetties. This indicates that most of the work of reconstruction of wharfs/jetties took place towards the closing of the TRP which was in December 2011. Another fact which emerges is that in the initial period after the Tsunami, focus was on reconstruction of wharfs/jetties of the Andaman Islands and not the Nicobar Islands, though it was the Nicobar Islands where all jetties were damaged.

The jetty damages were of various kinds. In northern part of A&N Islands, due to uplift, jetties were raised above the water level and thus

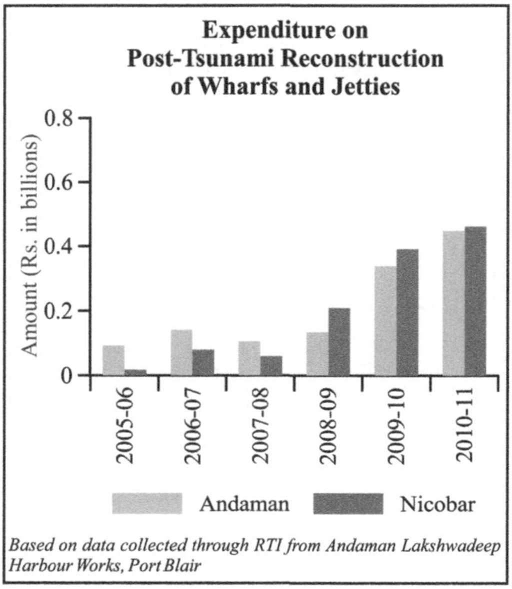

Figure 8.7 Expenditure on post-Tsunami reconstruction of wharfs and jetties

became unusable. The reconstruction involved extending the length of the jetties so that they could reach the sea; for example, the Gandhi Nagar jetty in North Andaman Island was extended by half a kilometre. The jetties of Middle Andaman Island and South Andaman Island were submerged due to subsidence. These had to be raised to the present sea level. The jetties of Little Andaman Island and Nicobar Islands were washed away and these had to be newly constructed. The Andaman Islands accounted for 51 per cent and Nicobar Islands accounted for 49 per cent of the total. The four islands, namely, Great Nicobar, South Andaman, Little Andaman, Car Nicobar accounted for 82 per cent of the total expenditure, sharing 34 per cent, 26 per cent, 17 per cent and 8 per cent, respectively. The remaining 18 per cent was shared between North and Middle Andaman, Kamorta, Nancowry, Chowra, Katchal and Teressa.

Spatial pattern of expenditure on physical infrastructure reconstruction

Reconstruction work was carried out in 12 islands, four in the Andaman Islands and eight in the Nicobar Islands. Four of the Nicobar Islands, namely, Bompoka, Trinket, Pulomilo and Kondul, were declared unfit for habitation, hence no reconstruction work was carried out in these four islands.

When all the expenditures of physical infrastructure reconstruction were mapped, it emerged that four islands held the lion's share (72 per cent) of the expenditure on reconstruction of physical infrastructure. Among the Andaman Islands, South Andaman accounts for highest expenditure followed by Little Andaman, North Andaman and Middle Andaman. South Andaman had the maximum expenditure for each of the infrastructures reconstructed.

Among the Nicobar Islands, Great Nicobar accounted for the highest expenditure followed by Car Nicobar, Kamorta, Katchal, Nancowry, Teressa, Chowra and Little Nicobar. Out of three dominant components of physical infrastructure, Great Nicobar had the highest expenditures for wharfs/jetties and roads and bridges, while for power and communications infrastructure Car Nicobar takes the lead.

The pattern of reconstruction follows the composite damage index (Figure 8.8). The damage index score was the highest for the four islands, namely, South Andaman, Little Andaman, Car Nicobar and Great Nicobar, and these four islands also account for maximum expenditure on reconstruction of physical infrastructure. The islands with moderate damage index score had lower expenditures. However, there is an exception among the Nicobar Islands. Between Katchal

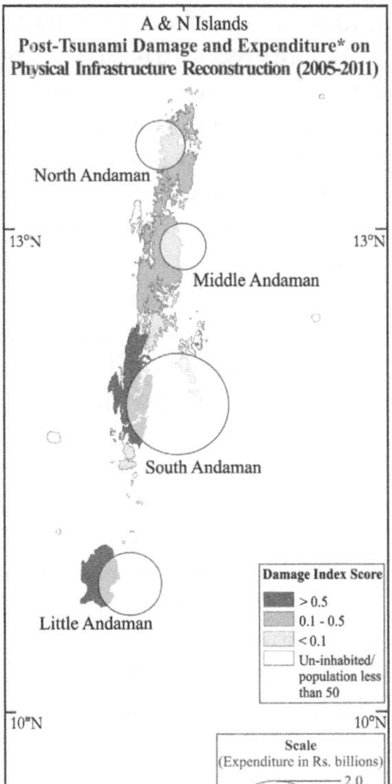

Figure 8.8 A&N Islands: post-Tsunami damage and expenditure on physical infrastructure reconstruction (2005–11)

and Kamorta, the former had a higher score of damage yet the expenditure on reconstruction of physical infrastructure was more in the latter. Except for this exception, expenditure on physical infrastructure was in sync with the damage pattern. Thus, it can be concluded that the status quo was maintained and wherever whatever was damaged was reconstructed. But in this process the tribal Nicobarese were once again neglected and their interest sidelined. There were no significant new additions to the pre-Tsunami infrastructure, especially in the Nicobar Islands. Once again, an opportunity was lost.

The reconstruction of infrastructure was concentrated in islands dominated by non-tribal population. All reconstruction of the Andaman Islands was for the non-tribal population as all of the Andaman Islands have more than 99 per cent of non-tribal population (Little Andaman is an exception where non-tribal population is 93 per cent). Among the Nicobar Islands, four islands with a high proportion of non-tribal population account for 87 per cent of the total expenditure on infrastructure reconstruction. Great Nicobar Island, with 87 per cent of non-tribal population, alone accounted for 47 per cent of the total expenditure of the Nicobar Islands on infrastructure reconstruction. In each of the infrastructure reconstructed, similar pattern was repeated. The four Nicobar Islands with dominant non-tribal population together account for 85 per cent of the total on power infrastructure reconstruction, 86 per cent of total on road reconstruction and 89 per cent of total on reconstruction of wharfs/jetties. Thus, the benefit of infrastructure reconstruction has been skewed in favour of the non-tribal population.

Apart from the physical infrastructure, the major reconstruction post-Tsunami was the houses.

Reconstruction of post-Tsunami houses

A major component of the TRP was reconstruction of houses. A sum of Rs. 12.21 billion was allocated for housing reconstruction, which was 46 per cent of the total allocation of Rs. 26.55 billion under the TRP. For the Tsunami-affected households two kinds of shelters (henceforth houses) were constructed; first were intermediate houses and second were permanent houses. Thus, reconstruction of houses became the most critical component of the TRP.

Under the norms laid out by the Eleventh Finance Commission of India (2000, 2005), the damaged houses were to be compensated in cash. For partially damaged houses of Middle Andaman Island and North Andaman Island, cash compensation of Rs. 2,000 was given for the repair. In the Nicobar Islands it was a different scenario. In the

Tsunami, coastal forests were damaged and markets of the Nicobar Islands were shattered. A meeting was held by 24 'experts' on 9 January 2005 at National Institute of Disaster Management to solve the shelter crisis of Tsunami-affected households of A&N Islands. This team of experts concluded that as there was a situation of 'unavailability of construction material and skilled labour', the households could not construct houses for themselves even with the cash compensation. Therefore, the A&N Administration should reconstruct the houses. The team also proposed that proper transition shelters were needed as the reconstruction process could exceed more than six months and the existing makeshift arrangements (relief camps) were harsh for living for longer time such as six months (A&N Administration 2006). Therefore, as per the recommendations of the experts there was a need to have two kinds of houses – first the intermediate houses and second the permanent houses.

For reconstruction of intermediate houses Rs. 0.97 billion was allocated under the RGRP. The actual expenditure was Rs. 1.15 billion, Rs. 0.18 billion more than the allocated fund, because the cost of each of the intermediate houses included the cost of clearing the land and for the amenities such as roads, street lights and a children's park, among others. The per unit cost of intermediate house was Rs. 120,000, even though it was a rudimentary one-room structure with walls and roof made of CGI sheets. The reconstruction of permanent houses also exceeded the allocated cost of Rs. 12.21 billion to Rs. 16.73 billion (RTI, APWD 2011). The average cost per unit of permanent house was approximately Rs. 140,000 and as 9,797 houses were reconstructed, the total cost of construction of the permanent houses was Rs. 13.72 billion (9,797 × Rs. 140,000). The additional cost of Rs. 3.01 billion was spent on providing community facilities at each of the settlement sites. In a Nicobar village, the community facility included a birth house, a death house, a community hall, a recreation hall and a shopping centre. In pre-Tsunami period, each of the traditional Nicobarese villages had a special house for the purpose of births, where the delivery of the baby took place and where the newborn with the mother would stay for 21 days or more before returning back to their home. This was the 'birth house'. Similarly, after death, the body of the deceased is kept at a separate house where the last rites take place. This is the 'death house'. 'Community hall' is used for functions such as marriage, a naming ceremony or any other celebrations. Therefore, in the post-Tsunami period each of these was reconstructed along with a 'recreation hall' and a 'shopping centre'; the latter two were alien to the traditional tribal life.

The Nicobar Islands accounted for 72 per cent and the Andaman Islands accounted for 28 per cent of the total expenditure on the reconstruction of intermediate houses. The intermediate houses were constructed at eight islands, two in the Andaman Islands (South Andaman and Little Andaman) and six in the Nicobar Islands (Car Nicobar, Great Nicobar, Teressa, Kamorta, Katchal and Nancowry). Out of the pre-Tsunami 12 inhabited Nicobar Islands, intermediate houses were constructed only in six islands because it was convenient for the administration to serve in a few locations as infrastructure had collapsed in most of the islands. In the absence of jetties, transportation of building materials and free ration was problematic. The A&N Administration provided free ration to the households during their stay in the intermediate houses for about two years. Therefore, it was convenient to deliver ration for a large population at a few sites in lesser number of islands. This was economical and saved both labour and time. Distribution of permanent houses followed a similar pattern.

The spatial pattern of expenditure, on reconstruction of permanent houses, was concentrated in the Nicobar Islands. Of the total expenditure, Rs. 16.73 billion, the Nicobar Islands accounted for 77 per cent, while the Andaman Islands accounted for 23 per cent. Among the Andaman Islands, permanent houses were constructed in two islands, Little Andaman and South Andaman. Reconstruction of permanent houses took place in eight out of the 12 Nicobar Islands. In four islands, namely, Bompoka, Trinket, Kondul and Pulomilo, permanent houses were not constructed because these islands were declared unfit for habitation by the Department of Science and Technology. Among the Nicobar Islands, Car Nicobar accounted for the highest expenditure followed by Great Nicobar, Kamorta, Teressa, Chowra, Katchal, Nancowry and Little Nicobar. The spatial pattern of houses reconstructed reflects the pattern of houses damaged. The two set of houses were different from each other.

Differences between intermediate houses and permanent houses

Two types of houses were reconstructed post-Tsunami – intermediate houses and permanent houses. The intermediate houses across the A&N Islands were constructed within six months, but the entire process of constructing and providing permanent houses to the Tsunami-affected households in A&N Islands took seven years. Not only in terms of time taken for reconstruction, both the intermediate houses and the permanent houses also differed in terms of size, number of rooms, floor plan, building materials, facilities and the services provided.

One block of intermediate houses consisted of six housing units aligned in a manner that two adjacent units shared a common wall. The distance between two blocks was 6.5 ft. These intermediate houses did not have any kitchen, toilet and bathroom. The toilet and the bathroom units were located at the end of each block that was shared by all the households of the block. The intermediate houses were makeshift arrangements to provide shelter to the households until the permanent houses were constructed. The size of each intermediate shelter was 209.3 sq. ft (length – 17.7 ft, width – 11.8 ft, height – 13 ft). One unit consisted of only a single room. The floor was either cemented or made of wood. No foundation was made as the structures had to be dismantled after the short period of use. In 2005, it was estimated that reconstruction of permanent houses would take two years to complete. Hence the intermediate houses were crude structures of CGI sheets and wooden posts which lacked the basic civic amenities. Though roads were constructed, street lights were erected and playgrounds were made, the construction materials were insubstantial and became non-functional within six months of use. The same kind of intermediate houses were built across the islands and there was no difference between the intermediate houses of the non-tribal and the tribal Nicobarese.

In contrast, the permanent houses for non-tribal and tribal Nicobarese differed (differences discussed in the next chapter). The floor area of a single unit of a permanent house is 450 sq. ft (length – 30 ft, width – 15 ft). The height of the unit from floor to roof is 10 ft. The houses for non-tribal households were constructed on ground and for tribal Nicobarese on stilts. The former is made on concrete platform of 3 ft height; the latter are mounted on stilts 8 ft above the ground. For walls and roofs, steel frames have been erected onto which the processed bamboo ply boards have been fastened for interior walls, wooden planks for external walls and CGI sheets for roofs. The bathroom and kitchen have been constructed using masonry walls made of aerated cement concrete blocks. The permanent house sites have been provided with basic amenities. Thus, the permanent houses were designed to be durable 'tsunami-proof' houses with 'model townships, catering to requirements of all facilities and amenities like schools, shopping, health centres, open spaces, government and semi-government establishments, etc.' (Kshirsagar and Kumar 2006). The process of reconstruction of houses was carried out by the Central Public Works Division (CPWD) under the Ministry of Urban Development and lasted seven years (Figure 8.9).

Before finalizing the design of the permanent houses, the team of experts visited the A&N Islands in February 2005 and interacted

Time Line of Post-Tsunami Sequence of Houses Construction and Allotment in A&N Islands

2004	December	The Tsunami
2005	January	Experts visit A&N Islands, assess damage, finalize design for intermediate houses
	February	Construction of intermediate houses begins
	March	The team submitted detailed project proposal to the Government of India
	April	Construction of prototype[1] of permanent houses by CPWD begins
	May	People shift to intermediate houses
	June	Construction of prototype of permanent houses by CPWD ended
	July	People reject the prototype. Experts visit A&N Islands to explain benefit of new house type; NGOs asked to put up prototypes of permanent houses
	August	Prototypes of permanent houses by NGOs completed
	September	Experts visit A&N Islands to finalize permanent house designs proposed by CPWD in July
	October	Ministry of Urban Development sends detailed report to the cabinet meeting for final approval
	November	
	December	Proposal accepted by Empowered Group of Ministers. NGOs allowed to participate only if they adopted the approved design and funded the endeavour
2006		Construction of permanent houses begins
2007		200 Permanent shelters allotted to people of Car Nicobar, 3 years after the Tsunami
2008		Permanent houses allotted to the people of Car Nicobar Island and South Andaman Island
2009		Permanent houses allotted to the people of following islands - Car Nicobar, Chowra, Teressa, Kamorta, Katchal, Nancowry, Little Andaman and South Andaman
2010		Permanent houses allotted to the people of Little Nicobar Island and Great Nicobar Island
2011		House allotment completed for all the islands

[1]Prototype is a proposed model of the structure of actual size.

A prototype was prepared by CPWD in June 2005, but was criticized by the islanders and the NGOs alike. NGOs requested to make prototypes which further delayed the reconstruction process. Finally, the prototype prepared by the CPWD was approved and reconstruction began by June 2006, which was 1½ years after the Tsunami. Therefore the households had to spend more time (average 4 years) in the intermediate houses than 2 years as envisaged earlier.

Based on Rawal, Desai and Jadeja (2006)

Figure 8.9 Time line of post-Tsunami sequence of house construction and allotment in the A&N Islands

with the tribal Nicobarese captains to understand the 'architecture of traditional houses'. In March 2005 the team of experts submitted a detailed project proposal to the Government of India and it was decided that several prototypes (i.e. a proposed model of the structure of actual size) would be constructed and among these the best would be selected. The reconstruction of prototype of the permanent house began in April 2005 and ended in June 2005. These prototypes were made as per the specifications of the Bureau of Indian Standards (BIS). BIS is a statutory organization under the Bureau of Indian Standards Act 1986. The BIS prepares building codes for construction of structures to make them strong enough to withstand the impact of natural hazards or to impose restrictions on land use so that the exposure of the society to the hazard situation is avoided or minimized.

Site selection: Land in the A&N Islands is scarce. The location of new sites for houses had to be carefully selected such that further tsunamis do not take a toll. Therefore, in order to reduce the vulnerability to the tsunami, the Department of Science and Technology (DST), New Delhi recommended an adoption of elevation based setback line for resettlement locations. Elevation of 10 m with 750 m distance from the coast was suggested for North Andaman Island, Middle Andaman Island and South Andaman Island and a 15 m elevation with 1.5 km distance from the sea was suggested for other islands. This new location consisted of reserved forests (as in the A&N Islands forests are 86 per cent of the total geographical area) which could not be cleared under the Supreme Court's orders of 2002. Therefore, the Forest and Environment Department of the A&N Administration appealed to the Central Empowered Committee of the Supreme Court for permission to clear 15,000 ha of land for rehabilitation.

The Supreme Court permitted the extraction of timber and non-timber produces for a period of six months under certain conditions. The order stated that – 'no felling will be done (i) up to 1,000 m from the sea shore, (ii) in national parks and sanctuaries and (iii) in mangrove forests'. The other conditions stated in the orders were:

> the felling will be exclusively for meeting the immediate requirement of rehabilitation work and not for any commercial purposes; proportionate reduction in the subsequent felling will be made while implementing the prescriptions of the working plan to adjust the present felling; and the location, quantity and actual utilization of felling/removal will be monitored by a committee not below the level of the Chief Conservator of Forests. The said Committee will also be responsible to provide monthly details of felling and utilization of timber and non-timber forest produce.

The Supreme Court also permitted conversion of round timber to sawn timber for the use in the reconstruction work. It also permitted use of forest land for restoring road and communication links and removal of sand from beaches for the purpose of buildings and road construction. The permissions were valid only for six months under certain conditions. Therefore, deforestation took place as a result of the Tsunami.

With the recommendation of the DST, 58 sites were identified across eight islands for reconstructing intermediate houses. Permanent houses were constructed at 84 sites across 10 islands (Table 8.8).

Most of the permanent houses were constructed next to the site of intermediate houses. Flat land was not always available, and therefore, the houses were built on hilltops and slopes as was the case in South Andaman Island. In other islands the new sites are 4–5 km inwards from the high tide line.

After the Tsunami, six small islands, namely, Chowra, Bompoka, Trinket, Little Nicobar, Pulomilo and Kondul were evacuated and the islanders were resettled in the neighbouring islands. Four of these had small geographical area – Pulomilo (1.3 km^2); Kondul (4.6 km^2); Chowra (8.2 km^2) and Bompoka (13.46 km^2) – and were inundated from all sides. Trinket, though, was of moderate size (36.3 km^2), and was divided into three smaller islands as its low lying areas were completely submerged. Little Nicobar had small population of about 350 individuals, hence it was convenient to evacuate the island rather than set up the relief camp and intermediate shelter there.

As the islanders from these six islands had to be relocated to the neighbouring ones, intermediate housing sites were set up in the 'host' island. Intermediate houses for the islanders from Chowra and Bompoka were set up in Teressa Island; islanders from Trinket were housed at Kamorta

Table 8.8 Type and number of houses reconstructed post-Tsunami

Island	*Intermediate shelters sites*	*No. of housing units*	*Permanent shelters sites*	*No. of housing units*
Car Nicobar	15	3,866	15	3,941
Little Andaman	5	1,945	4	1,973
Great Nicobar	7	1,154	8	995
South Andaman	4	729	9	823
Kamorta	10	507	15	518
Teressa	9	821	8	506
Chowra	0	0	5	346
Katchal	4	315	4	315
Nancowry	4	228	11	269
Little Nicobar	0	0	5	111
Total	58	9,565	84	9,797

Source: Based on A&N Administration (2005a, 2011a)

Island; islanders from Kondul, Pulomilo and Little Nicobar had their intermediate shelters at Great Nicobar Island. For the location of permanent houses, islanders from Chowra and Little Nicobar were reluctant to settle anywhere but on their own island and thus were allowed to return to their respective islands. The other four islands were declared unfit for habitation by the DST and villagers had to compromise and were accommodated on other islands. Households of Bompoka were settled in Teressa; of Trinket were settled in Kamorta; from Pulomilo settled in Little Nicobar; and from Kondul were accommodated on Great Nicobar. It can be concluded that sites for new houses were selected in such a way that a tsunami of the magnitude equal to the one of 26 December 2004 would not affect the new houses in future.

Numbers reconstructed: A total of 19,362 houses were reconstructed in the A&N Islands after the Tsunami. Of this 9,565 were intermediate houses and 9,797 were permanent houses. Of the total 9,565 intermediate houses, 3,866 houses were reconstructed in Car Nicobar Island accounting for 40 per cent of the total. It was followed by the islands of Little Andaman with 21 per cent (1,945 houses) and Great Nicobar with 12 per cent (1,154 houses); the islands of South Andaman, Teressa, Kamorta, Katchal and Nancowry accounted for the remaining 27 per cent (Figure 8.10).

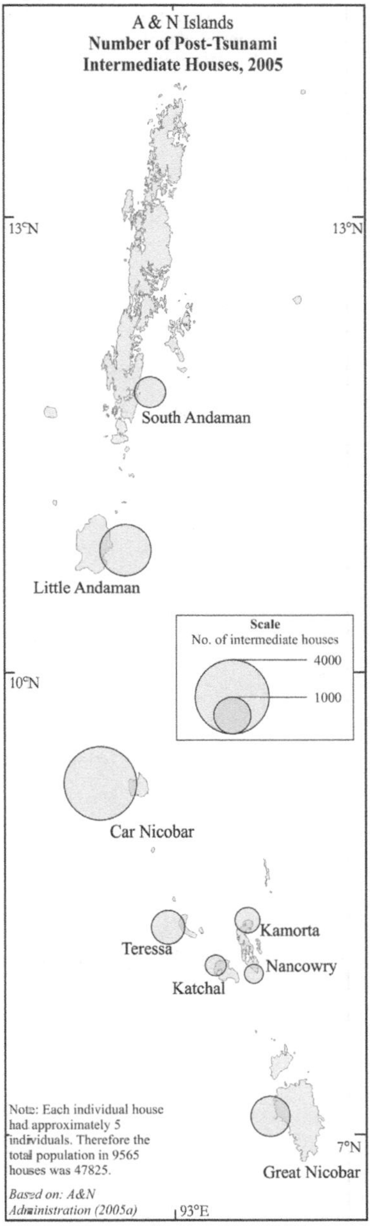

Figure 8.10 A&N Islands: number of post-Tsunami intermediate houses, 2005

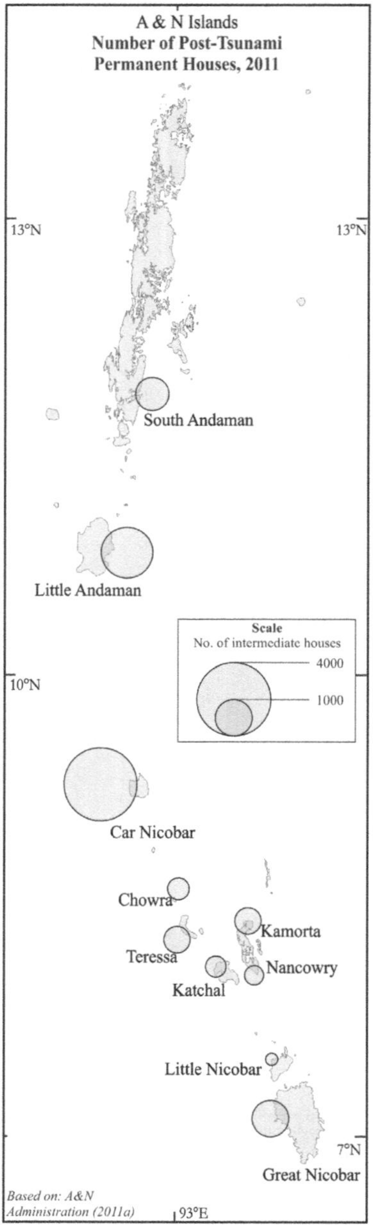

Figure 8.11 A&N Islands: number of post-Tsunami permanent houses, 2011

The spatial pattern of permanent houses followed the distribution pattern of the intermediate houses. The Nicobar Islands account for 71 per cent of the total 9,797 reconstructed permanent houses and the Andaman Islands account for the remaining 29 per cent (Figure 8.11). Permanent houses were constructed in 10 islands, two of which were in the Andaman Islands and eight in the Nicobar Islands. Of the total 9,797 houses reconstructed, islands of Car Nicobar, Little Andaman and Great Nicobar account for 71 per cent of the total. The spatial pattern of reconstructed houses, both intermediate and permanent, followed the pattern of completely damaged houses. Nicobar Islands accounted for one-third of the completely damaged houses and also of reconstructed houses. The reconstructed houses are concentrated in Car Nicobar Island.

Of the 9,565 intermediate houses, 3,850 (40 per cent of total) houses were reconstructed by eight NGOs for the non-tribal households. CRS-CARITAS India/ACANI constructed 1,000 houses in Campbell Bay in Great Nicobar Island; in Little Andaman Island 1,000 houses were constructed by Sustainable Environment and Ecological Development Society (SEEDS), 500 houses by Adventist Development Relief Agency (ADRA), 350 by DCC, 100 by Rotary Club and 50 by Hindustani Covenant Church (HCC).

In contrast the share of NGOs in the reconstruction of permanent houses was minimal. Of the 9,797 permanent houses, NGOs reconstructed only 706 houses (7 per cent). The NGOs engaged in reconstruction of permanent houses were Mata Amritanandamayi Math (MAM), CARE India, HCC and Association of Catholics A&N Islands (ACANI). In South Andaman Island, NGOs constructed 606 houses while in Little Andaman Island ACANI constructed 100 houses. The remaining 9,091 permanent houses were constructed either by APWD or CPWD. APWD constructed 217 houses in South Andaman Island and 908 houses in Little Andaman Island. CPWD constructed 965 houses in Little Andaman Island and all 7,001 permanent houses of the Nicobar Islands.

The quality of houses varied with the constructing agencies. Even though all houses appear similar, building materials used for the construction are different and the quality of construction also varies significantly. The details of this will be discussed in the next chapter. The new houses were provided free of cost to the Tsunami-affected households of the completely damaged houses, however, providing only the house cannot lead to rehabilitation. The TRP also provided for the rehabilitation of the Tsunami-affected households.

Restoration of livelihood

Livelihood restoration was initiated for four sectors, namely, agriculture and livestock, fisheries, tourism and environmental and coastal protection. For reviving livelihood Rs. 4.6 billion was allocated under the TRP. This was 17 per cent of the total TRP of Rs. 26.55 billion. Of the total Rs. 4.6 billion, Rs. 2.17 billion (47 per cent) was allocated for restoring agriculture and livestock, Rs. 1.48 billion (32 per cent) was allocated for fisheries, Rs. 0.67 billion (15 per cent) was allocated for revival of tourism and the remaining Rs. 0.28 billion (6 per cent) was allocated for environmental and coastal protection. Apart from these allocations, loans at low rates were offered to Tsunami-affected small traders for restoring their businesses. Compensation for the damage was major component of rehabilitation and for restoring livelihoods.

Compensation

Compensation was given for the loss of agricultural land, standing crop, livestock, boats and fishing gears, petty businesses, as well as marginally and partially damaged houses. Compensation for agriculture was sanctioned under three heads: (1) Rs. 10,000 per ha for permanently submerged land; (2) Rs. 5,000 per ha for reclamation of land; and (3) Rs. 24,231 per ha for damage of paddy, crops and vegetables and Rs. 94,465 per ha for damage of plantation crops such as coconut and areca nut.

For compensation of the loss of livestock, initially the A&N Administration decided to replace the livestock by bringing animals from the mainland India. However, it was soon realized that shipping of 50,000 animals consisting of pigs, goats and poultry from mainland was not possible. It takes minimum of three days for a ship to reach the A&N Islands from mainland India, and the ships are not equipped to carry so many animals. There was problem of health and hygiene not only of the animals but also the passengers aboard. Moreover, procuring animals from the mainland, transporting them first to Port Blair and then to the different islands, involved much higher cost than paying the compensation. This deterred the administration in procuring animals, thus cash compensation was paid instead. It was given to the heads of household who lost the livestock. The amount was deposited as a fixed deposit for a period of 10 years. The rate at which compensation was given was Rs. 12,000 per buffalo or cow, Rs. 4,000 per pig, Rs. 3,000 for per goat and Rs. 80 for each of the lost poultry birds (i.e. hen, cock and duck).

For fisheries, the proposal was to replace the damaged boats and lost fishing gears. Therefore, full subsidy was announced for the replacement/repair of country boats, in addition to the 50 per cent subsidy for upgrading of country boats, repair of mechanized boats and 100 per cent subsidy for the loss of nets and other fishing gears.

In case of damaged small business establishments of less than Rs. 10,000, the total amount of loss was compensated. These included small pan shops, STD booths, *dhabas* (local roadside eateries) and other shops with items worth not more than Rs. 10,000. For loss of businesses of more than Rs. 10,000 such as garages, bicycle shops and grocery shops, Rs. 10,000 was sanctioned as compensation and such businesses were entitled to avail a loan at a nominal interest rate.

Apart from providing compensation, the A&N Administration initiated several strategies to revive the livelihoods of the affected population. To revive agriculture among the non-tribal population, the administration undertook three initiatives – (1) reclamation of agricultural land affected; (2) creation of water bodies such as ponds, wells, check-dams and field bunds; and (3) distribution of farming implements, seeds and fertilizers. As the tribal Nicobarese were engaged in coconut plantation, schemes for the Nicobar Islands focused on restoring back the coconut plantation damaged in the Tsunami. The administration introduced the 'buy-back' scheme whereby, first, the tribal Nicobarese had to plant the coconut and areca nut seeds provided by the administration; and second they had to collect the nuts and the seedlings from both these plantations and sell it back to the administration.

For revival of animal husbandry, three primary schemes were sanctioned. First, cultivation of fodder on 100 ha of government land as fodder cultivation land was damaged. Second, animal feed was procured from mainland for immediate need. Third was regeneration and replacement of lost livestock. Cattle, goat, pig, poultry and duck units were supplied for regeneration of livestock. The livestock was supplied from the departmental farms and from owners of unaffected areas. To promote livestock rearing, the administration provided the livestock under a buy-back arrangement and the administration proposed to buy progeny of the supplied livestock. Revival of fisheries was also an important component of livelihood rehabilitation. For its revival, a live fish trade was promoted. Aquaculture was promoted in areas submerged after the Tsunami. For storing fishes, deep freezers have been installed in different islands – two in Great Nicobar and one each in Havelock, Neil Island, Little Andaman, Car Nicobar and Kamorta. Construction of 12 fish landing centres was sanctioned, 11 in Andaman Islands (South Andaman – 4; Middle Andaman – 3; North Andaman – 1; Little Andaman – 1; Neil – 1; and Havelock – 1) and one in Nicobar Islands (Campbell Bay in Great Nicobar). In the pre-Tsunami period, fishing for commercial purpose was carried out by the non-tribal population; hence all the fish landing centres have been situated in areas inhabited by the non-tribal population.

Even though no tourists were killed in the Tsunami, tourism in the A&N Islands suffered a setback. Therefore, to revive tourism in the A&N Islands several projects were approved such as refurbishment of government accommodations, creation of water sports facility, introduction of seaplanes and upgrading of tourist infrastructure.

For environmental and coastal protection two methods have been adopted. In settlement areas a sea wall has been constructed, while littoral forest species have been planted along uninhabited coasts. Both these efforts did not involve the affected population directly. The work was taken up by the administration.

The work done in A&N Islands under post-Tsunami response was based on a series of RTIs put at various departments provided by the A&N Administration. The RTIs provided the skeleton of the compensation given to the Tsunami-affected households. It provided the 'data' on all work done as a response to the Tsunami and gave the perspective of the government. But what is the ground reality? Did people receive support from A&N Administration? Did the people participate in the reconstruction process? What was the impact of this response on the affected people? Have the responses been effective in recovery of assets and resources of the Tsunami-affected population? The following chapter is an appraisal of the perception of the response of the Tsunami-affected households and the impact of this response on them. The next chapter will also bring forth the differences in response provided to the non-tribal households and the tribal Nicobarese tuhets and assess the recovery of both post-Tsunami.

Appendix

Table 8A.1 NGOs involved in relief and rehabilitation post-Tsunami in A&N Islands

Only Andaman		
Sl. No.	*Name of the NGO*	*Island of operation*
1	Bharat Seva Dal	Little Andaman
2	Calorex Foundation	Little Andaman
3	CINI-GOAL	Little Andaman
4	Dweep Peedit Sahayata Samity	Little Andaman
5	Institute for Mother and Child	Little Andaman
6	RC Church	Little Andaman
7	Rotary International	Little Andaman
8	Vishwa Jagriti Mission Trust	Little Andaman
9	West Bengal Gandhi Peace Foundation	Little Andaman
10	Tsunami Imdad Committee	North Andaman
11	ACANI	South Andaman
12	Action Aid	South Andaman
13	Art of Living – Vyakti Vikas Kendra	South Andaman
14	Association of Catholics	South Andaman
15	BAPS Swaminarayan Sanstha	South Andaman
16	Calcutta Urban Services	South Andaman
17	Friends Society in Social Service	South Andaman
18	Islands Development Organisation	South Andaman
19	Jamat-E-Hind	South Andaman

(*Continued*)

Table 8A.1 (Continued)

Sl. No.	*Name of the NGO*	*Island of operation*
20	The MATA Amritanandamayi Math	South Andaman
21	Mass Education	South Andaman
22	Pranab Kanya Sangha	South Andaman
23	Rajasthani Yuva Manch	South Andaman
24	Rajiv Gandhi Foundation	South Andaman
25	Ramkrishna Vivekananda Mission	South Andaman
26	Sant Nirankari Mandal	South Andaman
27	SEVAC Mental Hospital	South Andaman
28	SOS Children Village	South Andaman
29	Swami Vivekanand Yoga Anusandhan Sansthan	South Andaman
30	The Salvation Fellowship Trust	South Andaman
31	Thiruvalluvar Club	South Andaman
32	Vivekanand Kendra	South Andaman
33	Yuva Shakti	South Andaman
34	Adventist Development Relief Agency (ADRA)	Little Andaman and South Andaman
35	Ananda Marga Pracharak Sangha	Little Andaman and South Andaman
36	Bharat Sevashram Sangha	Little Andaman and South Andaman
37	Bharatiya Jeev Jantu Rakshan Samvardhan Parishad	Little Andaman and South Andaman
38	Chinmaya Mission	Little Andaman and South Andaman
39	EFICOR, DC, EHA – Disaster Co-ordination Committee (DCC)	Little Andaman and South Andaman
40	Hindustani Covenant Church (HCC) with CASA	Little Andaman and South Andaman
41	Islanders Sangathan Manch	Little Andaman and South Andaman
42	Ramkrishna Mission	Little Andaman and South Andaman
43	SEEDS – Sustainable Environment and Ecological Development Society	Little Andaman and South Andaman
44	Seva Bharathi	Little Andaman and South Andaman
45	Vanvasi Kalyan Ashram	Little Andaman and South Andaman
46	Voluntary Health Association of India	Little Andaman and South Andaman
47	West Bengal Voluntary Health Association	Little Andaman and South Andaman
48	World Vision, India	Little Andaman and South Andaman

Only Nicobar		
Sl. No.	*Name of the NGO*	*Island of operation*
49	Andaman Chamber of Commerce and Industry	Car Nicobar
50	Excel Industries	Car Nicobar
51	Church of North India (diocese of A&N Islands) in association with CASA	Car Nicobar, Chowra, Teressa, Katchal, Kamorta and Nancowry
52	Care, India	Chowra, Teressa, Katchal, Kamorta and Nancowry
53	Divine Youth Club	Great Nicobar
54	Guru Nanak Sarbat Sikh Sangat	Great Nicobar
55	INSANIYAT	Great Nicobar
56	United Sikhs	Great Nicobar
57	Dera Sacha Sauda	Katchal and Great Nicobar
Both Andaman and Nicobar		
58	Khalsa Aid	Little Andaman and Great Nicobar
59	PRAYAS	Little Andaman and Great Nicobar
60	Bharatiya Jain Sangathan	North Andaman, Middle Andaman, South Andaman, Little Andaman and Great Nicobar
61	GELC Tsunami Relief and Rehabilitation	North Andaman, South Andaman, Little Andaman and Great Nicobar
62	Church Auxiliary for Social Action (CASA)	South Andaman, Little Andaman and Car Nicobar
63	Nehru Yuva Kendra	South Andaman, Little Andaman and Car Nicobar
64	Save the Children	South Andaman, Little Andaman and Great Nicobar
65	CARITAS-CRS	South Andaman, Little Andaman, Car Nicobar and Great Nicobar
66	Oxfam	South Andaman, Little Andaman, Car Nicobar and Great Nicobar
67	Butterflies	All affected islands of A&N Islands
68	Child Line Foundation	All affected islands of A&N Islands
69	Red Cross	All affected islands of A&N Islands
70	Society for Andaman and Nicobar Ecology (SANE)	All affected islands of A&N Islands
71	UNICEF	All affected islands of A&N Islands
72	Unique Service Trust	All affected islands of A&N Islands

Source: Based on Digal (2006)

Table 8A.2 List of 70 items of the OXFAM package provided in A&N Islands

Items for personal hygiene and sanitation	*Clothes*	*Bedding items*	*Kitchen items*
1. Bathing soap	16. Hosiery Items	21. Bed sheet	28. Bowl
2. Clips, rubber bands	17. Sari, blouse, petticoat	22. Blanket	29. Containers
3. Comb	18. Shirt, trouser	23. Mattress	30. Cups
4. Detergent	19. Skirt, top	24. Mosquito net	31. Drum
5. Hair oil	20. Trouser	25. Pillow	32. Glasses
6. Razor		26. Pillow covers	33. Hatha (ladle)
7. Sanitary napkins		27. Sleeping mats	34. Kadai
8. Shampoo			35. Knife
9. Shaving brush			36. Measuring mug
10. Shaving cream			37. Plate
11. Soap case			38. Pressure-cooker
12. Tongue cleaner			39. Saucepan
13. Toothbrush			40. Spoon
14. Toothpaste			41. Stove
15. Washing soap			42. Tawa
			43. Water filter
First-aid kit	*School kit*	*Hardware*	*Other items*
44. Antiseptic cream	53. Colour pencils	62. Hammer	66. Bucket
45. Bandages	54. Crayons	63. Nail	67. Candle
46. Band-Aid	55. Eraser	64. Rope	68. Mugs
47. Burnol	56. Geometry box	65. Screwdriver	69. Radio
48. Cotton	57. Note book		70. Torch
49. Dettol	58. Pen		
50. Moov	59. Pencil		
51. Safety pins	60. Register		
52. Thermometer	61. Scissors		

Source: Based on fieldwork (2010)

9 Compensation and the Tsunami

The Tsunami that led to the devastation of the A&N Islands occurred on 26 December 2004. It was followed by an equally gigantic response in the A&N Islands which lasted till December 2011. In this duration of seven years, more than Rs. 30 billion was spent in the reconstruction of damaged infrastructures and houses and rehabilitation of the affected households of the A&N Islands. The armed forces, the A&N Administration and the Government of India worked together with 72 NGOs to bring back normalcy post-disaster. However, it is more essential to see what has been the recovery. How much have the affected households recovered? To get an overview of the response to the Tsunami and to assess the recovery of the affected households, fieldwork was undertaken three times – May to August 2010, January 2011 and January 2012.

The fieldwork

Fieldwork was carried to generate data on three aspects of the response. First, to assess the differences in response provided to the non-tribal and tribal Nicobarese tuhets; second, perception of the people regarding the permanent houses they were living in; and third, to assess the recovery of the affected households. Households were carefully selected keeping in mind the objectives of the study. Case studies were selected from South Andaman Island which is dominated by non-tribal households and Car Nicobar Island, which is inhabited by the tribal Nicobarese population.

An open-ended semi-structured questionnaire was used to generate information on various aspects of the post-Tsunami recovery. Head of the households were interviewed in all cases. The questions included their pre-Tsunami house location, house type, employment, number of family members, the immediate reaction after the Tsunami, the journey of the households from original house location to the permanent post-Tsunami houses, compensation received – heads of household, time of receiving compensation, process of claiming compensation, amount received as compensation and other questions. Questions were also asked regarding the perception about the house – its quality, cost, durability, effectiveness in case of next tsunami and other disasters.

Households residing in the post-Tsunami permanent houses were selected as case studies. These were vulnerable households and therefore were affected in the Tsunami. Moreover, these households experienced Tsunami firsthand and lost their houses and everything in them. The permanent houses were constructed post-Tsunami and were handed over to these households free of cost.

The houses were constructed in 10 islands; however, case study households were selected from two islands – South Andaman and Car Nicobar. Both these islands represent the contrasts of the Andaman Islands and the Nicobar Islands which offer scope for comparison. First, the permanent houses of South Andaman have the non-tribal households whereas the permanent houses of Car Nicobar have the tribal Nicobarese tuhets. The households of South Andaman belong to diverse linguistic and cultural background such as Tamil, Telegu, Malyalis, Bengalis, Ranchiwalas and others. Thus, the former is heterogeneous and segregated, while the latter is homogenous and united.

Second, the permanent houses of South Andaman also accommodated non-tribal households from the Nicobar Islands, especially Car Nicobar, Katchal and Teressa. These households were dislocated to a place 500 km away from their original house. On the other hand, Car Nicobar Island has the original tribal Nicobarese who were moved 5–6 km inwards from the sea, and remained on their 'own' island. Third, households of South Andaman have no access to common property resources which are under the A&N Administration, while in Car Nicobar Island the resources are managed by the tribal Nicobarese themselves. Fourth, the permanent houses of South Andaman are in and around an urban area, while tribal Nicobarese tuhets of Car Nicobar Island are in a rural area. The only feature these two islands share is that both have the district headquarters. Port Blair, the capital city of the A&N Islands, is situated within South Andaman, while Car Nicobar is the district headquarters for the Nicobar District.

A total of 9,797 permanent houses were constructed at 84 sites across the A&N Islands. Out of this, from four permanent house sites, 100 case study households accounting for 1 per cent of the total Tsunami-affected households of the A&N Islands were selected. Only 100 households were selected since the fieldwork was based on in-depth case studies, not on question schedules. Out of the 100 households, 65 are from South Andaman – 45 from Bambooflat and 20 from Austinabad; and 35 from Car Nicobar – 20 from Arong village and 15 from Kimius village (Figure 9.1).

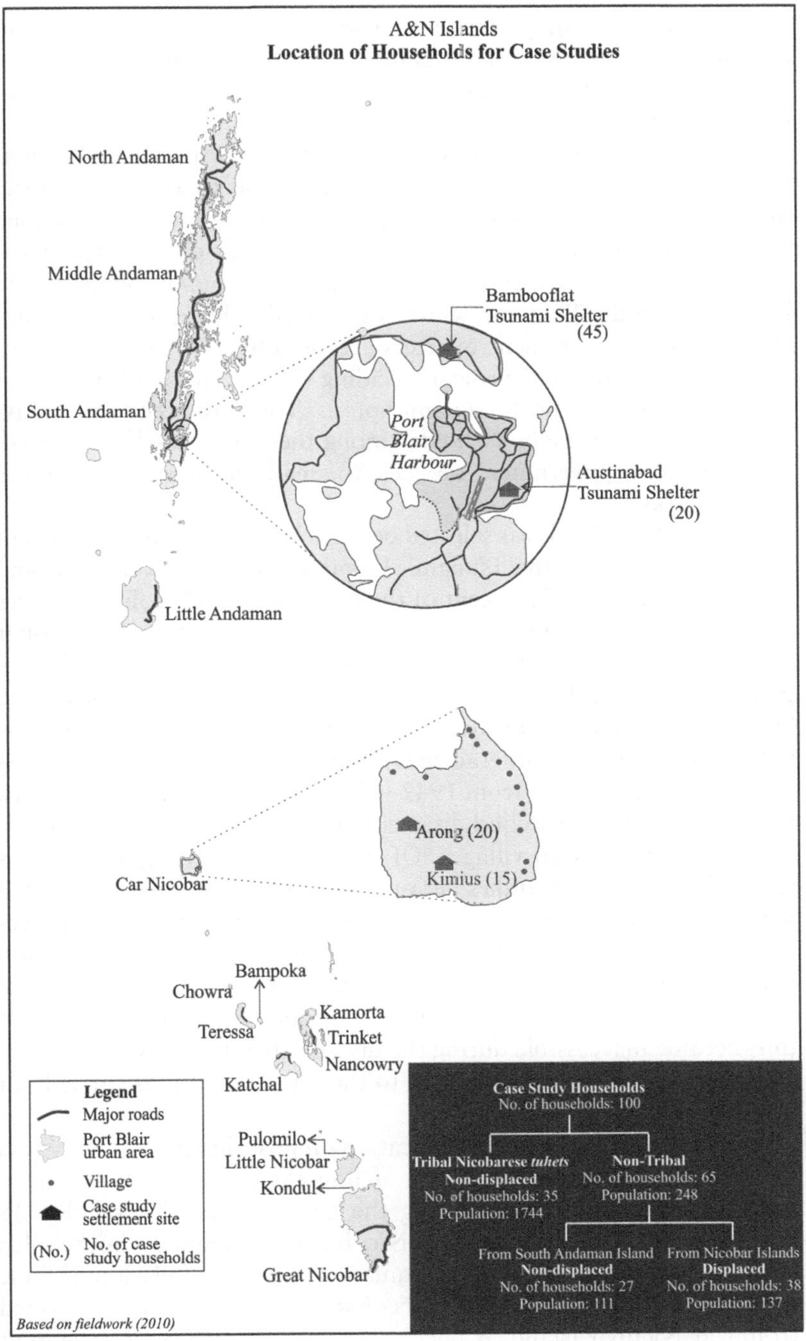

Figure 9.1 A&N Islands: location of households for case studies

In South Andaman, permanent house sites are at nine locations, out of which, two were selected – Bambooflat and Austinabad. Together these account for 65 per cent of the total 823 post-Tsunami permanent houses of South Andaman.

Again, these two locations had households dislocated from Nicobar Islands. As one of the objectives of the study was to assess the impact of dislocation on the households post-disaster, these sites provided the requisite case studies. Second, the households of these two locations belonged to various socio-economic backgrounds. Some owned houses before the Tsunami, while others lived either in rented accommodations or government quarters. These differences in the type of house pre-Tsunami had a role in deciding the perception of the households regarding the post-Tsunami houses. Third, the economic activity of the head of the households among these 65 households varied. It ranged from government services, to small businesses, fishing and labour works.

Car Nicobar Island was selected because it accounted for 28 per cent of the total killed in the Tsunami. It was second after Katchal Island that accounted for 37 per cent of the total. But Car Nicobar provides a unique situation which is not shared by any other of the Nicobar Islands: first, it accounts for 60 per cent of the total tribal Nicobarese population. Second, the Nicobarese of Car Nicobar are considered 'advanced' as it is the headquarters of the Nicobar District and the tribal Nicobarese have interacted with the outside world. Third, it was under the Japanese rule from 1942–45 during the World War II. Thus, the Tsunami was the second disaster experienced by Car Nicobar.

Car Nicobar has 16 villages. Of these villages, Arong and Kimius were selected for case studies because both villages together accounted for 55 per cent of the total population killed in Car Nicobar. Second, both these villages had 100 per cent tribal Nicobarese population before the Tsunami. Third, both these villages are located away from the headquarters (the administrative headquarters of the island), and thus became inaccessible during the disaster thereby delaying immediate relief. This gave the insight in to the relief operations carried out post-Tsunami.

As assessing the impact of dislocation on post-disaster recovery was the objective, near-equal number of households were selected from the above four locations. Among the 65 non-tribal households 38 households were dislocated from Nicobar Islands and 27 households were dislocated within South Andaman. The total population of these 65 households was 248. Tribal Nicobarese households were selected from Car Nicobar Island. A Nicobarese village consists of large joint families known as tuhets which consists of several households. Each tuhet functions as one unit and hence all tuhets were selected as case studies. The entire population of Arong village was covered under 20 tuhets and entire population of Kimius village was covered under 15 tuhets. Thus, even though the total number of tribal Nicobarese tuhets is 35, it has a total population of 1,744. All tuhets were dislocated within the pre-Tsunami island nearly 5–6 km inwards from the coast. Before one proceeds to analyze the response it is important to understand, what is a tuhet?

Tribal Nicobarese tuhets

A Nicobarese village consists of tuhets or joint families. A joint family is a mode of combining smaller families into larger family units through the extension of three or more generations including at least grandparents, parents and children. It is also defined as a group of people who generally live under one roof, who use a common kitchen and who hold common property. In case of a Nicobarese tuhet, fieldwork confirmed that the tuhets do not live under one roof nor have common kitchens, but there is common ownership of land and they are related by birth or by marriage, and they live together within a designated area in the village. This area is also known as tuhet. Thus a tuhet is akin to a locality. Consequently, the locality, its inhabitants and its houses are all known by the same tuhet name. In the pre-Tsunami period, the locality consisted of several houses for various purposes.

Each tuhet had a *ma-pati*, the large beehive house that was occupied by the tuhet head and the other members. Next to it was a cooking house called *talika*. There was the *sanra*, a small gable built on stilts for couples with small children. Separate dwelling houses and cooking houses came up to accommodate the increase in family members. These were set up within the designated area of the tuhets and hence became a part of the existing tuhet.

Every village had a *pati-yongni* (birth-house) and a *pati-kupah* (death-house), as both birth and death are considered 'unclean' processes. Before the introduction of hospitals, the birth-house was used for the birth. The expectant couple shifted to the birth-house closer to the date of delivery along with a senior female family member. The birth took place in this house and it was forbidden for the mother to the leave the house till about 21 days of the delivery. Only after the restriction period was over could the mother and baby rejoin their tuhets. This insured safety of the newborn and the nursing mother from infections. Similarly, in case of death, the body of the diseased was taken to the death-house for the last rites. Before introduction of Christianity, the Nicobarese were animistic. As per their animistic rituals, the person carrying out the last rites had to stay in this house for a couple of days. After a purification ceremony, the person could go back to his normal life. After Christianity was introduced, the death-house was used to keep the body of the diseased, where people from the village could come and pay homage to the deceased before being buried.

Every village also had an *el-panam*, which was the community hall used for gatherings, meetings and celebrations. Guests are never taken directly into a house, but stayed in the el-panam. Every village was an accumulation of several tuhets. The number of tuhets in a village varied. In Car Nicobar, the number of tuhets in villages range from 10 to 41. On an average a village has 21 tuhets (Table 9.1).

The number of people in a tuhet also varied from 10 to over 100 members (fieldwork, 2010). A tuhet is the basic economically functional unit of the Nicobarese society. For social occasions and other

Table 9.1 Tuhets of Car Nicobar Island

Village	*Number of tuhets*
Mus	41
Malacca	31
Sawai	29
Tamaloo	27
Perka	25
Small Lapathy	21
Arong	21
Kinyuka	20
Big Lapathy	17
Kinmai	16
Kimius	15
Tapoiming	13
Chukchucha	12
Teetop	12
Kakana	10
Average	**21**

Source: Based on fieldwork (2010)

financial needs, the tuhet acts as a single unit. The land is owned jointly by the tuhet. Each tuhet has a head, and if the tuhet is large then there may be assistants to help the head of the tuhet. The eldest of the family usually is the head of the tuhet.

The villages of Nicobar Islands are headed by a council of captains and tuhet heads, who take all decisions regarding the village and the villagers. The village council is a body of five elected members known as captains. The person with most votes is the first captain, followed by second captain, the third and so on. This system of chiefs or 'captains' was not a part of the traditional Nicobarese society. It became a necessity and was encouraged by the traders for their own purpose. It was started by the Portuguese in the seventeenth century and was carried on by Danes, Austrians, British and the present Indian government. The word captain comes from the trading ships where the head of the ships were known as captains (Imperial Gazetteer of India 1909).

A captain of a Nicobarese village is selected from among the people on the basis of qualities like communicative skills, good behaviour, sincerity, knowledge of traditional norms, wisdom, ability to make decisions and compassion for the fellow Nicobarese. The ownership right of the village land rests with the captain. This land consisting of forests and the coconut groves are divided among the tuhets of the village. The households (consisting of husband and wife and their children) of the tuhet have the rights to use the land. They look after the coconut groves and maintain the trees and have the ownership right on the produce from the land. They have to give some part of it to the captain as fees, or they contribute to the regular feasts organized by the village. The post of the captain is not hereditary. Therefore, the ownership of land is communal. However, fieldwork of this study confirmed that usually the eldest son of the captain succeeds his father, making the 'captainship' hereditary. But if the villagers feel that the son is incapable to be their leader, they do select someone else.

The village councils of all villages in the island form the tribal council. In the pre-Tsunami period there were 12 tribal councils, one in each island. Currently, there are eight tribal councils – Car Nicobar, Chowra, Teressa, Kamorta, Nancowry, Katchal, Little Nicobar and Great Nicobar – as four islands were declared unfit for habitation after the Tsunami. The tribal council is headed by the chairperson also known as the chief captain. He is assisted by a vice captain, the first, second and third captains from villages. The captains are the most respected persons of the island. They are never disobeyed and their decisions are accepted without protest.

In pre-Tsunami period, the tribal Nicobarese lived in partial seclusion with their own set of rules and regulations which were in contrast to the non-tribal population. Efforts were made by the A&N Administration to retain these differences. Thus, the differences in the response provided to the non-tribal households and the tribal Nicobarese tuhets were apparent in the immediate as well as in the long-term phase of response.

Differences during the immediate response phase

The differences between the response provided to the non-tribal households and the tribal Nicobarese emerged immediately after the Tsunami. In South Andaman Island, the affected parts included both rural and urban area. In rural area, the gram panchayat heads visited the affected area and requested the households to go to the nearest gram panchayat bhawans. In urban area, councillors of the wards visited the affected area and distributed plastic sheets and bamboo poles to the households so they could set up relief camps. The non-tribal households from Nicobar Islands were evacuated to Port Blair and were housed in the relief camps set up in school premises. Thus, the non-tribal households from Car Nicobar were evacuated to South Andaman. Non-tribal from other Nicobar Islands, especially Katchal and Kamorta, went back and forth between Teressa and Katchal before the relief camps were finally set up in Katchal. The households lived for six months in these relief camps and then were dislocated to South Andaman Island.

For the tribal Nicobarese tuhets, relief camps were set up on the pre-Tsunami island itself. The tribal Nicobarese were given an option to leave for Port Blair or remain in Nicobar Islands. Most of them preferred to remain on the island. Nearly 200 members (out of about 1,500) from the 35 case study tribal Nicobarese tuhets were evacuated to Port Blair and housed at the Island Tourism Ground. Some of them lived with their acquaintances in Port Blair. However, within a month they returned back.

Thus, the three groups ended up in three kinds of relief camps – first, the non-tribal households from South Andaman either found shelter in panchayat bhawans or set up relief camps with plastic sheets in the place of pre-Tsunami house location. The second group, the non-tribal households from Car Nicobar Island were housed in the school premises of Port Blair. The 24 schools (10 middle schools, seven secondary schools and 17 senior secondary schools) of Port Blair Municipal Area were converted into relief camps. For the third group of tribal

Nicobarese tuhets tents were set up as relief camps in Car Nicobar Island itself.

In the first two types of relief camps, i.e. in camps set up with plastic sheets in the pre-Tsunami location, gram panchayat bhawans and in the school premises, relief materials were distributed at the camp sites. It was convenient for the households to collect the relief items, food and cash compensation. But for the relief camps in Nicobar Islands, there were no such delivery mechanism since roads were damaged. As the villagers had taken refuge in the middle of the forest, they walked to the nearest base where ration and relief items were stocked. Helicopters were used to deliver these items to the base camp. Arong village itself became one of the base camps for the supply of relief material to neighbouring villages of Sawai and Teetop, while villagers of Kimius walked for about 5 km to Kakana to collect and carry back the ration.

Of the sample 100 households, 95 households were housed in relief camps after the Tsunami, while the other five stayed with families or lived in rented accommodations. The experience of living in the relief camp was described as a 'worst nightmare' by 85 per cent of the households. Unanimously, every household acknowledged that life in relief camp was one of the 'worst periods' after the Tsunami. The pain of losing family members, loss of house and everything that was dear was traumatic. Coupled with it, they were forced to live in the relief camps that were ill-equipped. Since households had lost everything, they had no other option but to live in these camps. The three important elements of human life – food, clothing and shelter – had become sparse. The households recalled that in the relief camps, hygiene was non-existent. Food was served from the common kitchen. The households had to wait in queue to 'collect' food, clothes and money distributed by the relief workers. The households said that they felt like 'beggars' as they were treated like them. There was no one to 'listen' to them. In the Nicobar Islands there was shortage of food and drinking water. A newspaper report of 5 January 2005 highlighted the plight of the tribal Nicobarese of the Teressa Island:

> stories unfold of lack of food, drinking water, officials 'deserting' the scene, and the hovering threat of a malaria outbreak. Stretched to the limit, the relief camp residents say they are now going to set up another camp on their own, near the only water source in the island – a stagnated pond. Says Harold, one of the 3,600 nameless faces stranded here, 'Babu, have you got something for us? There is no dal and edible oil. If you don't give us food, we won't be able to survive this'. The conditions are terrible. No district admin-

> istration official has bothered to come here and enquire about our needs. No effort has been made to restore water supply. The Assistant Commissioner of Nancowry arrived yesterday but he said his primary task is to evacuate people. But what's the point in thrusting more and more people into camps when there are no basic amenities?
>
> (Sanyal 2005)

Car Nicobar being the headquarters of the Nicobar district, it did not face such acute situation as the Teressa Island, however, its far-flung villages did face difficulty in collecting relief items.

Even though the households described their experiences of living in relief camps as 'hell', they were supportive of the work of the armed forces in the relief phase. The armed forces in this case refer to the army, air force, navy, coast guard, paramilitary forces as well as the A&N Islands Police and Fire Services. A non-tribal head of the household dislocated from Car Nicobar Island commented that:

> given the magnitude of the damage, the armed forces did a good work. At that time it was the armed forces which rescued us and brought us to Port Blair in aeroplane. When they found us, they gave us clothes and food. They also provided shelter to us. They were our saviours.

Even the Captain of Arong Village said that 'it was the armed forces that made the helipad in the forest in a day's time and brought food, clothes, medicines and other relief items. We were saved because of them'.

As for the efforts of the administration, the households unanimously agreed that 'the officers of the administration just came to visit; the actual work was done by the armed forces'. The difficulties of the households did not end with the closure of the relief camps and the end of the immediate phase of response. It lingered on in the long-term phase of response too.

Differences during the long-term response phase

The phase of long-term response began after the households moved to intermediate houses during May–June 2005, i.e. six months after the Tsunami. This phase was characterized by three aspects: (1) receiving ex-gratia compensation; (2) involvement in the livelihood schemes and (3) the reconstruction and relocation to the new houses. This chapter deals with the first two aspects while reconstruction and relocation of houses forms the next chapter.

Ex-gratia compensation

The Finance Commission of India sets up norms for the compensation given after a disaster. In 2004, the norms set by the Eleventh Finance Commission of India (2000, 2005) were in force. However, considering the scale of damage, the ex-gratia compensation provided, especially for persons killed and missing and for agriculture damage, exceeded the norms of the Eleventh Finance Commission. As per the norms, ex-gratia payment to families was fixed at Rs. 50,000 per deceased person, this was increased to Rs. 200,000 in case of the Tsunami. Similarly, for immediate relief Rs. 1,000 per household was the norm, in this case Rs. 13,000 was given. Again, for agricultural, horticultural and annual plantation crops, a compensation of Rs. 1,000 per ha was fixed for rain-fed area and Rs. 2,500 per ha in areas with assured irrigation. But in case of the Tsunami, Rs. 24,231 per ha for paddy, vegetables and other crops and Rs. 94,465 per ha for plantation crops (coconut and areca nut) was sanctioned. The revised rates varied from that of the Finance Commission's norms because the Government of India and the A&N Administration were of the opinion that the amount under the norms was insufficient to aid the households in rehabilitation as they were living in the remote islands.

The revised amount was also based on the damage assessment figures provided by the expert committee that visited the A&N Islands after the Tsunami to assess the damage. This assessment was based on the interaction of members of the expert committee with the panchayats in case of Andaman Islands and the tribal councils in case of the Nicobar Islands. Out of the 27 non-tribal households from South Andaman Island, 19 were under the panchayats. Damages of these households were assessed by the members of the panchayats, who in turn, prepared the damage list and handed it to the administration. The panchayat members assessed the damage of standing crops, cropped area affected, submerged land, livestock lost, damage to houses and other damages if any for each of the affected household. The remaining eight non-tribal households from South Andaman Island belonged to the urban area. The damage assessment of these households was carried out by the staff of the councillor's office who recorded the damage to the houses, loss of boats and fishing gears. The damages of the non-tribal households from Nicobar Islands were recorded by the officers in charge of the relief camps, who prepared the list of persons killed and missing, lost livestock, houses damaged, shops damaged and other details.

The case of the tribal Nicobarese was different. Immediately after the Tsunami, on the request of the administration, the village captain assigned the tuhet heads the task of drawing a list of persons killed and missing, injured, houses damaged, livestock lost, coconut groves destroyed, boats/canoes lost and other losses. The list thus prepared by the tuhet heads was compiled for the village and given to the tribal council. The tribal council after verifying the details with the captains of the village gave the final list to the A&N Administration.

The difference in assessment was that for the non-tribal households, the assessment was done by outsiders whereas in case of tribal Nicobarese, it was their own tuhet heads and village captains who assessed the damage. The structure of assessing and presenting the details of damage between panchayat and tuhet captain was very critical. The non-tribal households accuse the panchayat heads and officers in charge of relief camps of being partial and making a biased list. They are of the opinion that the list was drawn up by strangers who purposefully did not enumerate their damages. There were allegations that panchayat members included names of people not affected by the Tsunami after taking bribe. On the other hand, among the tribal Nicobarese the damage list was drawn by the tuhets themselves, there were no such grievances and allegations. The tribal Nicobarese lost more in the Tsunami, but with their method of assessment, the compensation received was high.

Compensation was given to the head of the households on the basis of the damage data provided by the panchayats, officers in charge of relief camps and the tribal council. Ex-gratia compensation of Rs. 2,000 per household as immediate relief was given in cash to each head of the affected household through the panchayats, officers in charge of relief camps and the tribal council for their respective cases. The other compensation was given through cheques issued by the Pay and Accounts Division of A&N Administration. For the non-tribal households, a notification regarding the issue of the cheque was published in *Daily Telegram* (government-run newspaper of A&N Islands) and the heads of the households were directed to collect the same from the deputy commissioner's office. For tribal Nicobarese tuhets, the cheques were handed over to the captains of the villages, who in turn distributed them to the members of the tuhets. Claiming and receiving compensation for the killed and the missing followed a different procedure.

In the Tsunami in the A&N Islands 432 persons were killed, while 2,164 were declared missing. The deceased was declared 'killed' if the body was found and identified while the 'missing' accounted for the bodies which were washed away and were never found. Claiming compensation for the persons killed did not involve any paperwork. For the non-tribal households a cheque was issued in the name of next of kin for the killed. Names of the heads of the households who were to collect the cheques were published in the *Daily Telegram*. For tribal Nicobarese tuhets, cheques for the persons killed were handed over to the village captains who distributed the same to the tuhet heads. Therefore, the non-tribal households had to visit the government offices to collect their cheques, while the tribal Nicobarese tuhets were given their cheques in their homes.

Claiming compensation for the missing persons was different from that of the killed. A non-tribal head of household had to file a First Information Report (FIR) along with the photograph of the missing person to receive the compensation. In addition with the FIR, he/she had to publish a 'missing notification' in the newspaper with the same photograph. The head of households faced difficulty in obtaining the photographs of the missing person as everything, including photo albums had washed away. As most of the non-tribal households had relatives in other parts of the A&N Islands and on mainland India who had old photographs, these were retrieved and were used for filing the FIR and the newspaper notification. Thus the non-tribal households had to retrieve the photograph of the deceased and then run between the nearest police station and the newspaper office to get the paperwork in order before they could file the claim for compensation.

In contrast, for the tribal Nicobarese tuhets, a police officer was stationed at the Office of the Special Relief Officer who filed the FIR for missing person. Only the captain of the village and the tuhet heads reported the number of persons missing. Therefore missing information for the village was filed collectively. Moreover the tribal Nicobarese tuhets were exempted from filing the notifications of the missing in newspapers as Nicobar Islands do not have a newspaper office. Therefore, the process of claiming ex-gratia compensation for the non-tribal households was more cumbersome than that of the tribal Nicobarese tuhets. Of the 2,164 persons declared missing, 390 were non-tribal and 1,774 were tribal Nicobarese.

Total population of the selected 100 case study households was 1,992 (248 non-tribal and 1,744 tribal Nicobarese) in 2010 during the fieldwork. Making adjustments for the new births (247 individuals were less than seven years of age at the time of fieldwork) and deaths (23 incidences of deaths) post-Tsunami, this population was estimated to be close to 1,760 at the time of the Tsunami. Out of this, 421 persons were killed or were declared missing. All the cases of persons killed or declared missing were either from non-tribal households from Nicobar Islands or from tribal Nicobarese tuhets. None of the 27 non-tribal households from South Andaman Island reported killings or missing cases. Out of 38 non-tribal households dislocated from Nicobar Islands in six households, two persons were killed and

18 were declared missing. Out of the 35 tribal Nicobarese tuhets, in 29 tuhets, 98 persons were killed and 303 were declared missing. All six non-tribal heads of households and 29 heads of tribal Nicobarese tuhets had received compensation for the persons killed and missing in the Tsunami. The funds were not released immediately, but were put in as fixed deposit for five years with an option given to the households to withdraw interest or collect the principal and the interest at maturity. This was done to ensure that the funds were not misused.

Official formalities for receiving compensation by households

Before funds could be released to the heads of the affected households, it was imperative to have a bank account, as payments were made through cheques. Of the 100 case study households, 67 heads of households had bank accounts before the Tsunami. The remaining 33 did not have a bank account. Of the 100 households with a bank account, 24 out of 27 (88 per cent) were non-tribal households from South Andaman, 21 out of 38 (55 per cent) were from non-tribal households from Nicobar Islands, while the remaining 22 out of 35 (63 per cent) were tribal Nicobarese tuhets. For the remaining households the administration opened bank accounts in the names of the heads of affected households before compensation cheques were issued. The banks relaxed their norms for opening the account. No address proof was necessary. The only documents required were passport photographs and the certificate issued by the officer in charge of relief camp. The administration arranged photographers to take pictures for the affected households. The bank accounts were opened by the end of April 2005 after which the cheques for the compensation amount were issued. All the households agreed the compensation for the killed and missing was received in two instalments of Rs. 100,000 each. Out of the three groups, it was the non-tribal households that had to make extra effort in order to receive the compensation. Retrieving the photograph of the deceased and then filing the FIR was tedious. But once the papers were in order, the cheques were issued.

Compensation received by the case study households

Ex-gratia compensation was sanctioned under 14 categories – (1) immediate relief, (2) temporary relief, (3) household articles, (4) small business, (5) crop loss, (6) submerged land, (7) reclamation of land, (8) livestock loss, (9) house damage, (10) orphan children, (11) widows, (12) unmarried girls, (13) seriously injured and (14) persons killed or declared missing in the Tsunami. All 100 case study households were not eligible for compensation under all 14 categories (Table 9.2).

Table 9.2 Sanctioned heads and amount for compensation

	Non-tribal households from South Andaman Island		*Non-tribal households from Nicobar Islands*		*Tribal Nicobarese tuhets*	
	No. of eligible households	*No. of households receiving compensation*	*No. of eligible households*	*No. of households receiving compensation*	*No. of eligible tuhets*	*No. of tuhets receiving compensation*
1. Immediate relief @ Rs. 2,000 per household						
	27	27	38	38	35	35
2. Temporary relief @ Rs. 3,000 per household						
	27	27	38	31	35	35
3. Household articles @ Rs. 10,000 per household						
	27	27	38	31	35	35
4. Small business @ not exceeding Rs. 10,000						
	0	NA	6	0	0	NA
5. (a) Crop loss of Paddy vegetable/root crop @ Rs. 22,231 per ha						
	0	NA	0	NA	0	NA
5. (b) Crop loss of Plantation crop (coconut and areca nut) @ Rs. 90,465 per ha						
	0	NA	0	NA	35	35
6. Submerged land @ Rs. 10,000 per ha						
	0	NA	0	NA	0	NA
7. Reclamation of land @ Rs. 5,000 per ha						
	0	NA	0	NA	0	NA

(*Continued*)

Table 9.2 (Continued)

	Non-tribal households from South Andaman Island		*Non-tribal households from Nicobar Islands*		*Tribal Nicobarese tuhets*	
	No. of eligible households	*No. of households receiving compensation*	*No. of eligible households*	*No. of households receiving compensation*	*No. of eligible tuhets*	*No. of tuhets receiving compensation*
8. Each livestock						
Cattle @ Rs. 12,000	0	NA	0	NA	10	10
Pigs @ Rs. 4,000	0	NA	2	0	35	35
Goats @ Rs. 3,000	5	0	20	0	35	35
Poultry @ Rs. 80	12	0	30	0	35	35
9. Each House damage*						
Severely damaged (permanent) @ Rs. 10,000						
Severely damaged (temporary) @ Rs. 6,000						
Marginally damaged @ Rs. 2,000						
10. Compensation for orphans @ Rs. 0.2 million for each orphan	0	NA	2	2	12	12
11. Compensation for widows @ Rs. 0.1 million for each widow	0	NA	0	NA	10	10
12. Compensation for unmarried girls @ Rs. 0.1 million for each unmarried girl	4	0	8	0	20	0
13. Compensation for seriously injured @ Rs. 5,000 for each severely injured person	0	NA	2	2	5	5
14. Compensation for persons killed or declared missing @ Rs. 0.2 million for each killed or missing person	0	NA	6	6	29	29

* Even though all case study households lost houses in the Tsunami, they did not receive cash compensation instead were given a new permanent house. Therefore compensation for houses was given for damages in Andaman Islands, who were not part of the case study households.

Source: Based on fieldwork (2010)

Compensation received by the non-tribal households from South Andaman Island: The selected non-tribal households from South Andaman Island were not eligible to receive cash compensation under nine categories that were not applicable to these households. These were: (1) small business loss, (2) crop loss, (3) submerged land, (4) reclamation of land, (5) house damage (they got a permanent house instead of cash compensation for the same), (6) orphans, (7) widows, (8) seriously injured and (9) persons killed or declared missing, therefore were eligible to receive compensation under five categories, namely, immediate relief, temporary relief, household articles, compensation of unmarried girls and for the livestock lost.

All 27 non-tribal households from South Andaman received compensation under three categories: immediate relief, temporary relief and compensation for household articles, but not for unmarried girls and the livestock lost. Out of the 27 households, four households had unmarried girls in the age group of 18–25 at the time of the Tsunami, who were eligible to receive compensation of Rs. 100,000 each. Yet they did not receive the compensation. Again, five of these households reared goats and 12 reared poultry birds. Yet, they too were denied the compensation. These households put allegations that the panchayat members who drew the list of affected households did not include their names in the list of 'cases eligible for the compensation'.

Compensation received by the non-tribal households from Nicobar Islands: The non-tribal households from Nicobar Islands were eligible to receive compensation under nine out of 14 categories. Of the 38 non-tribal households from Nicobar Islands, all received compensation for immediate relief; but, seven households did not receive compensation for temporary relief and for household articles even though they were eligible. This is because they did not stay in relief camps. Among these 38 households from Nicobar Islands, six had small businesses, and most households reared goats, pigs and poultry for domestic consumption (total households who reared pigs: 2; goats: 20; poultry: 30). However, none of these households received any compensation for either small business or for livestock loss. This is because as per the ANATPR 1956, entry, stay and trading by non-tribal in the tribal area is prohibited. Trading by the non-tribal is allowed after obtaining a licence. However, as these households were staying and trading 'illegally' in the area, they were not entitled to receive the compensation either for business damage or for loss of livestock. These households were not given compensation for unmarried girls as they were not enumerated. There were eight such households.

The eligible households among these 38 non-tribal households from Nicobar Islands received compensation under three categories, namely, orphaned children, seriously injured and compensation for the persons killed or declared missing. There were two households with orphan children (number of children orphaned was three), two households with seriously injured persons (number of injured was two) and six households with persons killed or declared missing in the Tsunami (number of persons killed or declared missing was 20). The non-tribal households from Nicobar Islands were one of the most vulnerable groups. As they were living in the Nicobar Islands 'illegally', they were not entitled for any compensation. However, the administration did provide compensation for some heads, but not for other. Thus, even though they were equally affected in Tsunami as the tribal Nicobarese, they were not provided with compensation under all heads. This was also the reason that post-Tsunami they were allotted houses in South Andaman and not in their original island of habitation.

Compensation received by the tribal Nicobarese tuhets: The 35 tribal Nicobarese tuhets were eligible to receive compensation under 10 out of 14 categories. All eligible tuhets received compensation except in one category, namely, assistance for unmarried girls. The tribal Nicobarese do not differentiate between the girls and the boys, and hence, did not claim compensation for unmarried girls. The tribal Nicobarese tuhets received compensation for 1,872 ha of land under coconut and areca nut plantations as well as compensation for livestock. A total of 98 cows, 5,040 pigs, 1,727 goats and 5,672 poultry birds were killed in the Tsunami among the 35 tuhets of the two case study villages of Car Nicobar Island. The tuhets were also compensated for orphaned children, grievously injured persons and for persons killed and declared missing in the Tsunami.

Therefore, it can be concluded that tribal Nicobarese and the non-tribal households from South Andaman Island were financially compensated. For the non-tribal households from Nicobar Islands, it was their pre-Tsunami vulnerability because of which they missed out on the compensation. This led to slower recovery and an accentuated vulnerability in the post-Tsunami period.

Amount received by the case study households as compensation

There was a stark contrast in the amounts of compensation received by the non-tribal households from South Andaman Island, the non-tribal households from Nicobar Islands and the tribal Nicobarese tuhets. Total amount received by all case study households together

Table 9.3 Ex-gratia compensation received by the case study households

Permanent shelter	*Amount received as ex-gratia (Rs. in millions)*	*Per cent of total*
Non-tribal households from South Andaman Island	0.4	0.14
Non-tribal households from Nicobar Islands	5.1	1.72
Tribal Nicobarese tuhets	290.8	98.15
Total	**296.3**	100

Source: Based on fieldwork (2010)

was Rs. 296.3 million (fieldwork 2010). Of this amount, the non-tribal households from South Andaman received 0.14 per cent, non-tribal households from Nicobar Islands received 2 per cent, while the tribal Nicobarese tuhets received the remaining 98 per cent (Table 9.3).

Thus, the 35 tribal Nicobarese tuhets accounted for lion's share of the total compensation received by the total case study households.

There are two reasons for this skewed distribution. First, the amount of compensation corroborates with data of the killed, declared missing, and damage to agricultural land; and Nicobar Islands accounted for largest share in both categories. Rs. 0.2 million per person killed/missing and Rs. 90,465 per ha for damage to plantation crops amplified the amount of compensation received by the Nicobarese tuhets. Added to this was the compensation given for the orphans, widows and for loss of livestock.

A second reason for the skewed distribution of the compensation amount is that among the tribal Nicobarese tuhets the compensation was paid 'family-wise' and not tuhet-wise. The Finance Commission prescribes that compensation is to be given to the head of the households. In this case a 'household' refers to a unit as found in mainland India. The Nicobarese tuhets though is one household, yet it does not abide by the concept of a household prevalent in mainland India. The administration requested the tribal council to provide list of affected 'households'. As a consequence, tuhets were broken down into 'families' for the purpose of receiving compensation. During the fieldwork it was assessed that the 35 tuhets were broken down into 430 families. Therefore, the amount of compensation received under relief, temporary relief and for household articles was given to 430 families and not 35 tuhets. This increased the compensation amount many times. The skewed distribution resulted in spatial concentration of the amount of compensation received in the Nicobar Islands as the tribal Nicobarese tuhets received 98 per cent of the total amount received by the selected 100 households.

The cash compensation for 27 non-tribal households from South Andaman Island was Rs. 405,000; for the 38 non-tribal households from Nicobar Islands it was Rs. 5,089,000; and for the 35 tuhets (i.e. 430 families) it was Rs. 290,761,240 (fieldwork 2010). Therefore the per capita (approximate) amount of compensation received by the non-tribal households in South Andaman was Rs. 15,000, meanwhile, the per capita compensation for the non-tribal households in Nicobar Islands was Rs. 133,921. For the tribal Nicobarese, it was Rs. 8,307,464 per tuhet but was Rs. 676,189 per 'family' (fieldwork 2010). It was the non-tribal households from Nicobar Islands that needed more money as they were relocated to an urban area which posed challenges and competitions. Cost of living in the urban area of Port Blair and Bambooflat is more than the rural Car Nicobar Island. Thus, the non-tribal households from Nicobar Islands were dissatisfied with the amount of compensation given to them as it was too paltry to begin new life to gain livelihood.

One of the reasons for dissatisfaction of the non-tribal households from South Andaman Island was the inadequate compensation for the damaged houses. Households belonging to Sippighat, Teylorabad, Biblitan and Burman Nallah (six households out of 27 non-tribal households from South Andaman Island) locality had their own houses. Some of them had also leased out the house on rent. Post-Tsunami, all households, irrespective of the house owner and the tenant, received compensation of Rs. 15,000 (Rs. 20,000 for immediate relief and Rs. 13,000 for temporary relief and household articles) and a new permanent house. The six households from South Andaman Island claimed that in the pre-Tsunami period they had property worth Rs. 2.5 million to 3 million and the land on which their house was built. Since what they received was a meagre sum of Rs. 15,000 and a semi-permanent house away from the main city, they were dissatisfied with the compensation received. The average size of their property was 3,068 sq. ft (285 m^2). However they did not receive compensation for the houses lost, and instead were allotted permanent houses. The differences of the size and value of property were not taken into consideration while giving compensation. These households claim that their houses were constructed after taking proper permissions from the authorities. They demanded higher compensation as they feel that the administration has cheated them by not pointing out that the location was unsafe. They also assert that administration could have denied them the permission to construct houses in the first place. Therefore, the households grudged that administration should compensate according to actual loss incurred.

What has added to the discontent is that equal compensation was given to the tenants as well as the landlords. Out of the 27 non-tribal households from South Andaman, six were landlords while the remaining 21 were the tenants. Now, the landlords felt that tenants did not lose property, yet were allotted permanent shelters. This equal compensation was criticized by the landlords. However, the same decision was appreciated by tenants as they maintained that though they did not lose a property, yet they did lose their 'homes'. The tribal Nicobarese tuhets, on the other hand, received near equal amount of compensation as the damage and the loss incurred by all tuhets was the same. Hence there was no dissatisfaction with the compensation received.

Again, the non-tribal households were not involved in the process of damage assessment. It was done by the relief camp in charge, and households alleged that being an outsider, the camps in charge did not draw the damage comprehensively, and thus the deserving households missed out on the compensation. The non-tribal households also alleged that there was bias towards tribal Nicobarese even though the non-tribal households were equally affected. They complained that the amount of compensation received was inadequate to cover and recover the losses. The case was different for the tribal Nicobarese. Here tuhet heads and the village captains drew the list and hence people were in the process, albeit, indirectly. Thus, there was no grievance of missing out on compensation.

Free ration: compensation in kind

Ration and vegetables were provided free of cost to the households as long as they were in intermediate houses. This time period was almost four and a half years for the non-tribal households and three and a half years for the tribal Nicobarese tuhets. Though ration was distributed with good intensions, the method of distribution was flawed because of three reasons: first, no ration shops were set up in close proximity to the non-tribal intermediate house sites. Second, ration for several months was provided at once. A truck load of ration came every two to three months. Third, the intermediate houses consisted of one room only. As there was no storage space therefore, some of the households sold the ration to earn some ready cash.

Among the tribal Nicobarese tuhets case was different. In Car Nicobar Island, there were only three ration shops for total 15 villages. The roads connecting these villages were damaged. Each of these shops catered to the surrounding four or five villages. Therefore, the villagers from the far-flung villages had to carry the ration on foot. One of the ration shops was set up in Kakana village and villagers from Kimius village had to walk for more than 10 km to collect ration from here. In contrast, Arong village did not face such problem because one of the ration shops was set up at Arong itself.

Critique of ex-gratia compensation

The A&N Administration ensured that the affected people do receive the cash compensation under the RGRP and the TRP. But during fieldwork it emerged that the process had several loopholes and was marred with delays and insensitivities.

Lack of verification

First, was the problem of verification of losses. After the Tsunami, the situation was that of chaos. The entire infrastructure and houses of Nicobar Islands had collapsed and washed away, people were stranded in the far-flung islands, there was no communication with the remote islands, people were killed in large numbers and their bodies were strewn across the islands. The administration was unprepared to handle this crisis. During fieldwork, the affected households recalled that there were several individuals who were not affected by the Tsunami but entered the relief camps at Port Blair and got enumerated as 'Tsunami-affected' households. There was no means to prove this fact, because at that point after the Tsunami there were no verifications. All who came to the relief camps were 'assumed' as victims and were given compensation and relief items. After second week of the stay at relief camps certificates were issued on the basis of which compensation was distributed. There were allegations that relief camp in charge took bribes and included names in the list, even though several individuals were not affected in the Tsunami. Such individuals received compensation, relief items as well as permanent houses. Therefore, the number of 'affected households' increased. Quite many individuals took advantage of the situation and benefitted in the Tsunami at the cost of the actual victims.

A second problem created in the absence of documented record was that several non-tribal households split up to receive more compensation. The compensation was paid 'household-wise'. The non-tribal households living in a joint-family registered themselves as two different households. By splitting, they could claim more money as well as relief. Out of the 65 non-tribal households, 22 had split up to receive more compensation and separate house. As there was no verification by the administration, they took advantage of the situation and benefitted in the disaster.

As there was no documentation of the assets of the households before the Tsunami, it was impossible to verify the claims. In the Nicobar Islands, there were no records of land ownership by the tuhets or the number of animals they owned. Therefore, an average estimates were provided by the tuhet heads to the captains of the village. Some of the tuhet heads gave greater figure than the actual damage. Some of the non-tribal households recounted that in order to claim more compensation, some tuhet heads included the names of people those who had died earlier. As there was no verification and as most of the bodies were 'missing', they took this as an opportunity to claim more funds. However, the tribal Nicobarese refuted such claims. In fact, they blamed the non-tribal households for such practice. The root cause of this problem was that compensation was provided based on what was said was lost rather than the actual loss.

Compensation not according to vulnerability

The administration did not provide compensation according to vulnerability of the households. The non-tribal households from Nicobar Islands were dislocated to a new place and needed greater assistance. Instead, they were denied the compensation under several categories because they were living 'illegally' in Nicobar Islands. They were the most vulnerable group, yet were given compensation less than the tribal Nicobarese tuhets as well as the non-tribal of South Andaman Island. Similarly, no distinction was made between age group or working status, be it a six-month-old baby, 35-year-old working men and women or a 70-year-old householder. The administration adopted a policy of 'equal value of life'. But, it is clear that the loss of a working member of the household is not same as the loss of a baby or a retired non-working person. Clearly, the household that lost the only earning member was more vulnerable than the others.

Delay in distributing compensation

All compensation was not received at one time. Rather, it was spread over the period of TRP. The compensation was received in instalments (Figure 9.2). An immediate relief of Rs. 2,000 cash was given within

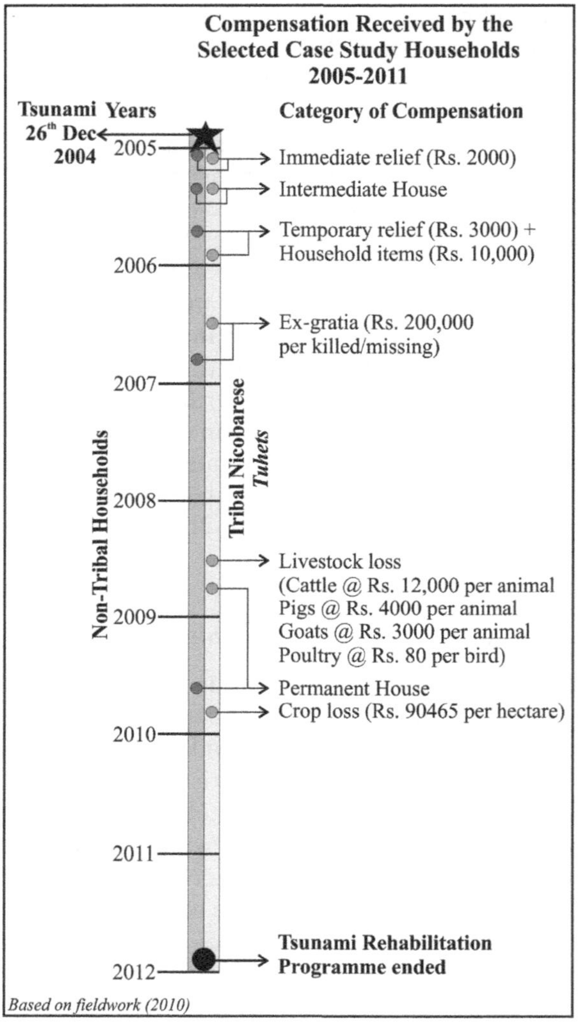

Figure 9.2 Compensation received by the selected case study households (2005–11)

seven days of the Tsunami. Temporary relief and compensation for the household articles, totalling Rs. 13,000 was given during July–September 2005, more than six months after the Tsunami. It was released after the households moved into the intermediate houses. Compensation for the killed, though sanctioned in March 2005, was given to the households much later. The non-tribal households received it only by September 2006 and the tribal Nicobarese received it in November 2006. The compensation for the killed/missing was not given either in cash or cheques. Rather it was deposited as a fixed deposit for a period of five years with a locking period of two years. This ensured that the fund was not misused and there would be a saving for the affected households.

Among the 100 case study households, only the tribal Nicobarese received the compensation for the livestock and for plantation. Compensation for the livestock damage was released in June 2008 while funds for plantation loss were released by October 2009. This amount too was put in fixed deposit for 10 years. However, the households were free to withdraw the interests of the same as per their needs. The permanent houses were allotted in 2008 and 2009 to tribal Nicobarese tuhets and non-tribal households respectively.

A pattern emerges in the distribution of funds. First, small amounts were released soon after the Tsunami and after six months, when the households moved into the intermediate houses. After which the households had to wait not only for permanent houses but also for compensation under different heads involving large sum of money such as ex-gratia for the killed and missing, compensation for livestock loss and for the loss of plantation crops. There was a time lag between the loss and the cash compensation received. However, compensation in kind was provided to the households after they moved into the intermediate houses. But, the compensation paid in kind was a case of misappropriate response.

Misappropriate response

After the Tsunami, the affected households were given relief items to help them build their lives again. The items given were useful for the non-tribal households, but it was a case of misappropriate response for the tribal Nicobarese (Table 9.4). The items provided to tribal Nicobarese, such as saris and blankets, reflect the unfamiliarity of the administration and the NGOs with tribal people. Saris were provided across Nicobar Islands even though Nicobarese women do not wear saris. In fact till about the early twentieth century, the Nicobarese did not wear clothes. In the historical documents, Nicobar Islands were referred to as 'the land of naked people'.

But after Christianity was introduced in the early twentieth century, people started wearing clothes. It was by the efforts of Bishop John Richardson, a Nicobarese himself, that Christianity spread in

Table 9.4 Misappropriate response – a case of tribal Nicobarese of Car Nicobar Island

Sector	*Items provided by Administration and NGOs*	*Misappropriate response*
Relief items	Saris	Tribal Nicobarese women do not wear saris
	Blankets	Nicobar Islands have no winter as it is situated in the equatorial region
	Cash	There was no shop to purchase goods from in Nicobar Islands after the Tsunami
	Radio	No electricity immediately after the Tsunami; no local radio station
	Uncooked food grains	No utensils or stoves immediately after the Tsunami to cook
	Bicycles	No roads immediately after the Tsunami
Agriculture	Vegetable cultivation introduced	Tribal Nicobarese do not cultivate vegetables and food grains. The land in Car Nicobar Island is not suitable for cultivation – high consumption and low return
	Pump sets, power tillers, organic manure (neem cakes)	Not required for coconut plantations; there is no paddy cultivation in Car Nicobar Island
	Farming implements	Poor quality
Fisheries	Diesel-operated fibre boat provided	Nicobarese prefer local canoes as it is easier to row in open sea Tribal not engaged in commercial fishing and diesel shortage in Car Nicobar Island
Livestock	Poultry rearing introduced	No commercial livestock produced Tribal rear pigs and poultry for domestic use
House	Intermediate House Permanent House	Contradictory to the pre-Tsunami houses; it broke down the cohesive living of the tuhets

Source: Based on fieldwork (2010)

the Nicobar Islands. Bishop John Richardson was trained in Burma (present-day Myanmar), therefore, the clothes that tribal Nicobarese took onto resembled the traditional dress of Burma. The men wore half-pants while the women wore long skirts with blouses. As saris were of no use to the Nicobarese women, they wore pants and shirts that were provided to the men. The saris were used as mosquito nets or to partition the one room of intermediate houses. Similarly, blankets were distributed that had no use for the Nicobarese as Nicobar Islands do not experience winter. Insensitivity of the administration was evident in the distribution of ration. During the fieldwork, the non-tribal households recalled that no ration shops were set up in the locality. A van with ration came and distributed ration for six months. As per the ration norms in A&N Islands, in a month an adult receives 5 kg rice, 2 kg dal and 1 kg sugar; while for a child (i.e. one below aged 12) 2.5 kg rice, 1 kg dal and 1 kg sugar was provided along with 2 kg potatoes per person per month. Therefore, for an average family consisting of two adults and two children, a total of 150 kg of ration and 48 kg of potatoes were provided at one time, despite the poor storing capacity of the intermediate shelters. The intermediate shelters did not had separate kitchen or any storage space. In the humid climate of A&N Islands, the ration rotted. Therefore, some of the households resorted to sell the ration given to them in the market while the others threw it away once it became unusable.

Another case of the misappropriate response was the 'agriculture revival package' for the tribal Nicobarese. The main economic activity of tribal Nicobarese pre-Tsunami was copra production. The coconut trees require minimum nurture. In a haste to implement post-Tsunami rehabilitation schemes, the Nicobarese were 'trained' to cultivate fruits and vegetables. As the tribal Nicobarese were not farmers, they were not inclined to pursue the training. Cultivation of vegetables requires skill and labour. Coconut plantation requires minimum maintenance and copra production requires a different set of skills. As the cultivation of vegetables and fruits was 'new', 'different' and 'difficult' for the tribal Nicobarese, they did not pursue it. Therefore, the power tillers, the pump sets and the organic manures that were distributed had no utility for the tribal Nicobarese. The farming implements did have some utility as these were used in digging land for planting traditional root crops, but, as the quality of these implements was poor these became unusable too soon. There were no returns from the cultivation even though the process demanded lot of time and effort, as the soils of Nicobar Islands being too sandy are infertile. The produce from the cultivation did not suffice even the needs of the tuhets, hence, it was easier purchasing the fruits and vegetables from the market. Added to this was an easily available compensation fund that discouraged the tribal Nicobarese to grow fruits and vegetables.

Similarly, under the agriculture revival strategies, water harvesting structures were constructed for irrigation in the Nicobar Islands. These structures lie unused today and have failed to have any impact on the livelihood of the tribal Nicobarese as the coconut plantation does not require ‘irrigation’. Similar was the fate with other livelihood initiatives and the fibre boats.

The case of revival of animal husbandry through introduction of poultry farming and cattle rearing among the tribal Nicobarese was also not successful. Tribal Nicobarese prefer to raise pigs rather than poultry, as traditionally pigs are considered a valuable asset among the Nicobarese tribe. The pigs not only provide food, but were used in barter until money was introduced in the mid-1950s. Pigs are the ‘least maintenance’ animals, as they are scavengers. Nicobarese reared pigs for domestic consumption and raising animals for commercial purposes did not fit with the tribal way of life. Thus, the scheme failed. Similar to livestock rearing, ‘revival of fishing’ schemes for commercial fishing also failed. Tribal Nicobarese were given fibre motor boats, but an uninterrupted supply of diesel to run the boats was not ensured. Car Nicobar Island has only one petrol pump. Queuing up for the ‘costly’ diesel was very ‘tiring’ for the Nicobarese. Thus, instead of the motor boats they preferred the traditional canoes, known as hoodies, for fishing.

From the items distributed it clearly emerges that the aim of these distributions by the government and the NGOs alike was to meet the target of distributing set number of items rather than the actual need of the households. One such case of misappropriate response was the cash flow in Nicobarese society that has had dire consequences.

Sudden cash flow in tribal Nicobarese society

The impact of the sudden cash flow in a subsistence tribal Nicobarese society was not researched before providing the compensation. In the pre-Tsunami period the tribal Nicobarese economy was dependent on resources like coconut, fishes, hunting and gathering. Cash was introduced after Independence in a limited manner. Most of the tribal Nicobarese tuhets sold copra for cash with which they procured items of need. In the post-Tsunami period, the flow of cash through compensation, be it for the killed, orphaned, widowed or for livestock and coconut plantation, has created a situation of ‘excess cash in the economy’. One of the early signs of the impact of sudden cash flow is reduced inclination to make copra. In the pre-Tsunami period copra was made and sold once a week, but post-Tsunami the production of copra has been reduced to once in three weeks. Ploughing money into gainful education, augmenting skills or savings has not happened.

The rehabilitation programmes failed to make an impact on the lives of the affected households. Thus, they have reverted back to their pre-Tsunami livelihoods and occupations. It can be concluded that distribution of ex-gratia funds was carried out mechanically without assessing either the needs or the impact of the fund flow among the affected people. The response was 'supply driven' and not 'need driven'. The well-intended exercise thus failed to reduce the vulnerability as the funds were 'consumed' rather than 'invested'. Another aspect of the Tsunami response directly involving the Tsunami-affected population was the post-Tsunami houses, which is dealt with in the next chapter.

10 Recovery and the Tsunami

A form of compensation was the free houses given to the households who lost their houses in the Tsunami. Post-Tsunami housing reconstruction was major segment of the TRP. Two types of houses, intermediate and permanent, were constructed. By the time of the fieldwork in 2010, all households had moved to the permanent houses and the intermediate houses were demolished. Therefore, the following section assesses the perception of, and, satisfaction with the permanent houses. 'Perception' is subjective and in this case it depended on the pre-Tsunami houses of the households. Therefore, an account of the pre-Tsunami houses is imperative before assessing the perception of the post-Tsunami permanent houses.

Pre-Tsunami houses of the case study households

The three groups, i.e. non-tribal households from South Andaman Island, non-tribal households from Nicobar Islands and the tribal Nicobarese tuhets from Car Nicobar Island were living in different types of houses in pre-Tsunami period. There was only one commonality among all case study households: all the houses were located on the coast, and hence were washed away in the Tsunami. Apart from the location, there were many variations of houses among the non-tribal households and the tribal Nicobarese tuhets. As there were no significant differences between the houses of the non-tribal households from South Andaman Island and non-tribal households from Nicobar Islands, in the following section the comparison is between two groups only, the 65 non-tribal households and the 35 tribal Nicobarese tuhets (Table 10.1).

There were variations in the pre-Tsunami houses. Therefore, for the purpose of this research the dominant (as described by the head of the households) house type has been considered. The houses of the 65 non-tribal households were mainly of two types – the permanent

Table 10.1 Difference between the pre-Tsunami and post-Tsunami houses

	Pre-Tsunami houses	*Post-Tsunami 'permanent' houses*
Location		
Non-tribal	On the coast	On a hill top, 10 km inward from the sea
Tribal	On the coast	4–5 km away from the coast
Building material		
Non-tribal		
Permanent	Concrete and CGI sheets	Steel frame, wooden planks, processed bamboo boards, concrete
Semi-permanent	Grass, ply board, plastic sheets	
Tribal		
Permanent	Concrete and CGI sheets	Steel frame, wooden planks, processed bamboo boards, concrete
Traditional	Wood and lallang grass	
Design		
Non-tribal		
Permanent	2–4 rooms Attached kitchen and toilet On ground	3 rooms Attached kitchen and toilet On ground
Semi-permanent	1 room Attached kitchen and common toilet On ground	
Tribal		
Permanent	1–2 rooms Separate kitchen Separate toilet On ground	Two room Attached kitchen Separate toilet On stilts
Traditional	One room Separate kitchen Separate toilet On stilts	

Source: Based on fieldwork (2010)

houses of households from Sippighat, Teylorabad, Biblitan and Burman Nallah (all in South Andaman Island) and the semi-permanent houses of households from Junglighat, Dairy Farm, Bambooflat, Panighat (in the Port Blair Master Plan, 2028 prepared by the Andaman Public Works Department (2009) these localities have been classified as 'slums') and households from Nicobar Islands. The permanent houses had floor and walls made of concrete. The roofs were made of sloping CGI sheets. The number of rooms in the permanent houses ranged from two to four. It had attached kitchen, toilet and bathroom. As the average size of the plot was 3,068 sq. ft, these houses had large rooms. In contrast, the semi-permanent houses consisted of single room with attached kitchen, but shared toilet and bathroom. Though the floor and the walls were of concrete, the roofs were made of plastic sheets, grass, tiles, CGI sheets or a combination of all. The average size of these houses did not exceed 170 sq. ft. Out of the 27 case study non-tribal households from South Andaman, six had permanent houses, while 21 had semi-permanent houses in the pre-Tsunami period. Among the non-tribal households from Nicobar Islands, 10 households had government jobs out of which only two lived in the government quarters that were permanent houses. The remaining 36 households lived in semi-permanent houses either constructed by the tribal Nicobarese or in makeshift shelters which they made themselves on land given to them by the tribal Nicobarese. In these houses, the floors were cemented. The walls were made in combination of two materials. From the ground, approximately 0.5 m was made of brick and cement. The upper part of 1.5 m was made of plywood board. The roof was made of the CGI sheets. For rooms, partition was created with the ply boards. These houses did not have solid foundation, or a sturdy superstructure.

The tribal Nicobarese tuhets also had two types of houses – the traditional houses which were completely made of forest products and the permanent houses. The traditional houses were huts constructed on stilts. These huts were circular with a diameter of approximately 60 ft and height of 15–20 ft that rested on stilts of 7 ft above the ground. For the main structure wooden posts were erected which was then covered with leaves called the lallang grass. Kloss (1903) described these huts as 'inverted bee-hive on posts' as these huts did not have separate wall and roofs. The roof was constructed in a manner that it covered the sides and served the dual purpose of roof as well as the walls. The grass is waterproof and was laid in an overlapping manner as a thatch for the roof so that rain water did not leak in. A narrow opening at the base of these huts was used for entry and exit. These huts have evolved over millennia and were adapted to the ecology of the island (Table 10.2).

These houses were best suited for climatic condition of the Nicobar Islands. The Nicobar Islands have equatorial climate with high temperature and rainfall for eight months. The roof provided protection

Table 10.2 Traditional huts: a design for mitigating disasters

Aspect of the hut	*Feature*	*Resistance to disaster or protection against disaster*
Shape	Circular	The shape acted as a wind break. The wind moved swiftly around the hut without facing an obstruction, therefore the huts do not collapse during high velocity winds during storms and cyclones.
Structure	Beehive	The rainwater does not stagnate on the roof, therefore there is no leakage even though rains occur on an average 20 days in a month.
	Stilts	The stilts prevent dampness in the house. It also protects the residents from insects that crawl on the ground. The space below the stilts is used for keeping the pigs and other livestock.
	Wooden frames	This gives stability to the structure. The islands are prone to frequent earthquakes. The wooden structure on which the house rests is dug deep in the earth. This protects the house from toppling over. In case it does fall, then there is no fear of injury as wood and grasses are light material. Within the huts, battens of areca palms are placed horizontally. These are used as attic for storing horticultural products.
	No nails	The house is made without any screws and nails. Instead ropes are used for tying up the structure. Thus there are no joints of weakness.
	Planks with leaves	This traps the air between the planks and the leaves and thereby insulated the huts. This maintains a comfortable temperature within the huts.
	Fireplace	Every hut has a fireplace built of clay. The smoke from the hearth strengthens the wood and the planks and also protects it from termites and other insects. This increases the life of the structure.
Building materials	Wood	The longevity of the wood such as coconut palm, jackfruit and other hardwood used for the stilts and the main structure is almost 20 to 30 years.
	Split bamboos	These provide ventilation to the house as there are no windows and only a small opening in the floor is used as a gate.
	Lallang grass (Dhanipati)	Is used for the roof and the walls. The grass is waterproof so it protects the house during the rains. The grass keeps the house cool on hot afternoons.

Source: Based on Kloss (1903); Imperial Gazetteer of India (1909); and Singh (2006) and fieldwork (2010)

from these elements of nature. The wooden frame of the hut provided stability during earthquakes. The circular design of the huts provided protection from high velocity winds as the structure allowed the winds to move around it rather than resisting it as in case of rectangular structures. The building material of hard wood from coconut tree trunks and jackfruit, cane and lallang grass was suited for the local weather conditions. The maintenance of the structure was easy and simple as building material was sourced from local forest and these structures could last for 25 to 30 years. The traditional Nicobarese 'inverted bee-hive' huts were more durable and safe to face earthquake, cyclone or tsunami. Yet, with time, these huts were modified and thus became susceptible to collapse in times of stress. These modifications range from change in the shape, structure as well as the building materials.

After 1950s, the traditional huts were modified and incorporated aspects of non-tribal houses. Most noticeable change was that the huts became rectangular and had walls made of ply board. The shape of the houses also changed to rectangular instead of circular as the construction of rectangular house was easier as it only needed four stilts which too were made of concrete. This led to the change in the basic structure of the houses. The height of the stilts was reduced to 1 m only instead of 2 m as the rectangular house could be supported at this height with the concrete blocks that were wider than the poles. The size of the house was also reduced and it was smaller than the traditional hut. Walls with door and windows were made in these houses, whereas the traditional houses neither had walls nor windows or doors. In the modified huts doors were made and thus instead of ladder there were steps. There was a change in building materials too. For the traditional huts the lallang grass was used for the roof that extended as walls too. In the new kind of houses, the grass was only used for the roof, in some of the houses, the grass was replaced with either CGI sheets or plastic sheets. As these new kinds of houses had walls, split bamboo mats and plastic sheets were used for walls. The number of rooms was increased to two in these huts. A later addition, after the non-tribal moved to the Nicobar Islands in the 1970s, to the house types of tribal Nicobarese were the second type of houses, the permanent houses. These had floor and walls made of bricks covered with concrete, while the CGI sheets were used for making sloping roofs. These houses resembled non-tribal houses with only one difference: cemented platforms were made in the rooms which were used as beds. The Nicobarese felt that this addition was more economical than making a cot. The traditional house of the Nicobarese did not have cots for sleeping as it was about 7 ft above the ground. The cots were laid out under these huts for resting during the afternoon. It was unsafe to sleep on the floor because the insects, crabs and snakes often crawl inside the house. The raised cemented platform acted as a cot and protected them from these animals. The raised platforms were also used for keeping items dry and safe during the rainy season. These platforms served the same purpose which the raised traditional houses performed, and, it was low maintenance structure and involved only one-time investment.

Both the non-tribal households and the tribal Nicobarese tuhets lived in different kind of houses largely because the two sets of population were culturally very different. To accommodate these differences, the permanent houses provided to both the groups varied even though they appear similar from the outside

Differences between non-tribal and tribal Nicobarese post-Tsunami houses

The households were allotted permanent houses nearly three to four years after the Tsunami. In this period both non-tribal households and the tribal Nicobarese tuhets had to change houses a number of times. On an average both the non-tribal households and the tribal Nicobarese tuhets changed houses twice: first, from relief camps to intermediate houses six months after the Tsunami; and, from intermediate houses to permanent houses in approximately three and a half years' time. However, some of the non-tribal households (especially, 11 households from Nicobar Islands) had to change houses four times in three years after the Tsunami (Table 10.3). The 11 households moved between different relief camps before finally arriving at relief camps at Port Blair. After which they moved to intermediate shelter within six months and then to the permanent houses in another three years. The non-tribal households lived in uncertainty as they were not aware of the location of the new houses; as site of relocation was not negotiated with them. The tribal Nicobarese had greater participation in the reconstruction process. There was no uncertainty as to where the new houses would be located. As the entire village is like a community, all tuhets moved to the intermediate houses and permanent houses at the same time.

Thus, all 35 tuhets changed location twice before being allotted the permanent houses.

This was not the only difference. Though the permanent houses across A&N Islands look similar, there are significant differences (Figure 10.1). There are four differences between the permanent houses of the non-tribal households and the tribal Nicobarese tuhets: in the naming of the house location, house structure, services provided and in the nature of ownership of the permanent houses.

Table 10.3 Number of times relocated before arriving at permanent house

Number of times relocated before allocation of permanent houses	*Non-tribal households*	*Tribal Nicobarese tuhets*	*Total*
One	12	0	12
Two	21	35	56
Three	21	0	21
Four	11	0	11
Total households and tuhets	**65**	**35**	**100**

Source: Based on fieldwork (2010)

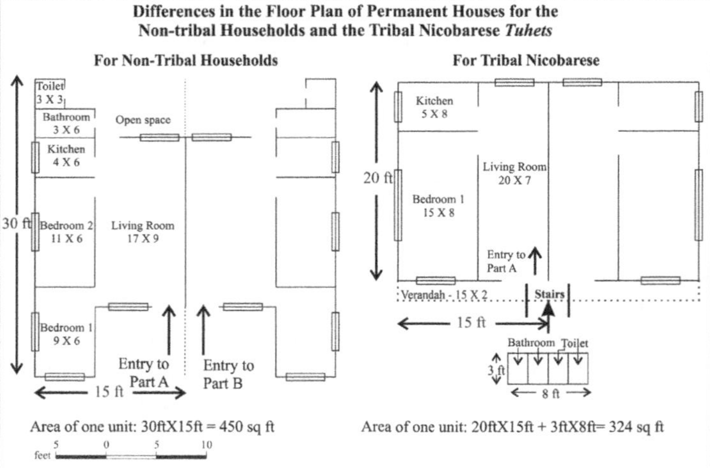

Aspect of the house	Non-tribal households	Tribal Nicobarese *tuhets*
Foundation	RCC	RCC
Super structure	Steel frame	Steel frame
Stilts	No stilts	8 ft above ground
Bathroom	Attached with the house	On ground, 2 ft away from the house
Walls External Internal	 Timber Processed bamboo board	 Timber Processed bamboo board
Floor Area (including toilet and bathroom)	2 units – 450X2 = 900 sq ft	1 unit – 324 sq ft 2 units – 648 sq ft 4 units – 1296 sq ft
Floor	Cement concrete	Processed bamboo board
Cost	140,000 for two units	120,000 for two units

Based on fieldwork (2010)

Figure 10.1 Differences in floor plan of permanent houses for the non-tribal households and the tribal Nicobarese tuhets

Difference in naming the settlement: There is a difference in the naming of the post-Tsunami house sites. The permanent house sites of the non-tribal households have a prefix 'Tsunami permanent shelter' to it. Thus, the houses are known as 'Tsunami permanent shelter – Bambooflat', 'Tsunami permanent shelter – Austinabad', 'Tsunami permanent shelter – Mithakhari', and so on. While for the tribal Nicobarese settlement, the names of villages have remained unchanged. Even though dislocated to a new site, the villages are known by their pre-Tsunami names. While referring to a pre-Tsunami village the prefix 'old' is used. Therefore, the erstwhile locations of villages are known as 'old Arong', 'old Kimius', and so on. The name of the settlement is a constant reminder of the Tsunami for the non-tribal households and is not the case with the tribal Nicobarese tuhets. One of the heads of non-tribal households commented: 'Tsunami will never leave us, even our address has Tsunami in it'. The name serves as a constant reminder of the Tsunami.

Difference of house structure: One of the most notable differences in the permanent houses of non-tribal households and tribal Nicobarese tuhets is that the former is built on ground, while the latter is built on stilts. However, the tribal Nicobarese from islands other than Car Nicobar Island preferred houses on ground, similar to houses of the non-tribal households. The pre-Tsunami houses of tribal Nicobarese tuhets were built on stilts. The space below the house was used as a pen for the pigs. Tribal Nicobarese, both men and women spent most of their time sitting under the house to rest during the day. The space below acted as a shade during summer as well as during rains. Therefore, to accommodate the needs of the tribal Nicobarese tuhets the post-Tsunami permanent houses were constructed on stilts.

The second difference is that all houses for non-tribal households have a floor area of 450 sq. ft (including the bathroom and toilet), while the permanent houses of tribal Nicobarese tuhets are of three sizes – a single unit, double unit and a quadruple unit. The floor areas of these units (without bathroom and toilet) are 300 sq. ft, 600 sq. ft and 1,200 sq. ft, respectively. The tribal Nicobarese tuhets had this variety to accommodate the large joint families.

The third distinct difference is in the floor plans of the permanent houses of the non-tribal households and the tribal Nicobarese tuhets. The permanent houses of the non-tribal households have one living room, two bedrooms, a kitchen, a bathroom and a toilet. There are two doors, five windows and open spaces in front and at the back of the house. Two units make one house. Thus each non-tribal household has a common wall. On the contrary, tribal Nicobarese tuhets have an option for a single unit house, a double unit or a quadruple unit. A single unit house of the tribal Nicobarese tuhets has one living room, one bedroom and one kitchen. The toilet and bathroom are on the ground, little away from the main house. A double unit has two variations. One model is like attaching two single unit houses. In the second type, the common wall joining two units has been removed to make single, but large living room. This unit has one living room, one kitchen and two bedrooms. In a quadruple unit four single units have been joined together to make one unit. However, a wall has been made in the middle of the house to support the ceiling. A door has been made in this wall. A quadruple unit has one kitchen, one living room and five bedrooms. All tribal Nicobarese tuhets permanent houses also have a 2 ft wide veranda and a staircase.

A fourth difference is that toilet and bathroom are attached with the house for the non-tribal households, but for tribal Nicobarese tuhets bathroom and toilet are on the ground away from the main house. Tribal Nicobarese prefer toilet and bathroom to be away from the main housing unit.

The post-Tsunami houses vary significantly between the non-tribal households and the tribal Nicobarese tuhets. This difference is a response to the different 'culture' of the tribal Nicobarese. Apart from differences in the houses, there were differences too in the services provided at the site of the settlement.

Difference in services provided: The services provided to the permanent houses' sites of the non-tribal households and tribal Nicobarese tuhets varied. Both the permanent shelter sites, Austinabad and Bambooflat of the South Andaman Island are located up on the hill, away from the main city by about 10 km. In non-tribal settlements many amenities were missing. During the fieldwork, it was ascertained that primary health sub-centres were non-functional at both the post-Tsunami permanent houses' sites. At Bambooflat, it was under construction, while at Austinabad it was yet to be inaugurated. There was no anganwadi (pre-primary play school for toddlers) though it was proposed in the plan. Few women together operated a temporary anganwadi centre from their homes. Again, there was no school, either primary or secondary, at both the sites. At Bambooflat, a school (middle school) was under construction, while there was no provision for schools at Austinabad. However, there are two private schools, i.e. Mahatma Gandhi International School and Kamraj School, that are located in the vicinity of the Austinabad Tsunami shelter (Table 10.4).

There were no community halls, recreation halls, playground, market or even a provision for a grocery shop. Few households had set up

Table 10.4 Services provided to permanent house sites of non-tribal households and tribal Nicobarese tuhets, 2010

Services provided	*Non-tribal households permanent house site*	*Tribal Nicobarese tuhets permanent house site*
Primary health sub-centre	No	Yes
Anganwadi	No	Yes
School	No	Yes
Community hall	No	Yes
Recreation hall	No	Yes
Playground	No	Yes
Market	No	Yes
Ration shop	No	Yes
Bus service	No	Yes
Birth house	No	Yes
Death house	No	Yes

Source: Based on fieldwork (2010)

make shift grocery shops. But these were illegally set up in temporary sheds. There was also absence of any state bus services from both the settlement sites. The nearest bus stop from both the sites was 3 km downhill. At Austinabad, a private operator's mini-bus (a 15-seater bus) was started in March 2010. From 8 a.m. to 8 p.m., every two hours the bus comes up to the permanent house sites. Bambooflat permanent house sites did not even have this service. The householders walked all the way to the bus stop daily. Most difficult was to carry the ration as most of the households could not afford the autos, which are the only means of conveyance, apart from the private vehicles. In the absence of basic amenities such as a health centre, grocery shop/ market or a bus service, it is difficult for the households to carry out day-to-day activities.

On the other hand, each tribal Nicobarese permanent house site has been provided with a primary health sub-centre, anganwadi, primary school, community hall, recreation hall, playground, shopping centre, ration shop, bus stop, birth house and a death house. Except for a shopping centre, all other services were functional at the time of the fieldwork. The shopping centre was not needed as the villagers procure necessary items from 'head quarter', the area where all the services are located such as post office, bank, ship and helicopter ticket booking counter, grocery shop, bakery, other shops and Bishop John Richardson Hospital. The villages are connected with metalled road on which the state transport bus plies once every two hours. It connects villages with each other and with the head quarter. Some Nicobarese have set up small shops which provide daily need items. Thus, the houses of the non-tribal households had no access to basic services as compared to the houses of the tribal Nicobarese tuhets. Apart from this, the ownership of the houses also varied among the non-tribal households and tribal Nicobarese tuhets.

Difference in the nature of ownership: In terms of ownership of the permanent houses, non-tribal households only have 'user' rights while tribal Nicobarese tuhets have 'owner' rights of the houses (see Appendix). An allotment order was issued to each of the non-tribal households and tribal Nicobarese 'families' as per which the houses were allotted. This order gives details such as name of the beneficiary, pre-Tsunami location of the households, permanent house location, permanent house number and the house constructing agency. Though it is an allotment order, it has terms and conditions of use of the permanent houses. It has been duly signed by the deputy commissioner of the district and heads of the non-tribal households, whereas the tribal Nicobarese did not have to sign the terms and conditions. Thus, the terms and conditions are not binding for the latter. This is because the houses for the non-tribal households were built on government land. Therefore, to ensure proper use of the shelter, the non-tribal households had to sign the contract, whereas, the houses for the tribal Nicobarese tuhet were constructed on their own land, and hence the tribal Nicobarese were exempted from the contract. The most important implication of this contract is that the non-tribal households were simply the 'occupants' of the houses and not the 'owners', whereas the tribal Nicobarese were the 'owners' of their houses.

The terms and conditions for the use of the permanent houses for non-tribal households and the tribal Nicobarese tuhets varied. There are 17 clauses in the agreement for the non-tribal households. As per the first and second clauses, 'the ownership of the shelter vests in the administration', but in case of death of the beneficiaries, their son/daughter shall be entitled to continue the possession of permanent shelter. Another significant clause is seventh one which states that 'beneficiaries shall not incorporate any alterations, modifications or additions to the shelter structure'. But Clause 8 states that, 'on allotment, the permanent shelter shall be the asset of the allotte thereby any maintenance/repairs, etc., to be met by the allotte him/herself'. Again, as per the clauses 10 and 11, 'the allotte shall not transfer the ownership of the permanent house by way of lease, sublet, sale or through any other legal instrument without prior permission of the A&N Administration' and 'the allotte shall, in no case, permit other people to stay in the permanent/temporary shelter on lease/rented basis for any commercial purposes'. From these clauses it is evident that ownership of the house does not belong to the households as sale or rent of the same is prohibited. Moreover, the households cannot modify the structure to accommodate their needs. The non-tribal households had to accept the houses as they had no other alternatives.

The tribal Nicobarese have no such clause for the use of the permanent houses. They can modify these structures as they are the 'owners' of the houses. However, they are not permitted to lease the house for any commercial purpose. Also, they cannot permit 'any non-tribal to stay in his permanent/temporary shelter without any valid authority, in violation of PAT [Protection of Aboriginal Tribe] Rules'.

It can be concluded that in the post-Tsunami, needs of non-tribal households were not catered to, while for tribal Nicobarese the houses were made not only in accordance to their daily needs, but have also incorporated their cultural needs. Similarly, in case of provision of services, the non-tribal permanent house sites have not been provided with the basic services. Comparatively, the tribal Nicobarese settlements have more services that are functional. But most important difference is that the non-tribal households have only 'user' rights of the permanent houses, whereas, tribal Nicobarese have greater 'ownership' of the permanent houses. These differences are reflected and reinforced in the level of satisfaction felt by the non-tribal households and the tribal Nicobarese tuhets.

Satisfaction with the 'houses': a fieldwork enquiry

In the Tsunami most of the houses were washed away, thereby leaving no other option but to reconstruct the houses. About 10,000 permanent houses were reconstructed making the reconstruction of the post-Tsunami houses the largest sector of the TRP. The government of India, together with the A&N Administration and the NGOs, constructed these houses. The government declared that the 'new kind' of houses would be 'complete with all modern amenities'. It was the only consoling hope for the affected households and both the non-tribal households and the tribal Nicobarese tuhets eagerly waited for the permanent houses. Immediately after the Tsunami, households were promised that they could move into new houses within two years. But it took nearly three to four years. With each passing day, the expectations for permanent houses kept increasing.

But when they were handed over these houses, reality was different. Were the non-tribal households and tribal Nicobarese tuhets satisfied with the permanent houses? What was the difference in the level of satisfaction between the two? What was the reason for dissatisfaction? To investigate these questions, fieldwork was undertaken from May to August 2010. By this time households had spent one and a half to two years in the 'permanent houses' and formed their perceptions, about the houses. The duration of stay of the non-tribal households and tribal Nicobarese tuhets at the time of the fieldwork varied.

Out of the 65 non-tribal households, 11 households had lived in the houses for more than two years; 49 households had spent six months to a year as they were allotted the houses in August 2009 and the remaining five households had spent less than six months in these houses. This difference in the duration of stay was because of the speed of the construction of houses by different agencies. The NGOs were quick in construction than the APWD. The NGO named MAM completed the houses and handed over the first. The second were houses by HCC and CARE India. The duration of stay of the tribal Nicobarese tuhets did not have this variation. The houses were allotted at the same time to all tuhets of the village. The tuhets of Kimius village were allotted the houses in July 2008 while the tuhets of Arong village were allotted the houses in October 2008. During fieldwork, the tuhets of Kimius had spent slightly over two years while the tuhets of Arong had spent one year seven months in the houses. Thus, 95 out of 100 case study households had spent more than a year in the permanent houses during the fieldwork and had formed their perceptions about the permanent houses (Table 10.5).

Table 10.5 Duration of stay in the permanent houses

Duration of stay	*Non-tribal households*		*Tribal Nicobarese tuhets*	
	Number	*Per cent of total*	*Number*	*Per cent of total*
More than 2 years	11	16.92	15	42.86
1 to 2 years	0	0.00	20	57.14
6 months to 1 year	49	75.38	0	0.00
Less than 6 months	5	7.69	0	0.00
Total	65	100	35	100

Source: Based on fieldwork (2010)

Assessment of satisfaction with 'houses'

To assess the level of satisfaction, the non-tribal households and the tribal Nicobarese tuhets were asked to assign scores on a scale of 1 to 5 to the level of satisfaction with different aspects of the permanent house, where, '1' indicated complete satisfaction, '5' complete dissatisfaction and the state of indifference was pegged at '3'. The case study households were asked to assign satisfaction scores to six important aspects. These were location of the house, house design, type of building materials, quality of construction of house, services provided and cost of the unit. The scores were then grouped into three categories – satisfied, indifferent and dissatisfied. Scores of 1 and 2 indicated satisfaction, 3 indicated indifference while scores 4 and 5 indicated dissatisfaction (Figure 10.2). The scores represent the perception of the non-tribal households and the tribal Nicobarese tuhets and not the actual conditions.

Satisfaction with 'location': In terms of location, of the total 100 case study households, 38 were dissatisfied, 27 indifferent and 35 were satisfied. The dissatisfaction was high among the non-tribal

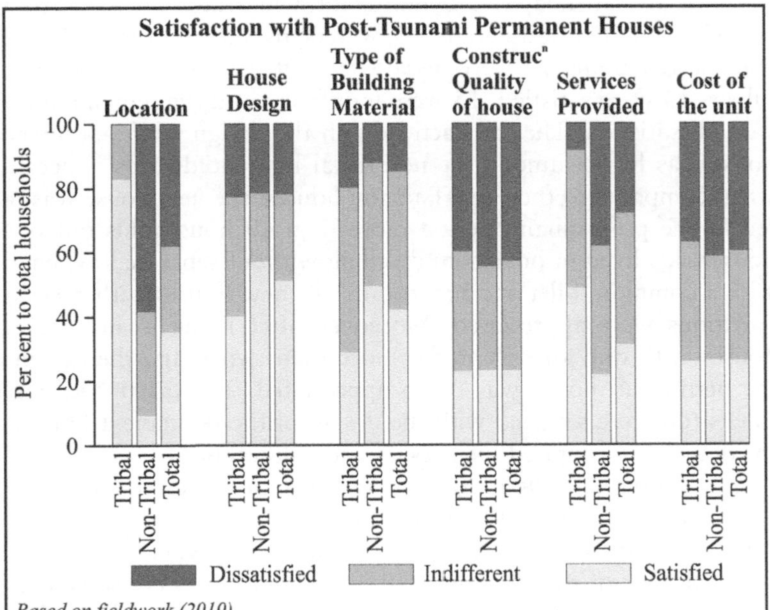

Based on fieldwork (2010)

Figure 10.2 Satisfaction with post-Tsunami permanent houses

households as compared to the tribal Nicobarese tuhets. Of the 65 non-tribal households, 38 were dissatisfied, 21 were indifferent and six were satisfied.

On the contrary, of the 35 tribal Nicobarese tuhets six were indifferent and 29 were satisfied. None of the tuhet heads said that they were dissatisfied with the location of the permanent houses. The reason for high dissatisfaction among non-tribal household and satisfaction among tribal Nicobarese tuhets is that the former had location on a hilltop away from the main hub of the city while the location of the latter faced no such problem. The root cause of problem was that the non-tribal households were not involved in site selected for the permanent houses, on the contrary, the location of permanent houses for tribal Nicobarese tuhets was selected in consent with tuhet heads, village captains and the Island Tribal Council. Even though for tribal Nicobarese accessing the sea became difficult and now they have to carry canoes to and fro between the settlement and the sea which are 5 km apart, they were satisfied with the location of the new settlement as it has been selected with the consent of their elders and leaders.

Satisfaction with 'design': The second aspect, the design of the house draws a satisfaction of half the sample. Of the 100 case study households 51 are satisfied, 27 are indifferent, while 22 are dissatisfied. Of the 65 non-tribal households, 37 were satisfied, 14 were indifferent and the remaining 14 were dissatisfied. Of the 35 tribal Nicobarese tuhets, 14 were satisfied, 13 were indifferent and the remaining eight were dissatisfied. The satisfaction with the design of the permanent house was higher among the non-tribal households, this is because, when compared to their pre-Tsunami houses, the new house was better. In the pre-Tsunami period more than 40 households out of 65 households lived in one room dwelling without separate kitchen and with a common toilet and bathroom. The new houses with necessary partitions were appreciated. Moreover, after living in intermediate houses with only one room for more than two years, the design of the permanent house was much appreciated. The tribal Nicobarese tuhets too were satisfied with the design of the permanent houses as the design incorporated their social and cultural needs.

Satisfaction with 'building material': The third aspect was the type of building materials with which 42 out of 100 case study households were satisfied, 43 households were indifferent, while 15 were dissatisfied. Out of the 65 non-tribal households, 32 were satisfied, 25 were indifferent and the remaining eight were dissatisfied. Of the 35 tribal Nicobarese tuhets, 10 were satisfied, 18 were indifferent and the remaining seven were dissatisfied. Most of the non-tribal households liked the exterior wooden walls and bamboo ply interior walls. The ones dissatisfied felt that entire house could have been made of concrete. They felt that wooden houses were flimsy and would decay while concrete houses are durable. The tribal Nicobarese tuhets on the other hand were indifferent. The 18 tuhet heads who were indifferent and seven who were dissatisfied said that when these permanent houses would decay and break down, they will set up their traditional houses. The reason for their indifference and dissatisfaction with the building material was that it was not durable. Within a year of occupation, the houses had begun to show signs of decay and breakdown. For example, the nuts and bolts used to put the iron superstructure had fallen off from couple of corners, thereby making the structure 'unsafe'. Moreover, these special bolts were unavailable in the market of Car Nicobar Island; therefore the people could not fix them up themselves.

Satisfaction with 'construction quality': The fourth aspect, construction quality of the house included the quality of building material and the assembling of the fittings. Out of the 100 case study households, only 23 were satisfied with the quality of construction, while 34 households were indifferent and 43 were dissatisfied. Again, out of 65 non-tribal households, 15 households were satisfied, 21 were indifferent and 29 were dissatisfied. Out of 35 tribal Nicobarese tuhets, eight were satisfied, 13 were indifferent and 14 were dissatisfied. The dissatisfaction was high because of the poor quality of building materials which was verified during the fieldwork. Among the houses of the non-tribal households, within a year of stay in the house, the steel frame rusted after the first monsoon, bamboo ply board was eaten by the rats and the cemented floor had developed cracks as the cement and sand mixture used was not in right proportion. The PVC pipes for harvesting rainwater were broken, the water tanks had leakage, the electrical and tap fittings were non-functional in more than 24 households. Again, the door handles were rusted and the window frame fittings had gaps from which the rainwater seeped in the house. In one household the fan fell from the ceiling. Compounding the problem for the non-tribal households was that there were no workers nearby the settlements who could mend the parts. Electricians, plumber, carpenters if called from the main market charge extra money to visit these settlements as these are not connected with the main market. For tribal Nicobarese tuhets the problem is worse, as there are no trained workers in Car Nicobar Island who can fix these new kind of houses.

Satisfaction with 'services provided': The fifth aspect is services provided. Of the 100 case study households 31 were satisfied, 41 were indifferent and 28 were dissatisfied. There is a sharp contrast between non-tribal households and tribal Nicobarese tuhets. Out of 65 households of the former, 14 were satisfied, 26 were indifferent and 25 were dissatisfied. On the other hand out of the 35 tribal Nicobarese tuhets, 17 were satisfied, 15 were indifferent and only three were dissatisfied. This is because functional amenities and services, such as primary health sub-centre, anganwadis, school, market, ration shop, community hall and others was not provided to the non-tribal households. Again, roads were not constructed and ones constructed were of poor quality that developed potholes after the first monsoon. Settlements of the non-tribal households were located on hill tops and thus required soil retaining wall. But this was not provided. The slopes are devoid of vegetation and rain washes off loose soil and floods houses located at lower terrace. This is a persistent problem as the A&N Islands receive heavy rain for nearly seven months in a year. On the other hand, the tribal Nicobarese tuhets have functional amenities and services. All services like the primary health sub-centre, anganwadis, school, community hall, recreation centre, ration shop, birth house and a death house were provided in each village. A shopping complex too has been constructed in each of the tribal Nicobarese village, however, it is non-functional as the villagers prefer to go to the headquarters (the central point of the island) where goods are available at cheaper rate and they also get some variety. Moreover, there is a regular bus service which connects the village to the headquarters. The ease of travel also discourages villagers to open a shop in the 'shopping centre' provided by the administration at each village. Tribal Nicobarese prefer to shop from the market. Thus, the tribal Nicobarese tuhets were satisfied with the services provided.

Satisfaction with 'cost': The sixth aspect of the permanent house was its cost. The average cost of the permanent house was Rs. 1.4 million. Forty out of 100 case study households were dissatisfied with the cost and felt that better houses could have been built at lesser cost. Another 34 households were indifferent. For them receiving a house was important, not its cost. These were largely tenants who did not 'own' a house in the pre-Tsunami period. The remaining 26 households were satisfied. They opined that Rs. 1.4 million for the house was genuine as cost of transporting the materials from mainland India to the A&N Islands was high. Among the 65 non-tribal households, 17 were satisfied, 21 were indifferent and 27 were dissatisfied with the cost of the permanent houses. Thus, the non-tribal formed the bulk of dissatisfied households. On the contrary, among the 35 tribal Nicobarese tuhets nine were satisfied, 13 were indifferent and another 13 were dissatisfied with the cost of the house.

From the above discussion it can be concluded that satisfaction with the permanent houses varied among the households. A factor determining this variation was pre-Tsunami houses of the people. The non-tribal households, who lived in slums before the Tsunami, are satisfied with the new houses, while those who had their own houses are dissatisfied. The tribal Nicobarese tuhets are satisfied with the new permanent houses. The non-tribal households had improvised 'permanent houses' as per their requirements even though by doing so they have breached the terms and conditions of the permanent house agreement. Of the 65 non-tribal households 96 per cent had modified the permanent houses allotted to them, by extending overhang of the roofs and added a room by covering the backyard. The tribal Nicobarese tuhets, on the other hand, have not modified the house structure. However, they have plans to reconstruct new houses when the permanent houses fall apart in due time, of which they are very sure.

Critique of the post-Tsunami response to housing

The post-Tsunami housing reconstruction was the most important aspect of the TRP. The reconstruction of over 19,000 houses (9,565 intermediate houses + 9,797 permanent houses) across A&N Islands was a feat for A&N Islands' Administration. However, interviews conducted with 100 households reveal a low level of satisfaction. There are certain aspects on which the post-Tsunami housing reconstruction was unsuccessful.

Delay

First is the delay in the reconstruction of permanent houses. Intermediate houses came up within four months of the Tsunami, while the first set of permanent houses could be completed only after three years of the Tsunami. The entire process of reconstruction and shift to permanent houses took over seven years for the entire A&N Islands. The delay led to frustrations among the households living in the intermediate houses. The intermediate houses were made to endure for a maximum two years, but large part of the affected households had to spend more than three years in it.

Poor condition of houses

The permanent houses provided are not suited for the climatic conditions of A&N Islands. The overhangs are small, even though A&N Islands receives rainfall for more than seven months in a year. After the delay in reconstruction, the people expected sturdier houses; instead the houses provided have begun to fall apart within three years of habitation. The permanent houses do not cater to the needs of the non-tribal households. The houses are so weak that nails cannot be drilled in it. Small racks are a necessity, which was not provided. The households felt that the houses were made in hurry, and they also commented that 'why doesn't the administration builds houses like these for their employees, if these are the best kind of house'?

People not involved in the reconstruction

The plan and design for the permanent shelter was prepared by the engineers of the Central Public Works Department (CPWD), who had not lived in the A&N Islands and were not aware of the ground realities such as climatic conditions and everyday needs and requirements of the population. As the reconstruction process lacked people's participation, the houses that have been constructed, are not 'suitable' for the people as it does not take into account the needs of the people living in it. The design was prepared by the CPWD, a department under the Ministry of Urban Development. The irony is this that out of the total 9,797 permanent houses constructed only 541 (i.e. 5 per cent of the total) were in the urban area while the remaining 95 per cent of the houses were in rural area. Yet, instead of Ministry of Rural Development, it was the Ministry of Urban Development which was entrusted with the charge to reconstruct the permanent houses. Thus the houses reconstructed are ill-suited for its inhabitants.

The process of reconstruction was handed over to private engineering companies, thereby further alienating the people from the housing reconstruction project. All constructing agencies, be it the CPWD, the APWD or the NGOs, leased out the work to engineering companies on the mainland India. These engineering companies brought in their skilled and unskilled workers, materials and the know-how. Thus, people for whom the houses were reconstructed were left out from the reconstruction process. The non-tribal households were neither consulted nor engaged in site selection, house design, layout of the settlement, choice of building material or in any part of the construction process. The involvement of the non-tribal households was limited to providing labour for the construction only.

The involvement of the tribal Nicobarese was more than the non-tribal. The Tribal Council was engaged in selection of the site which was based on the recommendations of the Department of Science and Technology (DST). The prototypes were constructed solely by the

CPWD engineer. These designs were rejected by the tribal Nicobarese leaders. However, later they accepted the same design after the persuasion by the administration with minor modifications. One of the tribal leaders from Car Nicobar Island remarked that 'they (the administration) were taking so long in finalising a design, therefore, we accepted. We thought if houses were uncomfortable, we would make new houses as per our needs. As of now these houses are good'. Only the tribal leaders were involved in the process of site selection and approval of design of the permanent houses. Thus, at household level neither the non-tribal households nor the tribal Nicobarese were involved in the reconstruction of the houses.

New vulnerabilities

The permanent houses have been constructed in such a way that future tsunamis cannot affect them. The sites for the permanent houses were selected following the guidelines of the DST. The house designs too were prepared under the Bureau of Indian Standards guidelines. Thus, in principle, house structures are resistant to ground shaking during earthquake, high velocity wind during cyclones and to tsunami inundation of less than 5 km. But in reality the structures have become susceptible to earthquakes, cyclones, fire and water logging.

The first reason for this susceptibility is non-durable building material and lack of maintenance. The post-Tsunami houses differ from the traditional houses. The non-tribal households and the tribal Nicobarese tuhets are not familiar with these new structures and are unable to do minor repairs. Humidity in the A&N Islands is high, because of which the main iron structures, on which the bamboo ply boards are fastened, have rusted, so nuts and bolts have fallen, while rats have made holes in the bamboo ply walls. The modifications made to the houses such as extension of overhang of roof and addition of rooms have further weakened the houses. The weakened structure can collapse during an earthquake.

The building materials used for permanent houses are vulnerable to fire. The primary building materials of these houses are wood and bamboo ply board. Both are inflammable. Coupled with it the concealed electrical wiring system with poor quality cables add to the hazard. Within three years of construction of permanent houses, incidents of fire were reported from the permanent houses. Two incidents occurred in Katchal Island where the permanent houses caught fire and burned down within minutes. The first incident occurred at Japan Tikery on 7 August 2010 at 7:15 p.m., within six months of stay in the house. The second incident occurred at East Wall Kapanga on 20 November 2011 at 8:35 p.m. Another incident occurred in Car Nicobar Island in 2014. The incident took place at Lapathi Village, Car Nicobar Island, on 8 February 2014 at 7:26 p.m. Although no one was killed, property worth nearly Rs. 8 million was charred in these three incidents together. In all these three incidents of fire, the cause was short-circuiting. The fear of fire was expressed by the non-tribal households and the Nicobarese tuhets alike. They stated that perhaps the permanent houses can protect them in case of a tsunami, but these houses are vulnerable to fire, made of wood with poor electricity wiring. Out of the total 100 case study households, 92 felt that the structures are fragile and will not be able to withstand an earthquake of great intensity, which are common in A&N Islands.

Apart from these, there is also the threat of epidemics. Fieldwork reveals that there is no waste disposable system in any of the four sampled permanent shelter sites (Austinabad and Bambooflat in South Andaman Island and Arong and Kimius in Car Nicobar Island). All the four case study locations are surrounded by forest in which the households dump the biodegradable as well as non-biodegradable garbage together. This is not only polluting the environment, but also has become the breeding grounds for pests. Risk of epidemics exists at the permanent house sites. Thus, it can be concluded that even though permanent houses seem to protect from future tsunamis, as these are located inward from the coast, types of raw materials used and design have rendered them vulnerable to cyclones, fire, earthquake and epidemics. New houses have reduced vulnerability on one end but rendered it on the other dimension.

Recovery of the Tsunami-affected households

The dictionary defines 'recovery' as an 'act of regaining something lost or stolen'. As per this definition, recovery for this research has been defined as 'return to improved or similar living conditions that existed pre-Tsunami'. As the fieldwork was conducted five and a half years after the Tsunami, the length of time gave adequate scope to assess recovery.

Recovery is a word encompassing many tangible and intangible dimensions. While it is difficult if not impossible to recover from the trauma associated with intangible losses such as, grief of the loss of beloved family and friends, sentiments associated with the damage of the house, loss of value of the 'land', disruption of lives and livelihoods as a result of the damage of critical infrastructure or the deterioration of quality of life, to name a few, in this research the aspect of emotional recovery or recovery of intangible losses has not been taken into consideration. Tangible recovery is measurable as it can be gauged through recovered tangible items such as house, shop, electronic items such as television and refrigerator, furniture such as table, chair, sofa etc., or recovery of livelihood such as stable sources of income from diverse activities.

The assessment of recovery was based on the interviews with heads of the households among the non-tribal households and with the heads of the tuhets among the tribal Nicobarese. The drawback of this method is that there is no way to verify the data, especially of the pre-Tsunami period. Thus, this assessment is solely based on the data and information provided by the head of the households. As the assets and resources of the non-tribal households and the tribal Nicobarese households varied, therefore, the indicators selected to assess recovery differed for both of them. Based on pre-and post-Tsunami analysis of ownership of goods and resources, recovery score for non-tribal households and tribal Nicobarese tuhets was calculated.

Recovery of assets among non-tribal households

In the pre-Tsunami period, the non-tribal households possessed several assets. To assess the recovery of non-tribal households, a comparison between pre-and post-Tsunami ownership of 10 assets was sought. These assets were selected after a reconnaissance was done to find out the material assets owned by the households pre-Tsunami. These were (1) telephone; (2) mobile phone; (3) radio; (4) television; (5) refrigerator; (6) other electronic gadgets like DVD player, mixer-grinder, washing machine, computer/laptop and vehicles such as (7) bicycle; (8) scooter/motorcycle; (9) car/jeep/SUV and (10) auto-rickshaw. Another reason for selecting only these items was that these were not donated either by the NGOs or the A&N Administration. The ability to repurchase the lost and damaged assets indicates recovery as the households themselves had to purchase these.

To assess recovery, two basic assumptions were drawn. First, all items were purchased by the households and were not gifted by anyone. Second, the quality and the cost of purchase of items by all households was the same. Given this, four possible situations emerged for the non-tribal households and a score was assigned to each of the envisioned scenarios: (1) those that possessed the asset before the Tsunami and have repurchased it after the Tsunami was assigned score 1; (2) possessed the asset before the Tsunami but have not purchased it after the Tsunami was assigned score 0; (3) did not possess the asset before the Tsunami but has repurchased it after the Tsunami was assigned score 2; and (4) did not possess the asset before the Tsunami and has not purchased it after the Tsunami was not assigned any score (Table 10.6). A score of 2 was assigned to the third category because it reflects the case of relatively 'more recovery' as it did not possess the item pre-Tsunami. The fourth case, the households did not own the asset pre-Tsunami nor did they own it post-Tsunami, therefore, no score was assigned as the case was not an indicator of recovery in this case.

The cost of these 10 selected items ranged from a few hundred rupees to thousands of rupees. For example, a simple music system (the cheapest one) was for Rs. 200 while a small car costs slightly more than Rs. 300,000. The minimum costs of other items ranged within these extremes. Therefore, for each household, a minimum cost was assigned to each item which was multiplied with the assigned score for each of the households. To calculate the final recovery, the costs were added. The 65 non-tribal households were then grouped into three categories of costs of assets recovered – high, moderate and low recovery. The households that have repurchased more assets thus have high recovery of tangible assets.

Of the 65 non-tribal households, 20 (31 per cent) are in the category of high recovery, 16 (25 per cent) in moderate recovery and 29 (45 per cent) in low recovery. Therefore, 70 per cent of the non-tribal households are in moderate and low recovery categories. The scores are relative, i.e. a case of high recovery refers to relatively higher recovery of assets by the households than other households. The high incidence of moderate and low recovery of asset is an indicator that the large sections of the non-tribal households are still reeling under the impact

Table 10.6 Scores assigned to the recovery of items by the non-tribal households

Situation	*Score*
Possessed the asset before the Tsunami and have repurchased it after the Tsunami	1
Possessed the asset before the Tsunami but have not purchased it after the Tsunami	0
Did not possess the asset before the Tsunami but have purchased it after the Tsunami	2
Did not possess the asset before the Tsunami and have not purchased it after the Tsunami	NA

Source: Scores assigned by the author

of the Tsunami. These households have not even managed to reach the stage they were before the Tsunami. The case with tribal Nicobarese was different. The assets of tribal Nicobarese tuhets vary from that of the non-tribal households, the methodology for assessing the recovery too varied.

Recovery of resources of tribal Nicobarese tuhets

The wealth of the tribal Nicobarese tuhets is their resources. They are dependent on natural resources such as sea, forest, coconut groves, pigs and chicken. Therefore, to assess the recovery among the tribal Nicobarese, present-day resources as percentage to pre-Tsunami total resources was calculated. In the Tsunami, coconut groves were destroyed, canoes lost and livestock were killed in large number. Hence, to measure recovery of tribal Nicobarese tuhet, three resources were included – number of coconut groves, number of canoes and livestock. As more than 80 per cent of the resources of the tribal Nicobarese were lost in the Tsunami, tuhets that had recovered 50 per cent of their pre-Tsunami resources were put in the category of high recovery. Tuhets with 30–50 per cent of their pre-Tsunami resources were put in moderate recovery and tuhets with less than 30 per cent of the pre-Tsunami resources were put in the category low recovery. When data were put together it revealed that of the 35 tuhets, 12 (34 per cent) were in high recovery category, while nine (26 per cent) and 14 (40 per cent) were in moderate and low category of resource recovered respectively. In this case too, two-thirds of the tribal Nicobarese have not been able to recover their pre-Tsunami resources. Recovery of resources is nature-dependent and the tribal Nicobarese hope that their resources stock would replenish in another 10 years.

Though among both non-tribal households and the tribal Nicobarese tuhets, large percentage is in moderate to low categories, yet there are subtle differences between the two (Table 10.7).

Lower percentage of non-tribal households are in the category of high recovery when compared to the tribal Nicobarese tuhets. Among the former, 31 per cent are in category of high recovery, while 34 per cent of the tribal Nicobarese are in the same category. In contrast, a higher (45 per cent) percentage of non-tribal households is in the category of low recovery than tribal Nicobarese tuhets (40 per cent). Thus, comparatively the non-tribal households have recovered less than the tribal Nicobarese tuhets. The following section assesses the recovery process and puts forth reasons for the differences between the recovery of the non-tribal households and the tribal Nicobarese tuhets.

Table 10.7 Repurchased assets/recovered resources as indicator of recovery

Category of recovery	*Non-tribal households*			*Tribal Nicobarese tuhets*		
	Cost recovered (in Rs.)	*Total households*	*Per cent of total*	*Per cent resources recovered*	*Total tuhets*	*Per cent of total*
High	More than 50,000	20	30.77	More than 50	12	34.29
Moderate	30,000–50,000	16	24.62	30–50	9	25.71
Low	Less than 30,000	29	44.62	Less than 30	14	40.00
	Total	**65**	100.00	**Total**	**35**	100.00

Source: Based on fieldwork (2010)

A comparison of recovery process

There is a difference between the recovery of the non-tribal households and the tribal Nicobarese tuhets. While the non-tribal households are still recouping from the disaster, the tribal Nicobarese have recovered better, even though the latter were the worst affected in the disaster. Among the 100 case study households, the tribal Nicobarese tuhets accounted for 98 per cent of the total amount received as compensation. There was also an indirect participation of the tribal Nicobarese tuhets in the house reconstruction project. The fieldwork also proved that the tribal Nicobarese have gone back to their original year-round activities. They have resumed their monthly and annual festivals. Their families have expanded by marriage and natural births. This indicates that the tribal Nicobarese population recovered in the face of Tsunami.

The recovery of the tribal Nicobarese tuhets can be attributed to their 'perception' of life. For tribal Nicobarese, the Tsunami was a part of life. One of the evidences of this perception is a wedding that was solemnized even in the face of tragedy. A wedding in Mus Village of Car Nicobar Island was scheduled for 4 January 2005. Even though the wedding was only nine days after the Tsunami, it was solemnized (Church of North India 2006). In the Tsunami, all was lost, there was no electricity and there was acute shortage of water, food, medicines and clothing. Even then, the wedding was held since it was scheduled before the Tsunami. Though the tribal Nicobarese were mourning the loss at that time, they thought moving ahead was equally important. In comparison, the non-tribal households did not and could not adapt well to the situation. A contrast is evident thus: a marriage in the non-tribal household from Nicobar Island was scheduled for February 2005, but it was postponed and solemnized in November 2006. Fieldwork brought forth that there are three main reasons that contributed to recovery of the tribal Nicobarese tuhets when compared to the non-tribal households – dislocation, the tuhet system and strong leadership.

Dislocation

The non-tribal households from Nicobar Islands had to move greater distance and to a different island when compared to either the non-tribal households from South Andaman Island or the tribal Nicobarese tuhets, who moved within the same island. The dislocation placed the non-tribal households from Nicobar Islands in new surroundings with new neighbours. The households who had no 'secured government jobs' had to scout for new sources of livelihood. In a new location with lack of funds to begin new businesses, the small traders were unable to resume their livelihood activities. In Nicobar Islands they had their niche customers and were successful in their small endeavours. Building the market again in competitive area was very difficult. The recovery of the 38 non-tribal households from Nicobar Islands was low as they have been uprooted and dislocated. The remaining 27 non-tribal households too were dislocated, but as these households remained within their pre-Tsunami islands, they remained in familiar surroundings. For these households only the location of the houses and the type of houses changed. But they pursued their pre-Tsunami activities.

The least affected by dislocation were the tribal Nicobarese tuhets as they were relocated on the same island, 4–5 km away from the coast. They remained in their 'own' habitat within their own families and resources. It was relatively lesser hardship for them once the jetties, roads and houses were reconstructed. All resources were not lost in the Tsunami. Coconut groves in the interior of the island were not affected while some pigs survived within the forests. Moreover, the tribal Nicobarese received ex-gratia compensation. Once the resources are regenerated, it will be easier to resume their pre-Tsunami activities of copra production, pig rearing and fishing. The predominant cause of better recovery among the tribal Nicobarese can be attributed to the strong tuhet system.

Tuhet system

The non-tribal households live in nuclear families, with five members on an average. Of these five, two to three are children. In these, the responsibility to look after the family rests mainly on one member.

From being enumerated in the list of the 'Tsunami-affected' to claiming compensation and the house, each head of household had to file an application individually. The security provided by a joint family was missing among the non-tribal households.

On the other hand, the tribal Nicobarese live in a joint family system called a tuhet. Fieldwork shows that a tuhet consists on an average 25 to 30 members. These large joint families are a social safety net which allowed better recovery. In the Tsunami some parents lost children, while some children were orphaned. Fieldwork confirms that in such cases children were adopted within the tuhet and the village. As all tuhets were equally affected in the Tsunami, the grief was shared by all members: 'cohesive community, family systems, social support, altruistic behaviour of the community leaders, and religious faith and spirituality were factors that helped survivors cope in disaster' (Math et al. 2008). Associated with the system of joint family was the strong leadership of the village captains.

Strong leadership

The non-tribal households lacked strong leadership in contrast to the tribal Nicobarese, where the tribal council and the village captains played a crucial role in the reconstruction of houses. There are two reasons for the lack of strong leadership among non-tribal households. First, the non-tribal households were diverse in occupation and in needs and also came from different islands – therefore, they lacked a leadership. Second, small groups were formed among the non-tribal households, each with a different set of problems. The personal differences are such that the households could not unite to put forward their needs. Thus, without leadership the non-tribal households were left out of the reconstruction process. On the other hand, the strong leadership was the reason for greater participation of the tribal Nicobarese.

In the post-Tsunami period the Tribal Council and the village councils played an important role in rebuilding Nicobar Islands. The crucial work of the village councils immediately after the Tsunami was to prepare the list of persons killed, houses damaged, livestock missing and coconut groves destroyed. Based on this list, the relief and rehabilitation was carried out. The Tribal Council was the key channel for the flow of relief items and compensation to the tribal Nicobarese tuhets. The Tribal Council was the mediator between the administration and the tribal Nicobarese tuhets. It ensured that no one was left out of the relief and rehabilitation process. Individual tuhets did not have to fend for themselves. For the reconstruction of houses too, the Tribal

Council was closely involved in site selection, clearing of land and distribution of houses. Approval of the house design was sought from the Tribal Councils before the construction could commence.

The non-tribal households had created no common representatives and each household had to seek the claim individually. Thus, a presence of a strong system of Tribal Council and the tuhet system allowed tribal Nicobarese to be involved in the reconstruction and rehabilitation processes, while non-tribal lacked a common representative and thus missed out not only on compensation, but were disengaged from the reconstruction process. Unfortunately, fieldwork points to the fact that a social system which provided safety net to the tribal Nicobarese shows signs of disintegration because of the response to the Tsunami. Following is the case study.

The ex-gratia compensation across India is paid to the head of the household of a nuclear family, who in most cases is a male. This decision is in conflict with the Nicobarese tuhet system, where the tuhet is the immediate family. All resources, such as coconut groves, vegetable gardens, canoes, pigs, traditional beehive houses, and other material possessions belong to the tuhet and all work, economic and domestic responsibilities were distributed among the tuhet members and all benefits too were for the entire tuhet. As the administration required a name list of nuclear families with a head of household and his wife and children for distribution of compensation, it requested for the same from the Tribal Council. The tuhet heads thus prepared the list with a male head of household, his wife and children as well as 'adopted' children. The total damaged resources were then divided according to the number of members in each of the nuclear families within the tuhet. The compensation and number of houses were received as per this list. The 35 case study tribal Nicobarese tuhets were segregated into 430 nuclear families.

In the post-Tsunami period, the financial resources and the material benefits are individually owned. As the nuclear households were given money, spending on material goods increased. The tribal Nicobarese have started to spend on purchasing alcohol, motorcycles, mobile phones and electronic appliances. Thus, most of the fund has been used up for consumption needs rather than for investments. Even the items of consumptions cannot be completely utilized as petrol and diesel is in short supply in the Nicobar Islands and hence there is no electricity for most part of the day (as electricity is generated using the diesel power generating sets). In the pre-Tsunami period, the expenditures were only on items required and used by the entire tuhet. But in the post-Tsunami period, it depends on the individual nuclear households. This has led to the breakdown of the age old tuhet system which helped the tribal Nicobarese in their recovery post Tsunami.

The impact of the Tsunami was so great that it also led to the change in the disaster management system of India.

Disaster management in India in the post-Tsunami period

The Tsunami-affected the disaster response mechanism of India, and therefore, that of A&N Islands. Disaster response post the IDNDR got a boost from the Hyogo Framework for Action taken up during the World Conference on Disaster Reduction organized by the United Nations International Strategy for Disaster Reduction (UNISDR) at the 10th Anniversary of the Kobe Earthquake of 1995 at Kobe, Japan. The framework for action had three components. The first was integrating disaster risk reduction (DDR) into sustainable development policies and planning. Second, developing and strengthening institutions, mechanisms and capacities to build resilience to hazards. Third, incorporating risk reduction approaches into emergency preparedness, response and recovery programmes. As India was a signatory, it abided by the framework. The decisions affected the disaster response mechanism of the A&N Islands. Tsunami was the catalyst which accelerated the institutionalization of the response mechanism. The Disaster Management Bill was introduced in the Rajya Sabha on 11 May 2005 and was passed by it on 28 November 2005. The Lok Sabha passed the bill on 12 December 2005 and it got the consent of the President on 12 January 2006 and became an Act of the Constitution.

As per the Act, the disaster management system came into force on 25 July 2008 in A&N Islands. The first measure was to set up the Union Territory Disaster Management Authority and District Disaster Management Authorities under the chairmanship of the lieutenant governor and district magistrates, respectively. There is one Union Territory Disaster Management Authority and three District Disaster Management Authorities, one each for (1) North and Middle Andaman, (2) South Andaman and (3) Nicobar. Emergency operation centres were also set up for dissemination of warning to the public and to monitor the disaster situation in their respective areas. The State Emergency Control Room has been set up in the secretariat in Port Blair.

The directorate has been entrusted with the responsibility of disaster management. To begin with, it has installed the disaster warning sirens. The first siren was put up at the secretariat in Port Blair in March 2010. During a disaster the siren blows at volume continuously at regular intervals. Two other sirens were installed on 1 May 2012: one at the district commissioner's office in Car Nicobar Island and the other at the assistant commissioner's office in Kamorta Island. The sirens were put to the test on 11 April 2012 when a powerful earthquake of magnitude 8.6 occurred off the coast of Sumatra. A tsunami warning was also issued. However, the people of A&N Islands complained that they were given no information whatsoever by the authorities as to what to do once the siren sounds. Their source of information was the national news channels which relayed that a tsunami alert has been declared for A&N Islands and people were requested to move away from the coast. Some people (in Port Blair) living on or near the coast in Port Blair did not hear the siren even though it was blown continuously for an hour. The siren is located at the secretariat which is far from the coast. People in Nancowry Island reported that the people from the administration arrived in the evening and asked them to stay away from the coast even after the tsunami warning was lifted. The non-coordination between Port Blair and remote islands was evident during the period. Another Disaster Management Plan for A&N Islands was prepared in 2012. It 'concentrates on the institutional set up and information flow for providing hazard specific responses along with the roles of primary agencies involved' (A&N Administration 2012). This plan also does not look at people's vulnerability.

The Tsunami: a missed opportunity

The Tsunami was an unprecedented global disaster directly affecting 14 countries, India being the third in terms of number of people killed. The Tsunami brought the A&N Islands to centre stage as the point of origin for the Tsunami was 300 km south-south-east of Great Nicobar Island. The world for the first time looked at A&N Islands from a different view other than 'pristine tropical islands with white sandy beaches and turquoise blue waters'. The event was highlighted by the world media which generated a response across the world. In India five states and union territories were affected. The A&N Islands were completely devastated and links with the remote islands were severed. The armed forces were called and they carried out the Operation Sea Wave, which to date is the largest peace time operation of the armed forces in which all three forces – army, navy and the air force – along with the coast guard and central para military forces participated. The armed forces transported nearly 2,000 million tons of relief materials to the affected islands of Andaman and Nicobar. Two hundred and seven relief camps were set up which housed 46,023 persons. Relief items were distributed. The Nicobar Islands were worst affected and the relief agencies were denied entry to the islands.

In the post-Tsunami period, the funds for the ex-gratia payments were received under the RGRP which was declared in January 2005 and for the reconstruction and rehabilitation under the TRP which was initiated in February 2005. The TRP was declared closed in December 2011, seven years after the Tsunami. It had two components – reconstruction and rehabilitation. Reconstruction of houses and infrastructure accounted for 82 per cent of the total funds allocated under the TRP. Tourism development, coastal protection and miscellaneous expenditures accounted for 4 per cent of the total. Thus, rehabilitation accounted for only 14 per cent of the total funds allocated under the heads of agriculture and fisheries. Ex-gratia compensation for loss of land, crop and compensation for boat damage was paid out of this 14 per cent, reducing the amount of rehabilitation further.

Spatially, Nicobar Islands received two-thirds of the total funds for reconstruction, while Andaman Islands received one-third of it. In case of housing reconstruction, Nicobar Islands received the lion's share. Seventy-one per cent of the total expenditure on reconstruction of housing was incurred in Nicobar Islands. As more houses were damaged in Nicobar Islands, more houses were constructed there. However, in case of infrastructure reconstruction, both the Andaman and Nicobar Islands received near equal allocations. This is because in the pre-Tsunami period Andaman Islands had the infrastructure that was damaged. In Nicobar Islands, the pre-Tsunami infrastructure was limited hence post-Tsunami infrastructure reconstruction too was limited.

From secondary data and fieldwork carried out five years after the Tsunami, this research concludes that reconstruction of infrastructure and houses has taken place in A&N Islands. However, the focus was on reconstructing what was damaged in situ instead of taking the Tsunami as an opportunity to rebuild islands equitably. Take the case of jetties of Nicobar Islands. Before the Tsunami, out of 12 inhabited islands of the Nicobar, cargo handling jetties were present only on two islands, while the remaining 10 islands had only ferry jetties. The status quo was maintained and the cargo handling jetties were reconstructed in islands, where they were damaged. The other jetties were reconstructed as ferry jetties and were not upgraded as cargo handling jetties. Similar pattern was followed for social infrastructure reconstruction. It included construction of schools, primary health centre and primary health sub-centres. The focus of reconstruction in Nicobar Islands was building back the infrastructure damaged rather than upgrading and expanding the services or improving the quality of the services provided. The reconstruction was focused on the islands with dominant non-tribal population such as Great Nicobar Island, Car Nicobar Island and Katchal Island. The other islands were not equally 'developed'.

One of the core components of the TRP was the housing reconstruction as 46 per cent of the total funds under TRP was allocated for housing reconstruction. Nearly 10,000 houses damaged in the Tsunami were reconstructed. The administration decided to construct intermediate houses before the onset of monsoons of 2005 as construction of large numbers of permanent houses would take more than two years period. Thus, a total of 19,362 (9,565 intermediate houses and 9,797 permanent houses) were constructed over a period of seven years.

Before being resettled in permanent houses, on an average, the households had to change location and type of shelter twice – first relief camp and second the intermediate shelter. However, the non-tribal households from Nicobar Islands had to shift between relief camps, and therefore, have changed location four times before being allotted the permanent houses. The intermediate shelters were put up in six months, but it was made of CGI sheets. In the fieldwork the households recalled that these intermediate houses were 'heat traps' and burned like a 'furnace' during summer while during the rains the roofs leaked. Though these houses were made to last only for two years, on an average the households spent more than four years in these intermediate houses. The permanent houses had other problems. Constructed of wood, iron poles and bamboo ply for living room and bedroom and concrete for kitchen and bathroom, the houses had different set of problems. The quality of construction of the houses is not good and the houses have developed several problems such as leaking roofs, rusted iron poles, cracks on the cemented floors, broken doors and windows, non-functional lights and fans, to name a few. The new houses though are protected from future tsunamis, yet have become vulnerable to different hazards such as fire, water-logging, cyclone and earthquakes.

A facet of response is recovery. Hence assets were selected as indicators to assess tangible recovery. For 65 non-tribal households the material assets were taken as indicators, while for tribal Nicobarese tuhets, resources were considered. Households are at different stages of recovery. However, more non-tribal households are in the category of low recovery than the tribal Nicobarese tuhets. A higher level of recovery of the tribal Nicobarese tuhets can be attributed to three factors – first, the tribal Nicobarese were not dislocated to different islands.

Remaining on 'their own' island gave them the freedom to rebuild as per their own needs. Second, the pre-Tsunami system of tuhets among the tribal Nicobarese acted as a safety net at the time of crisis. Third and most significant factor contributing to the recovery of the tribal Nicobarese was the leadership provided by the tribal council. Because of strong leadership provided by the tribal council and the village captains, receiving compensation was a smooth process for the tribal Nicobarese. Also, they had greater role in reconstruction of houses.

The lesson which one can draw from the Tsunami response in A&N Islands is that people require different types of assistance and in different forms. The one-size-fits-all approach does not aid in rehabilitating the affected people. The assistance provided or the schemes made for an area should be geographically rooted. Superimposition of mainland ideas adversely affects the vulnerability. Though A&N Islands were reconstructed after the Tsunami, it was a missed opportunity to reduce vulnerability. The vulnerability persists, though in a different form.

Appendix

Terms and conditions for the use and occupancy of the permanent houses

Andaman Islands

1. That the shelter is constructed by the Administration for the families affected by Tsunami of 26th December, 2004. The ownership of the shelter vests in the Administration.
2. The ownership of land on which permanent shelter is situated shall continue to rest with the A&N Administration.
3. The allotte shall occupy the permanent shelter immediately.
4. The allotte should apply for electricity connection from the Electricity Department at his/her expenses.
5. With allotment of permanent shelter, the allotte is requested to shift to this shelter from intermediate shelter in his or her occupation. Electricity Department will provide electric supply to the permanent shelter, disconnecting supply to the intermediate shelter without any notice or extension of time to the allotte.
6. The water charges etc. to the permanent shelter shall be borne by the allotte at his or her own expenses.
7. That the beneficiaries shall not incorporate any alterations, modifications or additions to the shelter structure.
8. On allotment, the permanent shelter shall be the asset of the allotte thereby any maintenance/repairs etc. to be met by the allotte himself or herself.
9. The allotte and his family members if any who are temporarily residing in the Govt. Accommodation shall be bound to vacate the Govt. Quarters on the allotment of permanent shelter. He or she shall furnish a declaration to this effect.
10. The allotte shall not transfer the ownership of the permanent house by way of lease, sublet, sale or through any other legal instrument without prior permission of the A&N Administration.

11. The allotte shall, in no case, permit other people to stay in the Permanent/Temporary shelter on lease/rented basis for any commercial purposes.
12. That the permanent shelter is allotted for the residential purpose and it should not be used for any purpose other than allotted for.
13. That the beneficiaries shall maintain the surrounding area neat and tidy and maintain a cordial relationship with the adjacent shelter owners.
14. That in case of death of both the beneficiaries, the son/daughter of the deceased shall be entitled to continue the possession of permanent shelter.
15. That in case of death of the beneficiary with no legal heirs, the permanent shelters will be taken over by the Administration.
16. That the administration reserves the right over the permanent shelter and it shall cancel the allotment at any time by giving one month notice, if the beneficiary had violated any of the terms and conditions specified above.
17. That in case of the beneficiary fails to vacate the permanent shelter after cancellation of the allotment, the possession of the permanent shelter shall be got vacated under the provisions of Public Premises (Eviction of unauthorized occupants) Act 1971.

The terms and conditions are signed by the beneficiary – it is then counter-signed by the DC of the South Andaman District.

Nicobar Islands

1. The allottee shall occupy the Permanent Shelter immediately.
2. The allottee should apply for electricity connection from the Electricity Department at his/her expenses.
3. With allotment of permanent shelter, the allotte is requested to shift to this shelter from intermediate shelter in his or her occupation. Electricity Department will provide electric supply to the Permanent Shelter, disconnecting supply to the intermediate shelter without any notice or extension of time to the allotte.
4. The water charges etc. to the permanent shelter shall be borne by the allottee at his/her own expenses.
5. On allotment, the Permanent Shelter shall be the asset of the allottee thereby any maintenance/repairs etc. to be met by the allotte him/herself.
6. The allottee and his family members if any who are temporarily residing in the Govt. Accommodation shall be bound to vacate

the Govt. Quarters on the allotment of permanent shelter. He/She furnish a declaration to this effect.

7. The allotte shall, in no case, permit other people to stay in the Permanent/Temporary shelter on lease/rented basis for any commercial purposes.
8. The allottee shall not allow any non-tribal to stay in his Permanent/ Temporary Shelter without any valid authority, in violation of P.A.T. Rules.

The terms and conditions were not signed by the beneficiary – it was only signed by the DC of the Nicobar District.

11 Conclusion

The vulnerable Andaman and Nicobar Islands

The A&N Islands have completely transformed after the British occupied it in 1858. Today the population of A&N Islands consist primarily of mainlanders who were 'settled' in A&N Islands over the last 159 years and the aboriginal tribal population. The mainlanders or the non-tribal constitute 92.66 per cent of the total population while the aboriginal tribes constitute only 7.34 per cent. Throughout its history, the A&N Islands were rendered vulnerable by the non-islanders. This vulnerability led to the disasters – the epidemics of syphilis, World War II and the Tsunami.

The three disasters experienced by A&N Islands were different from each other, yet all were the result of vulnerability. They were different in terms of time period, number of people killed, damage, spatial distribution and the vulnerabilities. The duration of each of these disasters varied. The epidemics of syphilis extended for 30 years; World War II was limited to three and a half years; and the Tsunami waves hit its coast three times within three hours. The number of people killed in each disaster varied too. In the epidemics 7,000 people were killed, in World War II 3,000–5,000 people, in Tsunami the killed were 3,513.

The impact of these disasters was spatially concentrated in few vulnerable islands only, i.e. the three disasters did not affect the entire A&N Islands at the same time. The impact of the epidemics of syphilis was concentrated in the Great Andaman Island. Similarly, the impact of World War II was on Port Blair in Andaman Islands and Car Nicobar in Nicobar Islands. In the case of the Tsunami, the waves were recorded in each and every island of A&N Islands, but the disaster was concentrated in parts of South Andaman Island, the entire Little Andaman Island and the Nicobar Islands. This concentration reflects the vulnerability of A&N Islands to different disasters at different point of time.

The only commonality in these three disasters was that, all were an outcome of vulnerability. These disasters were concentrated spatially and affected a particular group of people only. It was the vulnerable population that was killed in each of these disasters. It was the tribal Andamanese population in case of epidemics of syphilis; it was the local-born population and the tribal Nicobarese of Car Nicobar Island in case of World War II; and it was the tribal Nicobarese again in case of the Tsunami.

Another commonality was that in these three disasters, vulnerability was rendered by people who did not belong to A&N Islands. It was the British in case of the epidemics, who colonized A&N Islands and rendered the original tribal inhabitants vulnerable to diseases. It was the Japanese, who occupied the A&N Islands during World War II and rendered it vulnerable. In the Tsunami, it was the mainland Indians who were resettled in A&N Islands. They 'utilized' and 'exploited' the islands and in the process rendered the tribal Nicobarese vulnerable. Again, in case of all three disasters it emerged that those who rendered the vulnerability were less affected. In epidemics there is no record of killing of the British due to syphilis. Lesser numbers of Japanese soldiers were killed in A&N Islands in World War II and in the Tsunami, of the total killed the non-tribal accounted for 16 per cent.

In case of each of these disasters, the vulnerability was rendered differently – in epidemics the Andaman Home and Andaman Orphanage seeded the vulnerability. In the Andaman Home, the tribal Andamanese were treated as servants, kept in unhygienic conditions and provided nutritionally poor quality food, and sexually exploited which led to the introduction of syphilis. The tribal Andamanese had no resistance to this disease and they succumbed to it. Within 30 years of the contact with the British, the tribal Andamanese population declined by 90 per cent.

In World War II, the cause of vulnerability was lack of military defence and severing of the A&N Islands from mainland India. During the war, the British forces cut off all supplies by bombing the supply ships, which led to large-scale killing and famine.

In case of the Tsunami, the vulnerability can be traced to the Independence and the Partition of India. The A&N Islands were seen as a land, forest and sea resource-rich area where the Partition refugees from East Bengal (present-day Bangladesh) could be resettled. People were resettled across A&N Islands from 1949 to 1980. Along with resettled families, relatives and other job-seekers came to the islands. In order to develop the islands and cater to the needs of the families from mainland India, the original tribal inhabitants were left out. To protect the interest of the tribal population the Andaman and Nicobar Islands Protection of Aboriginal Tribes Regulation (ANPATR) was passed in 1956. It restricted entry of the non-tribal population the tribal areas thereby protecting the interest of the tribes. However, the regulation was amended 12 times in last 57 years. These amendments permitted regulated entry of non-tribal people in the protected areas. Thus, the tribal Nicobarese were neither left undisturbed to practice their traditional way of life nor were they assimilated in the development process. The tribal Nicobarese continued to pursue subsistence extractive economy which was dependent on coconut extraction and the sea and thus were settled on the coasts. In the Tsunami, the villages and the resources of the tribal Nicobarese were washed away and the disaster ensued.

Be it the epidemics, World War II or the Tsunami – or be it in the nineteenth, twentieth or twenty-first century – in each and every disaster the vulnerable population was affected and killed. The people of A&N Islands, especially the tribal population, has been rendered vulnerable time and again, and it is this vulnerability which has led to the disaster in every case. The research clearly proves that human-rendered vulnerability and not 'nature' is the cause for disasters.

Each of these disasters was followed by a response. Four characteristics of the response can be identified. First, response to disasters in the A&N Islands has always been imposed by outsiders. The islanders have been treated as passive victims and relief was given to them after the disaster. There was no involvement of tribal Andamanese in the response process in case of the epidemics. The British put the tribal Andamanese in hospitals against their wishes and were not given the opportunity to recoup themselves because the British felt that the Andamanese were incapable to treat the disease which was unknown to them. Moreover, the British were confident that hospital treatment was the best remedy. In World War II, the British brought in relief items post-war and sanctioned compensation and pension for the affected people. In this case too, the British imposed the relief, reconstruction and rehabilitation. In case of the resettlement post-Partition, it was the mainland Indian government which decided the scheme and its implementation in the A&N Islands. In case of the Tsunami, again the Government of India together with the A&N Administration imposed the relief and reconstruction on people without their consent or opinion. Post-Tsunami the tribal Nicobarese wanted to reconstruct their houses themselves and sought support of the A&N Administration for providing building materials and implements. Instead, new kinds of houses were handed over to them after a delay of three years.

Second, in case of all three disasters, the response was ad hoc and temporary. In case of the epidemics, when the British noticed the disease, they sent the infected persons to the hospital. The British were aware of the negative consequences of interaction between the tribal Andamanese population and the convicts yet the interactions and exploitations continued unabated. The response to the disease was to treat symptoms only and did not aim at restoring the health of the tribal Andamanese. Instead, the diet given in the Andaman Home weakened the constitution of the Andamanese and reduced their immunity system. Even after Independence no systematic effort was made to rehabilitate the tribal population. Though the Andaman and Nicobar Aboriginal Tribes Protection Regulation came in effect since 1956, the tribal needs were not addressed.

In case of World War II, the British response was to evacuate the islands. The Japanese, on the other hand, struggled to maintain a sufficient food supply but were unsuccessful. The Japanese also constructed defence infrastructure in A&N Islands such as the roads with bunkers along the coast, airstrips and jetties. However, even these were destroyed in the bombings. The British reoccupied A&N Islands, but their response too was limited to delivery of relief items, payment of ex-gratia compensation and reconstruction of damaged infrastructure. Even at this juncture, the defences of A&N Islands were not built.

The response to the Partition of India created vulnerability for A&N Islands. The policy to resettle the Partition refugees from East Bengal was unplanned, chaotic and neglected the needs of the tribal population, the original inhabitants of the islands. After the Independence, the disaster management of A&N Islands was the responsibility of the Government of India. Even though the administration had prepared two disaster management plans before the Tsunami, yet the response to the Tsunami was ad hoc. The administration was unprepared to handle the situation post-Tsunami and thus sought help of the armed forces. The disaster management plan did not focus on how to function in a situation in which all support systems such as transport and communications network, houses and markets collapsed.

The third common feature of the disaster response in A&N Islands was that it did not address the vulnerability of the people. Vulnerability was rendered to the people of A&N Islands which led to the disasters. The post-disaster response did not address this vulnerability as the response was provided by the outsiders who rendered the islands vulnerable. In case of epidemics, the tribal Andamanese were made vulnerable by the British when they forced them to live in Andaman Home and civilize them. However, the British did not address the vulnerability so that the remaining tribal Andamanese population could be saved. They simply closed the Andaman Home because there were no Andamanese left to stay in these Homes. The Andamanese who showed symptoms of syphilis were forcefully put in the hospitals. This left out the cases which did not show a symptom, thereby leaving substantial population at risk of infection. The response of the British to the epidemics did not reduce the vulnerability of the tribal Andamanese.

Similarly, the response of the British to World War II did not address the vulnerability of the people. In World War II, instead of defending the territory, the British evacuated the islands. Even after the war, the response aimed at compensating the loss and not strengthening the defence mechanism. The response of the British did not address the vulnerability of the people in any way.

The response of Government of India to the Partition of India created further vulnerability. It was lack of self-sufficiency of the local born population which led to famine conditions during War. Yet, immediately after Independence, the Government of India transplanted and resettled the Partition refugees from East Pakistan, thereby putting more pressure on the scarce resources. Thus, instead of creating self-sufficiency to reduce vulnerability, the Government of India created further dependency of A&N Islands on the mainland India without creating strong infrastructure for transport and communications. The non-tribal population became vulnerable, but the vulnerability of the tribal population increased manyfold.

The response to the Tsunami too did not address the vulnerability of the tribal Nicobarese. The subsistence economy and the resultant coastal location of tribal Nicobarese settlement were responsible for large-scale killing, yet, the tribal Nicobarese still continue with their subsistence economy even though post-Tsunami the settlement location has been shifted 5–6 km inwards from the sea. One other evidence of a well-meaning response leading to vulnerability is the case of post-Tsunami permanent houses. In A&N Islands, 9,797 new permanent houses were constructed 5–6 km away from the coast in order to provide shelter to the people who lost their houses in the Tsunami. Though the new houses constructed are protected against the Tsunamis because of their location inward from the coast, the new houses have become vulnerable to fire, heavy rains and waterlogging. The fieldwork confirmed that these houses are vulnerable to fire. Therefore, response to the Tsunami has created new forms of vulnerability. The poor quality building material and lack of maintenance has made these structures vulnerable to high intensity cyclones and high magnitude earthquakes.

This research also proves that post-Tsunami response aimed at 'developing' the tribal Nicobarese with cash compensation. Cash was given as compensation for the loss of livelihood resources such as coconut trees, boats and livestock and for persons killed. But sudden and surplus cash has had several drawbacks. The cash flow has reduced the inclination of the Nicobarese to work. The excess cash has not been invested in building economic security rather it is being spent recklessly on the purchase of material assets. Further, post-Tsunami ex-gratia compensation across India is paid to the head of the household of a nuclear family. This trend was in conflict with the Nicobarese tuhet system which was broken up for receiving the compensation. Thus, the tuhet that was fundamental in recovery of the tribal Nicobarese post-Tsunami has been fragmented thereby creating new vulnerabilities. The Tsunami was an opportunity to build back strong and vulnerability-free islands, yet it was missed. Vulnerability and response are not independent of each other as both are intricately linked and both need to be addressed simultaneously if one aims at a disaster-less if not a disaster-free world. Till then the vulnerable A&N Islands shall remain vulnerable!

Bibliography

Alongi, Daniel M. 2008. 'Mangrove Forests: Resilience, Protection from Tsunamis, and Responses to Global Climate Change', *Estuarine, Coastal and Shelf Science* 76 (1): 1–13, www.researchgate.net/publication/222540308_Mangrove_forests_Resilience_protection_from_tsunamis_and_responses_to_global_climate_change (accessed 10 March 2011).

Andaman and Nicobar Administration. 1996. *Disaster Management Plan for Andaman and Nicobar Islands*. Port Blair: A&N Administration.

Andaman and Nicobar Administration. 2004. *Disaster Management Plan for Andaman and Nicobar Islands*. Port Blair: A&N Administration.

Andaman and Nicobar Administration. 2005a. *Intermediate Shelter: Tsunami Rehabilitation Operation-GO-NGO Participation*, www.japanplatform.org/area_works/sumatra/0518adra/adar 02.pdf (accessed 3 June 2011).

Andaman and Nicobar Administration. 2005b. *Islandwise Damage Details*, www.tsunami.and.nic.in/killer.htm (accessed 20 February 2008).

Andaman and Nicobar Administration. 2006. *Andaman and Nicobar Islands: Towards a Brighter Tomorrow*. Port Blair: A&N Administration.

Andaman and Nicobar Administration. 2009. *List of Ex-Gratia Distributed for the Killed*, www.and.nic.in/ex-gratia/ex-gratiapayment.htm (accessed 1 April 2010).

Andaman and Nicobar Administration. 2011a. *Permanent Shelter Allotment List*, www.and.nic.in/shelter_SA/Permanent%20Shelters.htm (accessed 3 December 2011).

Andaman and Nicobar Administration. 2011b. *Relief and Rehabilitation Post Tsunami*, www.and.nic.in/Announcements/visit/sep/district_admn.pdf (accessed 3 December 2011).

Andaman and Nicobar Administration. 2012. *Disaster Management Plan for Andaman and Nicobar Islands*. Port Blair, www.indiaenvironmentportal.org.in/files/file/A&N%20Islands%20Disaster%20Management%20Plan%202012.pdf (accessed 12 February 2014).

Andaman and Nicobar Gazette. 1956. *The Andaman and Nicobar Protection of Aboriginal Tribes Regulation, 1956*. Port Blair: A&N Administration.

Andaman Public Works Department. 2009. *Draft Master Plan for Port Blair Planning Area, 2028*, Volume II. Port Blair: A&N Administration, www.indiaenvironmentportal.org.in/files/Volume%20II-%20Full.pdf (accessed 10 November 2010).

Anderson, Clare. 2008. *Refugee Resettlement in the Andaman Islands: Colonial Precedents, Post-Colonial Predicaments*, Research Report, ESRC End of Award Report, RES-067–27–0001. Swindon: ESRC, https://s3-eu-west-1.amazonaws.com/esrc-files/.../gCehpGU4EES8g7JaOJlrow.pdf (accessed 4 August 2011).

Ansari, M. T. 2008. *Malabar: Desheeyathayude Idapedalukal*, Malabar: Involvement in Nationalism. Kottayam: DC Books.

Arya, A. S. 2009. *Construction of Intermediate Shelters at Andaman and Nicobar Islands*, PowerPoint presentation at IIT Roorkee, http://bhagirathi.iitr.ac.in/ (accessed 5 December 2011).

Asian Development Bank, United Nations and World Bank. 2005. *India: Post Tsunami Recovery Program Preliminary Damage and Needs Assessment.* New Delhi, www.tnrd.gov.in/externallyaidedprojects/Tsunami_rehabilation/india-assessment-full-report[1].pdf (accessed 14 August 2007).

Asian Disaster Preparedness Centre. 2006. *Regional Analysis of Socio-Economic Impacts of the December 2004 Earthquake and Indian Ocean Tsunami.* Bangkok, www.adpc.net/igo/category/ID626/doc/2014-uyp0TG-ADPC_Regional_Analysis_of_Tsunami_Impacts.pdf (accessed 21 June 2012).

Banghar, A. R. 1987. 'Seismotectonics of the Andaman-Nicobar Islands', *Tectonophysics* 133 (1–2): 95–107.

Bhagat, R. 2005. 'Tsunami: A Blow to Andaman Business Community', *Hindu Business Line*, 22 February.

Bharatiya Jain Sangathan. 2005. *Andaman and Nicobar Islands: A Year after the Tsunami: A Report on Activities of Bharatiya Jain Sangathan, 26th December 2004 to December 2005.* Pune, www.bjsindia.org/oldwebsite/PDFs/DM_Reports/A_Year_After_tsunami.pdf (accessed 15 July 2013).

Bilham, R. 2005. 'A Flying Start, Then a Slow Slip', *Science* 308 (20 May 2005): 1126–1127.

Bilham, R., E. R. Engdahl, N. Feldl and S. P. Satyabala. 2005. 'Partial and Complete Rupture of the Indo-Andaman Plate Boundary 1847–2004', *Seismological Research Letters* 76 (3): 299–311.

Biswas, S. K. 2009. *Colonization and Rehabilitation in Andaman and Nicobar Islands.* New Delhi: Abhijeet.

Blaikie, P., T. Cannon, I. Davis and B. Wisner. 1994. *At Risk: Natural Hazards, People's Vulnerability, and Disasters.* London: Routledge.

Burton, I., R. W. Kates and G. F. White. 1993. *The Environment as Hazard.* London: Guilford Press.

Census of India. 1901. *Volume 3: Andaman and Nicobar Islands, Report on the Census*, Richard C. Temple (ed.). Calcutta: Office of the Superintendent of Government Printing, India.

Census of India. 1921. *Volume 1 and 2: The Andaman and Nicobar Islands, Report*, R. F. Lowis (ed.). Calcutta: Office of the Superintendent of Government Printing, India.
Census of India. 1931. *Volume 1 and 2: Andaman and Nicobar Islands, Report*, M. C. C. Bonington (ed.). Calcutta: Government of India Central Publication Branch.
Census of India. 1941. *Andaman and Nicobar Islands: General Report: Part I*. New Delhi: Government of India.
Census of India. 1951. *Andaman and Nicobar Islands: General Report: Part I*. New Delhi: Government of India.
Census of India. 1961. *Andaman and Nicobar Islands: General Report: Part I*. New Delhi: Government of India.
Census of India. 1971. *Andaman and Nicobar Islands: General Notes*. New Delhi: Government of India.
Census of India. 1981. *Andaman and Nicobar Islands*. New Delhi: Government of India.
Census of India. 1991. *Andaman and Nicobar Islands*. New Delhi: Government of India.
Census of India. 2001a. *Primary Census Abstract: Andaman and Nicobar Islands*. New Delhi: Government of India.
Census of India. 2001b. *Village Directory: Andaman and Nicobar Islands*. New Delhi: Government of India.
Census of India. 2001c. *Implication of Terms Used in Indian Censuses*, http://censusindia.gov.in/Data_Products/Library/Indian_perceptive_link/Census_Terms_link/censusterms.html (accessed 10 July 2008).
Census of India. 2001d. *Concepts and Definitions*, http://censusindia.gov.in/Metadata/Metada.htm#2l (accessed 10 July 2008).
Census of India. 2001e. *Housing Tables: Andaman and Nicobar Islands*. New Delhi: Government of India.
Census of India. 2011a. *Figures at Glance: Provisional Population Totals*, www.censusindia.gov.in/2011-prov-results/data_files/ani/PPT_Press_Release_Final.pdf (accessed 1 June 2011).
Census of India. 2011b. *Primary Census Abstract: Schedule Caste and Schedule Tribe, Andaman and Nicobar Islands*, www.censusofindia.gov.in/2011census/PCA/SC-ST/PCA-A11-Appendix/ST-35-PCA-A11-APPENDIX.xlsx (accessed 5 January 2017).
Chandramouli, R. 2005. 'Indifference in Aid, Allege Residents of Car Nicobar', *Times of India*, 2 January.
Chatterji, Joya. 2007. *The Spoils of Partition: Bengal and India (1947–1967)*. New Delhi: Cambridge University Press.
Chattopadhyay, Raghavendra and Subir Bhattacharya. 2007. *Half Yearly Monitoring Report of Indian Institute of Management, Calcutta on Sarva Siksha Abhiyan for Union Territory of Andaman and Nicobar Islands for the Period 1st October 2006 to 31st March 2007*. Calcutta: Indian Institute of Management, http://ssa.nic.in/monitoring-documents/mi/state-wise-

reports-of-monitoring-institutions/andaman-nicobar/1st%20Oct%20 06%20-%2031st%20Mar%2007%20.pdf (accessed 18 August 2012).
Chaudhry, Shivani and Enakshi Ganguly Thukral. 2006. *Battered Islands: Report of a Fact-Finding Mission to Tsunami Affected Areas of the Andaman and Nicobar Islands*, Fact-Finding Mission Report 9. New Delhi: Housing and Land Rights Network, www.hic-sarp.org/documents/Fact%20 Finding%20Report%20Mission%209.pdf (accessed 4 June 2009).
Cipriani, L. 1955. 'On the Origin of the Andamanese', *Census of India, 1951, Appendix E* 17 (1): 66–71, New Delhi: Government of India.
Coomar, P. C. 1997. *Migration and Social Change: A Study of Bhantus of Andaman Islands*, Memoir No. 98. Calcutta: Anthropological Survey of India.
Cooper, Z. 2002. *Archaeology and History: Early Settlements in Andaman Islands*. New Delhi: Oxford University Press.
Correspondent. 1978. 'Tourism to Boost Island Economy', *Daily Telegram*, 2 September.
Correspondent. 1979. '1980 Declared as "India Tourism Year" ', *Daily Telegram*, 3 August.
Correspondent. 1998. 'Fade to Black', *Down to Earth*, 21 August.
Correspondent. 2004. 'Asian Sea Quake Has Given Globe a Jolt', *Daily Telegram*, 30 December.
Correspondent. 2005. 'Earthquake "Redraws the Map" ', *Daily Telegram*, 3 January.
Curray, Joseph R. 1991. 'Possible Greenschist Metamorphism at the Base of a 22 km Sedimentary Section, Bay of Bengal', *Geology* 19 (11): 1097–1100.
Cutter, S. 1996. 'Vulnerability to Environmental Hazards', *Progress in Human Geography* 20 (4): 529–539.
Dam Roy, S. and P. Krishnan. 2005. 'Mangrove Stands of Andamans vis-à-vis Tsunami', *Current Science* 89 (11): 1800–1804.
Das, A. K. 2005. 'GIS Based Mapping of Tsunami Induced Land Use/Cover Change in Nancowry Group of Islands, Andaman and Nicobar Islands', *OSTI Newsletter* 10 (October): 1–4.
Das, D. P., S. S. Ghosh, D. Chakraborty and K. Dharmanik. 2007. 'Analysis of Satellite Data for Detection of Changes in Coastal Geomorphology of A&N Islands Due to 26 December 2004 Earthquake', in Surjit Dasgupta (ed.) *Sumatra-Andaman Earthquake and Tsunami 26 December 2004*, Geological Survey of India, Special Publication No. 89. Kolkata: Geological Survey of India, 51–62.
Dasgupta, J. 2002. *Japanese in Andaman and Nicobar Islands: Red Sun over Black Waters*. New Delhi: Manas.
Dass, F. A. M. 1937. *The Andaman Islands*. Bangalore: Good Shepherd Convent Press, https://archive.org/details/andamanislands035497mbp (accessed 20 September 2007).
Deendayal Upadhyaya Gram Jyoti Yojna. 2004. www.ddugjy.gov.in/mis/portal/definition_electrified_village.html (accessed 20 February 2012).

Department of Environment and Forest. 2006. *State of Environment: Andaman and Nicobar Islands*. Port Blair: A&N Administration.

Department of Ocean Development and Space Application Centre. 1997. *Coral Reef Maps of India*. Ahmedabad.

Dhanuja, Richa. 2003. 'History of the Andaman Trunk Road', *SANE Newsletter*, April/May, www.andaman.org/BOOK/Sane/sanetext/Newsletter_April (accessed 10 November 2012).

Dhar, A. 2005a. 'Andaman Tribal Want Mainlanders to Leave', *Hindu*, 17 February.

Dhar, A. 2005b. 'Katchal: Now a Picture of Devastation', *Hindu*, 18 February.

Dhar, A. 2005c. 'Political Bickering over Tsunami Relief', *Hindu*, 20 February.

Dhingra, K. 2005. *The Andaman and Nicobar Islands in the Twentieth Century: A Gazetteer*. New Delhi: Oxford University Press.

Digal, J. 2006. *Andaman and Nicobar Islands: Post Tsunami Livelihood Measures*. Port Blair: Andaman and Nicobar Administration.

Directorate of Economics and Statistics. 1965. *Basic Statistics*. Port Blair: Andaman and Nicobar Administration.

Directorate of Economics and Statistics. 2006. *Islandwise Statistical Handbook 2006*. Port Blair: Andaman and Nicobar Administration.

Directorate of Economics and Statistics. 2008a. *Basic Statistics 2007*. Port Blair: Andaman and Nicobar Administration.

Directorate of Economics and Statistics. 2008b. *Economic Survey of Andaman and Nicobar Islands*. Port Blair: Andaman and Nicobar Administration.

Directorate of Economics and Statistics. 2011. *Basic Statistics 2010–11*. Port Blair: Andaman and Nicobar Administration.

Directorate of Economics and Statistics. 2015. *Basic Statistics 2015*. Port Blair: Andaman and Nicobar Administration.

Distant, W. L. 1877. 'Our Present Knowledge of the Nicobarians', *Journal of the Royal Anthropological Institute of Great Britain and Ireland* 6: 209–213, https://ia601407.us.archive.org/31/items/journalroyalant10irelgoog/journalroyalant10irelgoog.pdf (accessed 6 April 2012).

Dorairaj and Soundararajan. 1997. 'Status of Coral Reefs of Mahatma Gandhi Marine National Park, Wandoor, Andamans', in *The Proceedings of the Regional Workshop on the Conservation and Sustainable Management of Coral Reefs*. Chennai: M. S. Swaminathan Research Foundation and Bay of Bengal Programme (FAO), www.fao.org/docrep/x5627e/x5627e0h.htm (accessed 2 May 2012).

Dutta, P. C. 1978. *The Great Andamanese: Past and Present*. Calcutta: Anthropological Survey of India.

Equitable Tourism Options (EQUATIONS). 2009. *Who Really Benefits from Tourism: EQUATION's Critique on Tourism Development*, Working Paper Series 2009–10, Section-I. Bangalore, www.equitabletourism.org/readfull.php?AID=1402 (accessed 23 February 2011).

Equitable Tourism Options (EQUATIONS), Indian National Trust for Art and Culture Heritage, Society for Andaman & Nicobar Ecology, Kalpavriksh, Jamsetji Tata Centre for Disaster Management of Tata Institute of Social

Sciences, ActionAid International India. 2008. *Rethink Tourism in the Andamans: Towards Building a Base for Sustainable Tourism*. Bangalore, http://equitabletourism.org/files/fileDocuments631_uid13.pdf (accessed 13 December 2014).

Finance Commission of India. 2000. *Report of the Eleventh Finance Commission of India*. New Delhi: Government of India.

Forest Survey of India. 2003. *State of Forest Report 2003*. Dehradun.

Forest Survey of India. 2005. *State of Forest Report 2005*. Dehradun.

Gabor, T. and T. K. Griffith. 1980. 'The Assessment of Community Vulnerability to Acute Hazardous Materials Incidents', *Journal of Hazardous Materials* 3 (4): 323–333.

Geological Society of India. 2005. *Great Tsunami: 26 December, 2004, Sumatra Region*, Memoir No. 64. Bangalore: Geological Society of India.

Ghosh, T., P. Jana, T. S. Giritharam, S. Bardhan, S. R. Basir and A. K. Ghosh. 2007b. 'Tsunami Survey in Andaman Nicobar Group of Islands', in Surjit Dasgupta (ed.) *Sumatra-Andaman Earthquake and Tsunami 26 December 2004*, Geological Survey of India, Special Publication No. 89. Kolkata: Geological Survey of India, 153–170.

Ghosh Roy, A. K., S. Bardhan, P. Jana and S. R. Basir. 2007a. 'Macroseismic Survey in Andaman and Nicobar Islands in the Aftermath of the Great Earthquake of 26 December 2004', in Surjit Dasgupta (ed.) *Sumatra-Andaman Earthquake and Tsunami 26 December 2004*, Geological Survey of India, Special Publication No. 89. Kolkata: Geological Survey of India, 17–50.

Goel, D. 2002. *Impact of Infrastructure on Productivity: Case of Indian Registered Manufacturing*, Working Paper No. 106, Centre for Development Economics. New Delhi: Delhi School of Economics.

Goel, P. 2006. *Andaman & Nicobar Islands through Tsunami: A Saga of Courage*. Port Blair: A&N Administration.

Gonsalves, C. 2005. 'The Deadly Bureaucracy in Andamans', *Indian Express*, 26 January.

Goodman, P. S. 2005. 'Tsunami's Unpredictable Outcome: Few Injuries', *Washington Post*, 3 January, www.washingtonpost.com/wp-dyn/articles/A58761-2005Feb2.html?nav=rss_world/issues/tsunami (accessed 15 September 2011).

Granger, K. 1997. 'Drought: Issues for the Twenty-First Century', in *United Nations: International Decade of Natural Disaster Reduction and QUIPUNET Internet Conference on Floods*, www.quipu.net:1997/English/Welcome.htm (accessed 28 March 2009).

Gupta, M. and A. Sharma. 2006. 'Compounded Loss: The Post Tsunami Recovery Experience of Indian Island Communities', *Disaster Prevention and Management* 15 (1): 67–78.

Gutenberg, R. and C. F. Richter. 1944. 'Frequency of Earthquakes in California', *Bulletin of the Seismological Society of America* 34: 185–188, http://authors.library.caltech.edu/47734/1/185. full.pdf (accessed 10 August 2008).

Hewitt, K. 1997. *Regions of Risk*. Harlow: Longman.

Hoon, Vineeta. 1997. 'Coral Reefs of India: Review of Their Extent, Condition, Research and Management Status', in *Proceedings of the Regional Workshop on the Conservation and Sustainable Management of Coral Reefs*. Chennai: M. S. Swaminathan Research Foundation and Bay of Bengal Programme (FAO), B1–B26, www.fao.org/docrep/x5627e/x5627e06.htm (accessed 2 May 2012).
Hunter, W. W. 1909. *Imperial Gazetteer of India: Provincial Series: Andaman and Nicobar Islands*. Calcutta: Office of Superintendent of Government Printing.
Ibrahim, H. M. and Nazery Khalid. 2007. *Growing Shipping Traffic in the Strait of Malacca: Some Reflections on the Environmental Impact*. Kuala Lumpur: Maritime Institute of Malaysia, www.mima.gov.my/mima/wp-content/uploads/GMIL/ (accessed 5 May 2016).
Indian Meteorological Department. 1881–2011. *Earthquakes ≥ 5 within 5–15°N and 90–95°E from Earliest Recorded Till 2011*. New Delhi.
Indian Meteorological Department. 1977–2007. *Disastrous Weather Events*. Pune.
Indian Meteorological Department. 1979. *Tracks of Storms and Depression in Bay of Bengal and Arabian Sea, 1877–1970*. New Delhi.
Indian Meteorological Department. 1996. *Tracks of Storms and Depression in Bay of Bengal and Arabian Sea, 1971–1990: An Addendum to the Storm Track Atlas, 1877–1970*. New Delhi.
Indian Meteorological Department. 2011. *Earthquakes ≥ 5 within 5–15°N and 90–95°E from Earliest Recorded Till 2011*. New Delhi.
Indian Meteorological Department. 2013. *Tracks of Storms and Depression in Bay of Bengal and Arabian Sea*, www.rmcchennaieatlas.tn.nic.in (accessed 25 January 2013).
International Coral Reef Initiative. 2016. *Benefits of Coral Reef*, www.icriforum.org/about-coral-reefs/benefits-coral-reefs (accessed 12 August 2016).
International Monetary Fund. 2005. *World Economic Outlook, April 2005*. Washington, DC, www.imf.org/external/pubs/ft/weo/2005/01/ (accessed 23 September 2009).
Ishii, M., P. M. Shearer, H. Houston and J. E. Vidal. 2005. 'Extent, Duration and Speed of the 2004 Sumatra-Andaman Earthquake Imaged by the Hi-Net Array', *Nature* 435: 933–936.
Jaiswal, Arvind. 2007. *Shelter Report: Intermediate Shelters in Andaman and Nicobar Islands*, Report No. 1, World Housing Encyclopaedia: An Encyclopaedia of Housing Constructions in Seismically Active Areas of the World, Earthquake Engineering Research Institute and International Association for Earthquake Engineering, www.world-housing.net/shelterreport1view.php?id=70 0008 (accessed 3 December 2011).
Jayasuriya, Sisira and Peter McCawley (eds.). 2010. *The Asian Tsunami: Aid and Reconstruction after a Disaster*. Cheltenham, UK: Edward Elgar and the Asian Development Bank.
Jeyakumar, S., Jai Sunder, M. S. Kundu, A. Kundu and T. P. Swapna. 2014. 'Traditional Pig Rearing Practices among the Nicobari Tribes of Nicobar

Group of Islands, India', *IOSR Journal of Agriculture and Veterinary Science* 7 (12): 35–41, www.iosrjournals.org (accessed 16 July 2016).

Justin, Anstice. 1990. *The Nicobarese*. Calcutta: Seagull Books on behalf of Anthropological Survey of India.

Kanapathipillai, V. 1995. 'The Repatriation of Indian Tamil Plantation Workers from Sri Lanka to India', in R. Cohen (ed.) *The Cambridge Survey of World Migration*. Cambridge: Cambridge University Press.

Kapur, Anu. 2008. *On Disasters in India*. New Delhi: Cambridge.

Kapur, Anu. 2010. *Vulnerable India: A Geographical Study of Disasters*. New Delhi: Sage.

Kapur, Anu, Neeti Chopra, Meeta Ajit, Deeptima Massey, Roshani Devi and Debanjali Chakraborti. 2005. *Disasters in India: Studies of Grim Reality*. New Delhi: Rawat.

Kaul, Rahul and Vivek Menon. 2005. *The Ground beneath the Waves: Post-Tsunami Impact Assessment of Wildlife and Their Habitats in India*, Volume 2. New Delhi: Wildlife Trust of India, www.wti.org.in/publications/ground-beneath-the-waves-vol2.pdf (accessed 26 December 2008).

Kayal, J.R. 2008. *Microearthquake Seismology and Seismotectonics of South Asia*. Dordrecht: Springer.

Khosa, Aasha. 2005. 'Centre Sees Andamans as Ideal Island Township: Rebuilding Project Would Need about Rs 3000cr', *Indian Express*, 2 March.

Kloss, Boden. 1903. *In the Andaman and Nicobars: The Narrative of a Cruise in the Schooner 'Terrapin', with Notices of the Islands, Their Fauna, Ethnology, Etc.* London: John Murray Press.

Krebs, Carola. 2007. *Befriending the Jarawa: Indian Tribal Policy Illustrated in the Case of a Hunter Gatherer Group in the Andaman Islands*. Translated in English by George Weber, www.andaman.org/BOOK/originals/Krebs/art-jarawa.htm (accessed 12 May 2008).

Kshirsagar, J. B. and Pawan Kumar. 2006. *Permanent Settlements in Andaman and Nicobar Islands*. New Delhi: Town and Country Planning Organization, Ministry of Urban Development, Unpublished PowerPoint Presentation.

Kudiasya, Gyanesh. 1996. 'Divided Landscape Fragmented Identities: East Bengal Refugees and Their Rehabilitation in India, 1947–79', *Singapore Journal of Tropical Geography* 17 (1): 24–39.

Kumar, Krishna. 1997. 'The Coral Reef Ecosystem of the Andaman and Nicobar Islands: Problems and Prospects and the World Wide Fund for Nature: India Initiatives for Its Conservation', in *The Proceedings of the Regional Workshop on the Conservation and Sustainable Management of Coral Reefs*. Chennai: M. S. Swaminathan Research Foundation, www.fao.org/docrep/x5627e/x5627e0f.htm (accessed 2 May 2012).

Lal, D. S. 1989. *Climatology*. Allahabad: Chanakya Publishing House.

Lall, B. 1992. *A Regime of Fears and Tears: History of the Japanese Occupation of the Andaman Islands*. New Delhi: Farsight.

Majumdar, Diptosh. 2005. 'Car Nicobar KV Teachers can Choose any Posting', *The Indian Express*, 6 January.

Majumdar, R. C. 1975. *Penal Settlement in Andamans*. New Delhi: Gazetteers Unit, Department of Culture, Ministry of Education and Social Welfare, Government of India.

Malik, Javed N., C. V. R. Murty and Durgesh C. Rai. 2006. 'Landscape Changes in the Andaman and Nicobar Islands, India, after the December 2004 Great Sumatra Earthquake and Indian Ocean Tsunami', *Earthquake Spectra* 22 (S3): S43–S66, http://earthquakespectra.org/doi/abs/10.1193/1.2206792?code=eeri-site (accessed 13 August 2009).

Man, E. H. 1883 (reprint-2001). *The Aboriginal Inhabitants of the Andaman Islands*. London: Anthropological Institute of Great Britain and Ireland; New Delhi: Mittal (reprint).

Mann, R. S. 2005. *Andaman and Nicobar Tribes Restudied: Encounters and Concerns*. New Delhi: Mittal.

Martínez, Rodrigo and Andrés Fernández. 2010. *The Social and Economic Impact of Illiteracy: Analytical Model and Pilot Study*, OREALC/2010/PI/H/12. Santiago: Economic Commission for Latin America and the Caribbean/United Nations Educational, Scientific and Cultural Organization, http://unesdoc.unesco.org/images/0019/001905/190571E.pdf (accessed 10 March 2017).

Math, Suresh Bada, John P. John, Satish Chandra Girimaji, Vivek Benegal, Biju Sunny, K. Krishnakanth, Uday Kumar, Ameer Hamza, Shweta Tandon, Kavitha Jangam, K. S. Meena, Biju Chandramukhi and D. Nagaraja. 2008. 'Comparative Study of Psychiatric Morbidity among the Displaced and Non-Displaced Populations in the Andaman and Nicobar Islands Following the Tsunami', *Prehospital and Disaster Medicine* 23 (1): 29–34, www.ncbi.nlm.nih.gov/pub med/18491658 (accessed 30 September 2011).

Mathur, K. N., S. K. Ray, S. Sengupta, Prabhas Pande and Surjit Dasgupta. 2007. 'A Preliminary Report on Investigation of Effects of the Sumatra-Andaman Earthquake of 26 December 2004 in Andaman and Nicobar Islands', in Surjit Dasgupta (ed.) *Sumatra-Andaman Earthquake and Tsunami 26 December 2004*, Geological Survey of India, Special Publication No. 89. Kolkata: Geological Survey of India, 5–16.

Ministry of Agriculture. 2001. *High Powered Committee on Disaster Management Report*. New Delhi: Government of India, http://nidm.gov.in/PDF/pubs/HPC_Report.pdf (accessed 12 September 2007).

Ministry of Environment and Forest. 1991. *Coastal Regulation Zone Notification*. New Delhi: Government of India, www.moef.nic.in/legis/crz/crznew.pdf (accessed 11 November 2011).

Ministry of Home Affairs. 2003. *Report of the Expert Committee on Jarawas of Andaman Islands*. Port Blair: Directorate of Tribal Welfare, Andaman and Nicobar Administration.

Ministry of Home Affairs. 2005a. *Tsunami Impact (26th December 2004) in A&N Islands, UT of Pondicherry and Other Coastal States of Tamil Nadu, Andhra Pradesh and Kerala: Relief, Rehabilitation and Other Follow Up Measures*, www.ndmindia.nic.in/Tsunami2004/sitrep35.htm (accessed 7 September 2007).

Ministry of Home Affairs. 2005b. *Prime Minister's-Tsunami: A Report to Nation*, www.alnap. org/pool/files/indian_gov_tsunami_report_nation_june_2005. pdf (accessed 5 June 2007).

Ministry of Home Affairs. 2006. *Report on Performance Audit Report on Tsunami Relief and Rehabilitation*, Report 20. New Delhi: Government of India, www.www.cag.gov.in/html/reports/civil/2006_20_peraud/20%20 OF%202006/highlights.htm (accessed 3 December 2011).

Ministry of Power. 2011. *Rural Electrification*, www.powermin.nic.in/rural_ electrification/main.htm (accessed 5 October 2011).

Ministry of Rehabilitation. 1965. *Report by the Inter-Departmental Team on Accelerated Development Programme for Andaman and Nicobar Islands, Special Area Development Plan*. New Delhi: Government of India.

Mishra, O. P., G. K. Chakraborty and O. P. Singh. 2007. 'Aftershock Investigation of the December 26, 2004 Sumatra-Andaman Islands Earthquake', in Surjit Dasgupta (ed.) *Sumatra-Andaman Earthquake and Tsunami 26 December 2004*, Geological Survey of India, Special Publication No. 89. Kolkata: Geological Survey of India, 115–152.

Mittal, R. M. 2009. 'Role of Armed Forces in Disaster Management Post-Tsunami in Andaman and Nicobar Islands: An Overview', *National Seminar on Housing and Disaster Mitigation Efforts in India: Case Studies*, www.bhagirathi.iitr. ac.in/dspace/bitstream/123456789/936/1/2%20RM%20 Mittal.pdf (accessed 10 November 2012).

Mondal, G. and D. C. Rai. 2008. 'Performance of Harbour Structures in Andaman Islands during 2004 Sumatra Earthquake', *Engineering Structures* 30 (1): 174–182, http://home.iitk.ac.in/~dcrai/details/SumatraEQ2. pdf (accessed 10 July 2011).

Mouat, F. J. 1863 (reprint-1995). *The Andaman Islanders*. New Delhi: Mittal.

Murthy, R. V. R. 2005. *Andaman and Nicobar Islands: Development and Decentralization*. New Delhi: Mittal.

Myka, F. P. 1993. *Decline of Indigenous Populations: The Case of Andaman Islanders*. Jaipur: Rawat.

Nambath, S. 2004. 'Death Toll in Andamans Put at 10,000', *Hindu*, 30 December.

Nath, S. K., S. N. Bhattacharya, M. Vyas, I. Pal, P. Sengupta, S. Mitra, W. K. Mohanty and S. S. Ganguly. 2005. 'The Great Sumatra-Andaman Earthquake of 26 December 2004 as Perceived from the Broadband Seismograms of Indian Institute of Technology, Kharagpur Seismological Observatory', *Current Science* 88 (9): 1381–1383, www.iisc.ernet.in/currsci/may102005/1381. pdf (accessed 30 June 2007).

Negi, D. S. 2007. *Andaman and Nicobar Islands: Past, Present and Brutal Tsunami Waves*. New Delhi: Tushar.

Netobel, R. Sobi. 2004. An Economic Study of Tourism Industry in Kanyakumari District, unpublished PhD thesis, Post Graduate and Research Centre, Department of Economics, Scott Christian College (Nagercoil), Manonmaniam Sundaranar University, Tirunelveli, http://shodhganga.inflibnet.ac.in/ handle/10603/61560 (accessed 30 December 2013).

Nicholas, S. and P. R. Shergold. 1989. 'Transportation as Global Migration', in S. Nicholas (ed.) *Convict Workers: Reinterpreting Australia's Past*. Cambridge: Cambridge University Press.

Oldham, R. D. 1884. 'Note on the Earthquake of 31 December 1881', *Records of the Geological Survey of India* 17 (2): 47–53.

Ortiz, M. and R. Bilham. 2003. 'Source Area and Rupture Parameters of the 31 December 1881 Mw 7.9 Car Nicobar Earthquake Estimated from Tsunami Recorded in the Bay of Bengal', *Journal of Geophysical Research* 108 (2215): 1–16, http://onlinelibrary.wiley.com/doi/10.1029/2002JB001941/full (accessed 12 August 2011).

Pandit, T. N. 1986. 'Tribes of Andaman and Nicobar Islands', in K. S. Singh (ed.) *The Tribal Situation in India*. Simla: Indian Institute of Advanced Studies.

Pandya, V. 2002. 'Jarawas of Andaman Islands: Their Social and Historical Reconstruction', *Economic and Political Weekly* (14 September): 3830–3834.

Pandya, V. 2005. *When Land Became Water: Tsunami and the Onge of Little Andaman Island*, www.andaman.org/BOOK/originals/Pandya/pandya.htm (accessed 5 January 2008).

Partiyogita Darpan. 2015. *General Knowledge Digest 2015*. Agra: Upkar Prakashan.

Paul, Bimal Kanti. 2011. *Environmental Hazards and Disasters: Contexts, Perspectives and Management*. Oxford: Wiley-Blackwell.

Perry, A. 2004. 'Best Beaches You Can Go to: Best of Asia 2004, Best for Your Soul', *Time Magazine (Asia Edition)*, 22 November.

Planning Commission. 2007a. *Report of the Task Force on Islands, Coral Reefs, Mangroves & Wetlands in Environment & Forests for the Eleventh Five Year Plan, 2007–2012*. New Delhi: Government of India.

Planning Commission. 2007b. *Tsunami Rehabilitation Notes of the Tsunami Rehabilitation Programme Cell, Planning Commission*. New Delhi: Unpublished.

Planning Commission. 2008. *Andaman and Nicobar Islands Development Report*. New Delhi: Academic Foundation.

Portman, M. V. 1899 (reprint-1990). *A History of Our Relations with the Andamanese*, 2 Volumes. Calcutta: Office of Superintendent of Government Printing; New Delhi: Asian Education Service (reprint).

Quarantelli, Enrico L. 2005. 'A Social Science Research Agenda for the Disasters of the 21st Century', in R. W. Perry and E. L. Quarantelli (eds.) *What Is a Disaster? New Answers to Old Questions*. Philadelphia, PA: Xlibris.

Radcliffe-Brown, A. R. 1922. *The Andaman Islanders: A Study in Social Anthropology*. Cambridge: Cambridge University Press.

Raghavan, S. and S. Rajesh. 2003. 'Trends in Tropical Cyclone Impact: A Study in Andhra Pradesh, India', *Bulletin of the American Meteorological Society* 85 (5): 635–644.

Rai, Durgesh C., C. V. R. Murty, Sudhir K. Jain, Hemant B. Kaushik, Goutam Mondal, Suresh R. Dash, Alex Tang, Mark Yashinsky and Martin Eskijian.

2006. 'The Effect of the December 2004 Great Sumatra Earthquake and Indian Ocean Tsunami on Transportation Systems in India's Andaman and Nicobar Islands', *Earthquake Spectra* 22 (S3): S561–S579.

Rajendran, C. P., Anil Earnest, Kusala Rajendran, R. Dev Das and Sreekumari Kesavan. 2003. 'The 13 September 2002 North Andaman, Diglipur Earthquake: An Analysis in the Context of Regional Seismicity', *Current Science* 84 (7): 919–924, http://tejas.serc.iisc.ernet.in/~currsci/apr102003/919.pdf (accessed 10 February 2008).

Ramasamy, S. M., C. J. Kumanana, R. Sivakumar and B. Singh. 2006. *Geomatics in Tsunami*. New Delhi: New India Publishing Agency, Sponsored by Department of Science and Technology, Government of India.

Rastogi, B. K. and R. K. Jaiswal. 2006. 'A Catalogue of Tsunamis in the Indian Ocean', *Science of Tsunami Hazards* 25 (3): 128–143.

Rawal, Vivek, Rajendra Desai and Dharmesh Jadeja. 2006. *Assessing Post Tsunami Housing Reconstruction in Andaman and Nicobar Islands: A People's Perspective*. Bangalore: Books for Change, www.preventionweb.net/publications/view/2037 (accessed 14 March 2011).

Regional Medical Research Centre (RMRC), National Institute of Mental Health and Neuro Science (NIMHANS), Action Aid. 2008. *Alcohol Consumption: Prevalence and Pattern in Andaman and Nicobar Islands*. Port Blair: RMRC, NIMHANS and Action Aid, http://nimhans.ac.in/cam/sites/default/files/Publications/10.pdf (accessed 4 March 2012).

Rodríguez, Havidán, Enrico L. Quarantelli and Russel R. Dynes (eds.). 2007. *Handbook of Disaster Research*. New York: Springer.

Roy, P. and S. Choudhary. 2002. *The Lost Horizon: A Tale of Ross: The Deserted Island's Citadel*. New Delhi: Farsight.

Roy, Sanjib. 2012. 'Post-1942 Bengali Settlers Get OBC Status in Central List', *Andaman Sheekha*, www.andamansheekha.com/2012/02/14/post-1942-bengali-settlers-get-obc-status-in-central-list/ (accessed 14 February 2016).

Roychowdhury, R. 2004. *Black Days in Andaman and Nicobar Islands*. New Delhi: Manas.

Sankaran, R. 2005. 'Impact of the Earthquake and the Tsunami on the Nicobar Islands', in Rahul Kaul and Vivek Menon (eds.) *The Ground beneath the Waves: Post-Tsunami Impact Assessment of Wildlife and Their Habitats in India*, Volume 2. New Delhi: Wildlife Trust of India, 10–77, www.wti.org.in/publications/ground-beneath-the-waves-vol2.pdf (accessed 26 December 2008).

Sanyal, S. 2005. 'Wanted: A Mother Teressa', *Indian Express*, 5 January.

Sareen, T. 2001. *Sharing the Blame: Subhas Chandra Bose and the Japanese Occupation of the Andamans (1942–1945)*. New Delhi: S. S. Publishers

Saxena, A., R. Rajan and S. S. Choudhury. 2008. 'Status of Coral Reefs in Post-Tsunami Period in Andaman and Nicobar Islands', in *Proceedings of the 11th International Coral Reef Symposium*, Volume 18. Ft. Lauderdale, Florida, 854–858, http://nsuworks.nova.edu/cgi/viewcontent.cgi?filename=175&article=1000&context=occ_icrs&type=additional (accessed 23 August 2012).

Sen, Satadru. 1998. 'Rationing Sex: Female Convicts in Andamans', *South Asia: Journal of South Asian Studies* 21 (2): 29–59.

Sen, Satadru. 1999. 'Policing the Savage: Segregation, Labor and State Medicine in the Andamans', *Journal of Asian Studies* 58 (3): 753–773.

Sen, Satadru. 2000. *Disciplining Punishment: Colonialism and Convict Society in the Andaman Islands*. New Delhi: Oxford University Press.

Shankar, Malini. 2014. *Preparing for the Day After: A Multimedia Tribute to the Victims of Natural Disasters on the Occasion of the 10th Anniversary of the Asian Tsunami*, https://books.google.co.in/books?id=EbzkBQAAQBAJ&source=gbs_navlinks_s (accessed 20 July 2016).

Sherman, T. C. 2009. 'From Hell to Paradise? Voluntary Transfers of Convicts to the Andaman Islands, 1921–1940', *Modern Asian Studies* 43: 367–388, www.cambridge.org/core/journals/mod ern-asian-studies/article/from-hell-to-paradise-voluntary-transfer-of-convicts-to-the-andaman-islands-19211940/35BD8DF8294653CE2374B7A1146F2914 (accessed 2 February 2009).

Singh, Simron Jit. 2006. *The Nicobar Islands: Cultural Choices in the Aftermath of the Tsunami*. Vienna: Czernin Verlag.

Singh, Simron Jit. 2009. 'Complex Disasters: The Nicobar Islands in the Grip of Humanitarian Aid', *Geographische Rundschau International Edition* 5 (3): 48–56.

Sircar, P. 2006. *Water on Land: The Killer Tsunami in Andaman and Nicobar Island*. Port Blair: Sangeeta Publishing House.

Spalding, Mark D., Corinna Ravilious and Edmund P. Green. 2001. *World Atlas of Coral Reefs*. Berkeley: University of California Press and United Nations Environment Programme, https://archive.org/stream/worldatlasofcora01spal/worldatlasofcora01spal_djvu.txt (accessed 30 June 2009).

Special Correspondent. 2005. 'Cabinet Approves Rs. 2731 Crore for Tsunami Relief', *Hindu*, 20 January.

Srivastava, H. N. and H. M. Chaudhury. 1979. 'Seismicity and Focal Mechanism of Earthquake in Andaman Islands', *Mausam* 30 (2): 213–220.

Stein, J. 1966. *The Random House Dictionary of the English Language*. New York: Random House.

Strand, Carl and John Masek (ed.). 2007. *Sumatra-Andaman Islands Earthquake and Tsunami of December 26, 2004: Lifeline Performances*, Technical Council of Lifeline Earthquake Engineering Monograph 30. Reston, VA: American Society of Civil Engineers.

Survey of India. 1996a. *District Planning Map Series: Andaman District*. Dehradun: Survey of India.

Survey of India. 1996b. *District Planning Map Series: Nicobar District*. Dehradun: Survey of India.

Talbot, Ian and Gurharpal Singh. 2009. *The Partition of India*. New Delhi: Cambridge University Press.

Tang, A., Durgesh C. Rai, David Ames, C. V. R. Murty, Sudhir K. Jain, Suresh R. Dash, Hemant B. Kaushik, Goutam Mondal, Ganapathy Murugesh,

Graham Plant, John McLaughlin, Mark Yashinsky, Martin Eskijian and Rao Surrampalli. 2006. 'Lifeline Systems in Andaman and Nicobar Islands, India, after the December 2004 Great Sumatra Earthquake and Indian Ocean Tsunami', *Earthquake Spectra* 22 (S3): S581–S606, www.iitk.ac.in/nicee/RP/2006_ Lifeline_EQSpectra.pdf (accessed 12 June 2008).

Timmerman, P. 1981. *Vulnerability, Resilience and the Collapse of the Society*, Environmental Monograph 1. Toronto: Institute of Environmental Studies, University of Toronto, www.ilankelman.org/miscellany/Timmerman1981.pdf (accessed 28 March 2017).

Turner, J. R., D. Vousden, R. Klaus, C. Suryanarayana, D. Fenner, K. Venkataraman, P. T. Rajan and N. V. Subba Rao. 2001. *Coral Reef Systems of the Andaman Island, Report of Phase I: Remote Sensing and Rapid Site Assessment Survey, April 2001*. New Delhi: Government of India/United National Development Programme – Global Environmental Facility. GOI/UNEP GEF. Coral Reef Ecosystems of Andaman and Nicobar Islands, www.reefbase.org/resource_center/publication/pub_14284.aspx (accessed 20 June 2008).

UN/ISDR (United Nations International Strategy for Disaster Reduction). 2009. *2009 UNISDR Terminology on Disaster Risk Reduction*. Geneva, http://unisdr.org/files/7817_UNISDR_Terminology_English.pdf (accessed 3 January 2012).

United Nations Environmental Programme. 2003. *Assessing Human Vulnerability to Environmental Change: Concepts, Issues, Methods and Case Studies*. Nairobi, www.unep.org/geo/GEO3/pdfs/AssessingHumanVulnerabilityC.pdf (accessed 16 March 2008).

Weber, G. 2005. *Tsunami: The 2004 Indian Ocean Earthquake and Tsunami*, www.andaman.org/mapstsunami/tsunami.htm (accessed 4 January 2008).

Weber, G. 2008. *The Andamanese*, www.andaman.org/BOOK/text-groupBodyChapters.htm (accessed 3 December 2007).

World Development Report. 1994. *Infrastructure for Development*. Washington, DC: The World Bank, https://openknowledge.worldbank.org/handle/10986/5977 (accessed 7 April 2012).

Online resources

www.alhw.and.nic.in
www.andaman.org
www.and.nic.in
www.ids.nic.in
www.ourairports.com
www.tn.gov.in
www.usgs.gov
www.wikipedia.org

Right to Information (RTI)

RTI replies from the following departments in 2011:

(1) Andaman Lakshadweep Harbour Work
(2) Andaman Public Works Division
(3) Department of Electricity
(4) Department of Transport
(5) Department of Social Welfare
(6) Department of Education
(7) Department of Environment and Forest
(8) Department of Fisheries
(9) Department of Animal Husbandry
(10) Department of Tourism

Personal communication

Andaman and Nicobar Police Fire Services, 2014.
Indian Meteorological Department.
Planning Commission, 2007.

Index

Note: Page numbers in *italics* refer to figures and those in bold refer to tables.